Solitons in Molecular Systems

Mathematics and Its Applications (*Soviet Series*)

Volume 61

Solitons in Molecular Systems

Second Edition

by

A. S. Davydov

Institute for Theoretical Physics,
Kiev, U.S.S.R.

Translated by Dr. Eugene S. Kryachko

SPRINGER SCIENCE+BUSINESS MEDIA, B.V.

Library of Congress Cataloging-in-Publication Data

Davydov, A. S. (Aleksandr Sergeevich), 1912-
[Solitony v molekuli͡arnykh sistemakh. English]
Solitons in Molecular systems / by A.S. Davydov. -- 2nd ed.
p. cm. -- (Mathematics and its applications (Soviet series) ; v. 61)
Translation of: Solitony v molekuli͡arnykh sistemakh.
Includes bibliographical references and index.
ISBN 0-7923-1029-2 (alk. paper)
1. Solitons. 2. Exciton theory. 3. Molecules. I. Title.
II. Series: Mathematics and its applications (Kluwer Academic Publishers). Soviet series ; 61.
QC174.26.W28D3813 1990
541.2'2--dc20 90-48842

ISBN 978-94-010-5475-1 ISBN 978-94-011-3340-1 (eBook)
DOI 10.1007/978-94-011-3340-1

Printed on acid-free paper

This is an updated, expanded second-edition of the book:
Solitons in Molecular Systems
published by Springer Science+Business Media Dordrecht, Dordrecht, Holland © 1985

SERIES EDITOR'S PREFACE

'Et moi, ..., si j'avait su comment en revenir, je n'y serais point allé.'
Jules Verne

The series is divergent; therefore we may be able to do something with it.
O. Heaviside

One service mathematics has rendered the human race. It has put common sense back where it belongs, on the topmost shelf next to the dusty canister labelled 'discarded non-sense'.
Eric T. Bell

Mathematics is a tool for thought. A highly necessary tool in a world where both feedback and non-linearities abound. Similarly, all kinds of parts of mathematics serve as tools for other parts and for other sciences.

Applying a simple rewriting rule to the quote on the right above one finds such statements as: 'One service topology has rendered mathematical physics ...'; 'One service logic has rendered computer science ...'; 'One service category theory has rendered mathematics ...'. All arguably true. And all statements obtainable this way form part of the raison d'être of this series.

This series, *Mathematics and Its Applications*, started in 1977. Now that over one hundred volumes have appeared it seems opportune to reexamine its scope. At the time I wrote

> "Growing specialization and diversification have brought a host of monographs and textbooks on increasingly specialized topics. However, the 'tree' of knowledge of mathematics and related fields does not grow only by putting forth new branches. It also happens, quite often in fact, that branches which were thought to be completely disparate are suddenly seen to be related. Further, the kind and level of sophistication of mathematics applied in various sciences has changed drastically in recent years: measure theory is used (non-trivially) in regional and theoretical economics; algebraic geometry interacts with physics; the Minkowsky lemma, coding theory and the structure of water meet one another in packing and covering theory; quantum fields, crystal defects and mathematical programming profit from homotopy theory; Lie algebras are relevant to filtering; and prediction and electrical engineering can use Stein spaces. And in addition to this there are such new emerging subdisciplines as 'experimental mathematics', 'CFD', 'completely integrable systems', 'chaos, synergetics and large-scale order', which are almost impossible to fit into the existing classification schemes. They draw upon widely different sections of mathematics."

By and large, all this still applies today. It is still true that at first sight mathematics seems rather fragmented and that to find, see, and exploit the deeper underlying interrelations more effort is needed and so are books that can help mathematicians and scientists do so. Accordingly MIA will continue to try to make such books available.

If anything, the description I gave in 1977 is now an understatement. To the examples of interaction areas one should add string theory where Riemann surfaces, algebraic geometry, modular functions, knots, quantum field theory, Kac-Moody algebras, monstrous moonshine (and more) all come together. And to the examples of things which can be usefully applied let me add the topic 'finite geometry'; a combination of words which sounds like it might not even exist, let alone be applicable. And yet it is being applied: to statistics via designs, to radar/sonar detection arrays (via finite projective planes), and to bus connections of VLSI chips (via difference sets). There seems to be no part of (so-called pure) mathematics that is not in immediate danger of being applied. And, accordingly, the applied mathematician needs to be aware of much more. Besides analysis and numerics, the traditional workhorses, he may need all kinds of combinatorics, algebra, probability, and so on.

In addition, the applied scientist needs to cope increasingly with the nonlinear world and the extra mathematical sophistication that this requires. For that is where the rewards are. Linear models are honest and a bit sad and depressing: proportional efforts and results. It is in the non-linear world that infinitesimal inputs may result in macroscopic outputs (or vice versa). To appreciate what I am hinting at: if electronics were linear we would have no fun with transistors and computers; we would have no TV; in fact you would not be reading these lines.

There is also no safety in ignoring such outlandish things as nonstandard analysis, superspace and anticommuting integration, p-adic and ultrametric space. All three have applications in both electrical engineering and physics. Once, complex numbers were equally outlandish, but they frequently proved the shortest path between 'real' results. Similarly, the first two topics named have already provided a number of 'wormhole' paths. There is no telling where all this is leading - fortunately.

Thus the original scope of the series, which for various (sound) reasons now comprises five subseries: white (Japan), yellow (China), red (USSR), blue (Eastern Europe), and green (everything else), still applies. It has been enlarged a bit to include books treating of the tools from one subdiscipline which are used in others. Thus the series still aims at books dealing with:

- a central concept which plays an important role in several different mathematical and/or scientific specialization areas;
- new applications of the results and ideas from one area of scientific endeavour into another;
- influences which the results, problems and concepts of one field of enquiry have, and have had, on the development of another.

This is the greatly modified and supplemented second edition of the first book on Davydov solitons and related matters which appeared in 1985 and is now practically sold out. The main differences are a large new chapter on the quantum theory of solitons, a thorough discussion of two component solitons and a great amount of attention paid to discussions of the physical consequences of the theory. These last two aspects have caused modifications and additions throughout the book (as compared with the first edition).

As I wrote in my preface to the first edition there is a good, albeit heuristic - there is mathematical work to be done here - argument that says that the next approximation after the linear one for (quasi) one-dimensional wave like phenomena will, inevitably, involve soliton equations. Developments since 1985 on quasi one-dimensional structures (such as protein chains), the topic of this book, have done much to bear this out, and soliton waves appear practically all the time in such structures. Examples are the soliton waves of energy transport along protein chains; these are called Davydov solitons, after their discoverer - the author of this book.

It is a pleasure to welcome this totally-up-to-date new version of this work, which is, discounting the 1st edition, still the unique comprehensive volume on the topic.

The shortest path between two truths in the real domain passes through the complex domain.

J. Hadamard

La physique ne nous donne pas seulement l'occasion de résoudre des problèmes ... elle nous fait pressentir la solution.

H. Poincaré

Never lend books, for no one ever returns them; the only books I have in my library are books that other folk have lent me.

Anatole France

The function of an expert is not to be more right than other people, but to be wrong for more sophisticated reasons.

David Butler

Bussum, 7 October 1990

Michiel Hazewinkel

Table of Contents

Preface

The problem of electrons pairing in quasi-one-dimensional structures with the formation of bisolitons is discussed. A new bisoliton mechanism of high-T_c superconductivity of quasi-one-dimensional nonmetallic complex compounds with strong electron-phonon coupling is investigated by means of nonlinear equations. In such compounds, the coherence length is small ($\sim 15\text{Å}$) compared $10^{-4} cm$ inherent for metals described by the BCS theory. The conditions of the generation of solitons and bisolitons at the boundary of a soft quasi-one-dimensional structure are also studied.

In a new and rather large Chapter, the basis of a quantum theory of solitons is presented which enables heat vibrations to be taken into account. The material dealing with proton conductivity via molecular chains with hydrogen bonds between molecules has been extended. The role of proton conductivity in the near-surface layers of water is also discussed.

Pekar's theory of polarons in ionic crystals is modified. The importance of an account of the spatial dispersion of optical phonons characterizing polarization of a crystal is demonstrated, if this is neglected, as has been done in all previous theories, polarons cannot move in a crystal.

In conclusion, A.A. Eremko, V.N. Ermakov, A.V. Zolotariuk, N.I. Kislukha, G.M. Pestryakov, A.I. Sergienko, and V.Z. Enol'skii who participated in the development of the original studies of the theory of solitons involved in this book, E.M. Zaika for her help in the preparation of this book for publication, and E.S. Kryachko for his English translation.

Alexander S. Davydov

Preface to the Second Edition

The first English edition of the present book was published in 1985 (also by Kluwer Academic Publishers). However, since then the theory of solitons in molecular systems has been developed considerably so that a second edition is more than justified.

This second edition is greatly modified and supplemented in view of the modern state of theory of two-component solitons in quasi-one-dimensional molecular systems. Compared with the first edition, some mathematical details have been omitted. Great attention has been paid to a discussion of the physical results of the theory outlined and its possible applications.

In molecular systems, solitons are two-component formations arising due to the nonlinear interaction of two types of quasiparticles, or fields. One should emphasize that these nonlinear formations move only with velocities less than some limiting velocity. The reasons leading to such a constraint are analyzed in the present book.

A considerable part of the book is devoted to modern approaches in the study of energy transduction and electrons in quasi-one-dimensional molecular systems and of protons in macro-molecules with hydrogen bonds. The dominant role of nonlinear phenomena in biology leading to the formation of solitons is demonstrated, and based on these concepts, some problems of modern bioenergetics are discussed. An important role of solitons in elucidating the contraction mechanism of muscles of animals, the motion of bacteria plaits, and intracellular motions are also considered.

Introduction

For a long time linear equations have been used for describing diverse phenomena in physics, chemistry, biology, and related applications. However, these take into account only a linear response of a system to an external influence. Consequently, if the intensity of an external influence increases, the response increases at the same rate.

Fundamental to the linear equations of mechanics (Newton's equations), electrodynamics (Maxwell's equations), and quantum theory (Schrödinger's equation) is the superposition principle, which enables any physical quantity to be represented as a sum of its elementary components. For instance, white light may be considered as the sum of various monochromatic, or singe-frequency components.

Many properties of systems consisting of a large number of interacting molecules and atoms have been explained using linear equations. In particular, the introduction of the concept of collective excitations has been extremely fruitful. A collective excitation characterizes a mutual and consistent, or coherent, motion of a large number of particles. Acoustic waves in gases, liquids, and solids are examples of such collective excitations. An acoustic wave is an oscillation of particles of average density which propagates in a wavelike manner on a background of a huge number of disordered heat vibrations. Simple acoustic waves are monochromatic and are therefore characterized by a specific frequency and wavelength.

The concept of elementary excitations of different types is widely found in solid state physics, e.g. *phonons* are quanta of collective vibrations of atoms and molecules, *excitatons* are quanta of collective excitations of solids whose frequencies correspond to those of visible and ultraviolet light, and *magnons* are quanta of spin (magnetic) excitations. All these elementary excitations are described by monochromatic waves. A strongly monochromatic wave has an infinite spatial extension. For this reason, it is unable to carry energy and information. Hence, energy is transferred only by vibrational excitations which move with a given velocity and are localized within a rather small spatial region. Such localized excitations are called *wave packets* since they are formed by a large number of monochromatic waves.

In various media, the phase velocity of monochromatic waves, i.e., the velocity of motion of a constant phase, depends on the wavelength. These media are called *dispersive.* As far as the motion of a wave packet in such a media is concerned, the different monochromatic components move with different velocities, which results in an increasing spatial dimension of the wave packet. One can say that a wave packet 'runs' with a time. The effect of 'running' is one of the main difficulties connected with the transfer of energy by wave-packet-type excitations. A second noteworthy shortcoming is that a wave packet loses energy in motion. This energy is transferred to random vibrations of atoms and molecules, i.e., to heating of the medium.

It has recently been recognized that the ideal model of transport of vibrational excitations, electrons, and protons in a medium is their transfer in the form of solitary waves known as *solitons.* Unlike ordinary waves which represent a spatial periodical repetition of elevations and hollows on a water surface, or condensations and rarefactions of a density, or deviations from a mean value of various physical quantities, solitons are single elevations, such as thickenings, etc., which propagate as a unique entity with a given velocity. The transformation and motion of solitons are described by nonlinear equations of mathematical physics.

The word 'soliton' was first coined by Zabusky and Kruskal [278] in 1965 to designate briefly solitary waves corresponding to particular solutions of some nonlinear equations describing the propagation of excitations in continuous media with dispersion and nonlinearity. However, the first qualitative description of solitary waves was observed in 1834 on a surface in a shallow channel at Edinburgh, Scotland, by the naval engineer John Scott-Russell and described in 'The Report on Waves' [261]. Based on initial observations, Scott-Russell emphasized the extreme stability and auto self-organization of solitary waves. The great stability of solitons has led to numerous attempts in recent years to explain many novel phenomena in terms of solitons in different branches of physics and other sciences.

The mathematical description of solitary waves on a water surface in shallow channels was first given on a basis of the equation,

$$\left[\frac{\partial}{\partial t}+u\frac{\partial}{\partial z}+\beta\frac{\partial^3}{\partial x^3}\right]u = 0, \quad u = u(x,t). \tag{1}$$

proposed in 1895 by Korteweg and de Vries (KdV equation) [213].

Interest in solitary waves has considerably increased due to related studies in plasma physics. In 1958, Sagdeev [104,105], showed that solitary waves, similar those observed on a water surface, could propagate in plasma in a strong magnetic field. Kadomtsev and Karpman [73] have given a complete review of magnetic and ionic-acoustic solitons in plasmas, and have considered in particular the nonlinear Schrödinger equation whose one-dimensional analogue has the form

$$\left[i\hbar\frac{\partial}{\partial t}+\frac{\hbar^2}{2m}\frac{\partial^2}{\partial z^2}+G|\psi|^2\right]\psi(z,t) = 0,$$

where G is a nonlinearity parameter and where $\hbar^2/2m$ is associated with dispersion. This equation is used for describing the self-focus phenomena in nonlinear optics, one-dimensional self-modulation of a monochromatic wave in plasma, etc.

Some problems in the theory of superconductivity and ferromagnetism lead to the sine-Gordon equation (see Chapter X):

$$\left[\frac{\partial^2}{\partial z^2}-\frac{\partial^2}{\partial t^2}\right]\psi(z,t) = \sin\psi(z,t).$$

Besides the equations mentioned above, are known many other nonlinear equations involving dispersion, whose solutions have a form of stable solitary waves due to the mutual compensation of nonlinear and dispersion effects. The formation of waves in a continuum described by nonlinear equations is related to a spontaneous breakdown of the local symmetry of a homogeneous system, i.e., with self-localization of the excitation energy, electric charge density, or other physical quantities.

The exceptionally important peculiarity of nonlinear equations is that they cannot be studied by means of a linearization procedure, even with subsequent treatment of small nonlinearities via perturbation theory based on expansion into normal linear modes. Nonlinear equations lead to nonlinear phenomena, such as *solitons, kinks, breathers,* and others which are impossible to obtain within finite-order perturbation treatment. These nonlinearities are fundamental as quasiparticles of linear theories, they give important information relating to the properties of appropriate media, and they play an important role in energy processes, in energy and charge transfer phenomena, and in structural transitions.

Solitary waves possess some remarkable properties which make it possible to compare them with particles. For instance, the conservation of shape and velocity after interaction is a particle-like property. It is, in fact, such properties which prompted Zabusky and Kruskal to call them *solitons* [278].

Together with ordinary extended waves, the solitary-like excitations are inherent to many nonlinear dynamical systems. Their analytical description is, however, well developed for one-dimensional systems only. In the one-dimensional case the nonlinear equations mentioned possess an infinite number of conservation laws and admit exact solutions via the so-called *method of inverse scattering* for the auxiliary linear operator. This method originated in the work of Gardner et al. [229], and was further developed by Zakharov, Faddeev, Novikov, Calogero, Marchenko, and others. The book by Zakharov et al. [66] is devoted to a complete treatment of this method. In the mathematical literature, the description 'soliton' is used only with respect to the localized solutions of completely integrable one-dimensional systems. Localized excitations described by incompletely integrable nonlinear equations are usually called *solitary waves.*

In the description of real systems, one cannot restrict the treatment to completely integrable equations alone, since the latter correspond to idealized dynamical systems. This neglects phenomena relating to the existence of boundaries and other degrees of freedom, dissipation, and small physically relevant perturbations due to surrounding bodies (complete isolation is, in fact, impossible). Unstable solitary waves can be used for a description of realistic systems if their lifetimes are larger compared with the time inherent for the phenomenon under study.

Here, the term 'soliton' is applied in a broader sense, i.e., to describe any self-localized excitations propagating without substantial change of form and velocity, due to a dynamical balance between nonlinearity and dispersion.

In solid state physics, approximate ideas of elementary excitations, as quasiparticles described by plane waves or, strictly, by wave packets, are of a current interest. General solutions of many linear phenomena are expressed in terms of such linear or quasilinear modes. One can suggest that solitons should play the same role in nonlinear dynamics.

The present book is devoted to the study of nonlinear phenomena occurring in quasi-one-dimensional molecular structures and their explanation in terms of solitons. However, we would like to begin our treatment of nonlinear phenomena by the example of the classical solitary waves observed by John Scott-Russell [261] in 1834 on a water surface in a shallow rectangular channel. As mentioned before, surface waves in such a channel are described by a dimensional equation of the type (1) given by Korteweg and de Vries (for short, KdV equation).

The function $u(x,t)$ in eqn. (1) characterizes a deviation from an average magnitude of velocity density, or any other real classical quantity, which is useful for a description of diverse phenomena in physics. The KdV equation is applied particularly to the study of ion-acoustic [271], magneto-hydrodynamical waves in plasma [203,207], acoustic waves in an anharmonic lattice [265-270,275,277], and many other phenomena. A complete treatment of the KdV equation is presented in the reviews of Kadomtsev and Karpman [73] and Toda [267], in the monograph by Karpman [74], and in other works.

In a particular case, when the function $u(x,t)$ in eqn. (1) characterizes a velocity in a continuum, we redesignate it by $\Phi(x,t)$ and rewrite eqn. (1) as follows

$$\Phi_t + 6g\Phi\Phi_x + \beta\Phi_{xxx} = 0, \quad \Phi = \Phi(x,t). \tag{2}$$

Henceforth, we shall use subindices of functions for denoting the appropriate derivatives. The nonlinear equation (2) describes a perturbation which propagates in a medium with nonlinearity and dispersion,

$$\omega(q) = c_0 q - \beta q^3, \tag{3}$$

where c_0 is the phase velocity of small vibrations described by a plane wave with wavenumber q in the limit $q \to 0$.

Neglecting the nonlinear term, $g = 0$, one can transform eqn. (2) into the linear one,

$$\Phi_t + \beta\Phi_{xxx} = 0, \tag{4}$$

which is traditionally called the linearized KdV equation. It is noteworthy that eqns. (2) and (4) are written within the coordinate frame x,t where a phase velocity c_0, involved in the dispersion (3), vanishes. Within this coordinate frame moving with velocity c_0, eqn. (4) possesses a particular solution of a plane-wave type,

$$\Phi(x,t) = \Phi_0 \exp\{i(qx-\omega t)\} \tag{5}$$

with negative dispersion ($\beta > 0$)

$$\omega = -\beta q^3.$$

The phase and group velocities of the wave given by eqn. (5), are respectively,

$$v_p = \frac{\omega}{q} = -\beta q^2, \quad v_{gr} = \frac{d\omega}{dq} = -3\beta q^2. \tag{6}$$

Hence, the wave (5) moves towards a negative x axis, as should be expected for a negative dispersion law. The general solution of the linearized KdV equation has been completely analyzed by Karpman [74].

The KdV equation is invariant relative to Galilei transformations,

$$t \to t, \quad x \to x - ct,$$

where c is an arbitrary constant. It possesses an infinite set of integrals of motion which are discussed in detail in [66], [67]. Here we deal with the simplest ones which have a clear physical meaning. For this purpose, let us rewrite eqn. (2) as follows

$$\Phi_t + \left[3g\Phi^2 + \beta\Phi_{xx}\right]_x = 0.$$

Integrating the latter over x in the limits $-\infty$, $+\infty$, and, in addition, imposing a condition of periodicity or the vanishing of the function $\Phi(x,t)$ its derivatives at the boundaries, one obtains the first integral of motion,

$$S_1 = \frac{1}{2}\int \Phi(x,t)dx = M. \tag{7}$$

This gives the law of conservation of an excitation 'mass'. Multiplying eqn. (2) by Φ, one obtains the following equation

$$\frac{1}{2}(\Phi^2)_t + (2g\Phi^3 + \beta\Phi\Phi_{xx} - \frac{1}{2}\beta\Phi_x^2)_x = 0,$$

which results in the second integral of motion,

$$S_2 = \frac{1}{2}\int \Phi^2(x,t)dx = P. \tag{8}$$

which is a corollary of the translational invariance of the KdV equation. For this reason, this determines the momentum of excitation. One can demonstrate in a similar way that the following quantity becomes the next integral of motion,

$$S_3 = \frac{1}{2}\int(2g\Phi^3 + \beta\Phi_x^2)dx = E. \tag{9}$$

It can be interpreted as an excitation energy. One can prove that the quantity,

$$S_4 = \int x\Phi(x,t)dx = x,$$

is also conserved. This may be called the 'centre of inertia of excitation'.

The KdV equation provides a good example illustrating that linearization of a nonlinear equation leads to loss of the most interesting solutions. Let us study the solutions without any reduction to a linearized form. We seek a solution of the translational invariant KdV equation in the form of stationary running waves having a constant velocity V. For this purpose, we pass to a moving coordinate frame assuming $\Phi(x,t) = \Phi(\xi)$ with

$$\xi = x - x_0 - Vt. \tag{10}$$

Then one obtains

$$\Phi_t = -V\Phi_\xi, \quad \Phi_x = \Phi_\xi.$$

Equation (2) transforms into the ordinary differential equation

$$(6g\Phi - V)\Phi_\xi + \beta\Phi_{\xi\xi\xi} = 0.$$

Integrating this, one obtains

$$\beta\Phi_{\xi\xi} = V\Phi - gA - 3g\Phi^2, \tag{11}$$

where $A > 0$ is an integration constant. Multiplying both sides of eqn. (11) by Φ_ξ and integrating again, one obtains the equation

$$\Phi_\xi^2 = \frac{2}{\beta}\mathcal{G}_A(\Phi), \tag{12}$$

where

$$\mathcal{G}_A(\Phi) \equiv \frac{1}{2}V\Phi^2 - g(A + \Phi^2)\Phi.$$

The new integration constant is assumed to vanish due to the translational invariance. The general solution of eqn. (12) is expressed in terms of the elliptic integral,

$$\int_{\xi_0}^{\Phi(\xi)} \frac{d\Phi}{\sqrt{\mathcal{P}_A(\Phi)}} = S\sqrt{\frac{2}{\beta}}\xi, \quad s = \pm 1, \quad \Phi_0 = \Phi(0). \tag{13}$$

At the very start, we deal with localized excitations which correspond to functions satisfying the following boundary conditions,

$$\Phi(\pm\infty) = \Phi_\xi(\pm\infty) = \Phi_{\xi\xi}(\pm\infty) = 0. \tag{14}$$

According to eqn. (11), such excitations correspond to the vanishing integration constant A. In this case eqn. (12) transforms to the following one,

$$\Phi_\xi = s\sqrt{2/\beta}\Phi\sqrt{V/2 - g\Phi}, \quad s = \pm 1. \tag{15}$$

If $\Phi_0 = \Phi(0)$ corresponds to the maximum value of $\Phi(\xi)$, $\Phi_\xi(0) = 0$, and from eqn. (15) the relation between Φ_0 and the phase velocity of excitation follows:

$$V = 2g\Phi_0. \tag{16}$$

Solitons with a larger amplitudes move faster than those with smaller ones.

Using relation (16), one can transform eqn. (15) as follows

$$\Phi_\xi = s\sqrt{2g/\beta}\Phi\sqrt{\Phi_0 - \Phi}. \tag{17}$$

Integrating the latter from 0 to ξ, one obtains the solution

$$\int_{\Phi_0}^{\Phi(\xi)} \frac{d\Phi}{\Phi\sqrt{\Phi_0 - \Phi}} = s\xi\sqrt{2g/\beta}.$$

Evaluating this integral, one obtains finally the form of the localized solution of the KdV equation:

$$\Phi(\xi) = \Phi_0 \mathrm{sch}^2(\xi\sqrt{g\Phi_0/2\beta}). \tag{18}$$

With the help of eqn. (16), one can express eqn. (18) in terms of a phase velocity,

$$\Phi(\xi) = (V/2g)\mathrm{sch}^2\left[\frac{1}{2}\xi\sqrt{V/\beta}\right]. \tag{19}$$

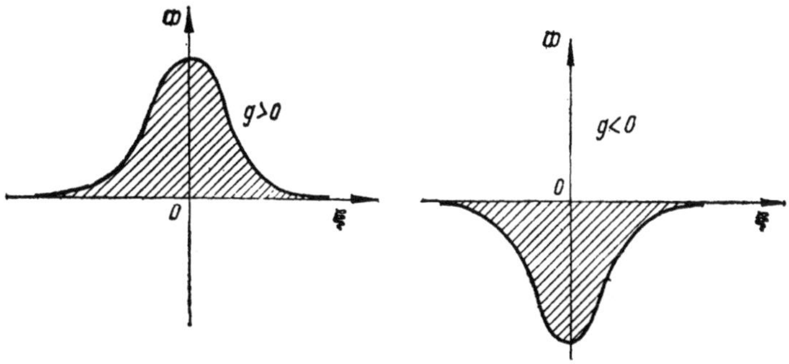

Figure 1. The amplitudes $\Phi(\xi)$ for the positive and negative nonlinearity coefficient g.

Negative values of the nonlinearity coefficient in eqn. (2) correspond to the negative values of $\Phi(\xi)$ (see Fig. 1). Therefore, for a positive nonlinearity coefficient, a soliton corresponds to an elevation on a water surface, or a thickening in other systems. This soliton carries a 'mass' of

$$M = \frac{1}{2}\int_{-\infty}^{\infty}\Phi(\xi)d\xi = 2\sqrt{\beta V^3}/g, \tag{20}$$

a momentum of

$$P = \frac{1}{2}\int_{-\infty}^{\infty}\Phi^2(\xi)d\xi = \sqrt{\beta V^3}/3g^2 \tag{21}$$

and energy,

$$E = \frac{1}{2}\int_{-\infty}^{\infty}(2g\Phi^3+\beta\Phi_{\xi}^2)d\xi = \frac{4V^{5/2}(2+\beta^2)}{15g^2\sqrt{\beta}}. \tag{22}$$

For a negative nonlinear coefficient, a soliton corresponds to a hollow, or a decrease of density in other systems. Then a soliton 'mass' becomes negative, and its momentum and energy are determined by the same equations (21) and (22). Both types of solitons move towards a positive direction of the x-axis of a coordinate frame moving with velocity c_0 appearing in eqn. (3). Relative to the coordinate frame linked with a system, the soliton velocity is c_0+V and hence, is always larger c_0. For a negative nonlinearity coefficient, the solutions of the KdV equation can be obtained from the appropriate solutions with positive g by the formal substitution $\Phi(x,t)\to-\Phi(x,t)$.

For a medium with positive dispersion, neglecting nonlinearity, one obtains

$$\omega(q) = c_0q+\alpha q^3. \tag{23}$$

Within the coordinate frame moving with velocity c_0, the KdV equation takes thus the following form

$$\Phi_t + 6g\Phi\Phi_x - \alpha\Phi_{xxx} = 0. \tag{24}$$

Its solutions are obtained from those corresponding to the KdV equation with a negative dispersion via the substitutions

$$x \to -x, \quad t \to t, \quad \Phi(x,t) \to -\Phi(x,t). \tag{25}$$

Relative to a coordinate frame moving with velocity c_0, such solitons propagate towards the negative direction of the x-axis. If their velocity is V within the coordinate frame chosen, then their velocity relative to the immobile coordinate frame of the medium, is $c_0 - V$. If the nonlinearity coefficient g is positive, the soliton movement corresponds to that of a 'rarefaction' region.

To study the periodic solutions of the KdV equation, one should ignore the boundary conditions (14) and preserve the integration constant in eqn. (11). In this case, the polynomial (13) takes the form

$$\mathcal{P}(\Phi) \equiv g\Phi(\Phi_0 - \Phi)(\Phi - \Phi_1), \tag{26}$$

where Φ_0 and Φ_1 correspond to the maximum and minimum values of $\Phi(\xi)$, respectively. The latter are defined via the relations

$$\Phi_0 + \Phi_1 = V/2g, \quad \Phi_0\Phi_1 = A > 0. \tag{27}$$

Introducing the ratios

$$\Phi/\Phi_0 = t^2 \text{ and } \Phi_1/\Phi_0 = k_1^2, \tag{28}$$

one can transform eqn. (13) into the following,

$$\int_1^{\sqrt{\Phi/\Phi_0}} \frac{dt}{\sqrt{(1-t^2)(t^2-k_1^2)}} = \xi\sqrt{2g\Phi_0/\beta}. \tag{29}$$

Using the definition of the Jacobian elliptic function $dn(u,k)$ through inverting the integral

$$\int_1^{dn(u,k)} \frac{dt}{\sqrt{(1-t^2)(t^2-k_1^2)}} = u, \quad k = \sqrt{1-k_1^2},$$

one can rewrite eqn. (29)

$$\Phi(\xi) = \Phi_0 dn^2(\xi\sqrt{2g\Phi_0/\beta},k), \tag{30}$$

where k is the modulus of the elliptic function related with the complementary modulus $k_1 = \sqrt{1-k^2}$.

Applying relations (27) and (28), one can express the amplitude Φ_0 through the phase velocity V and the complementary modulus,

$$\Phi_0 = V[2g(1+k_1^2)]^{-1}. \tag{31}$$

Using the latter formula, one obtains the expression of the wave function (30) in terms of the phase velocity:

$$\Phi_d(\xi) = \frac{V}{2g(1+k_1^2)}dn^2\left[\sqrt{\frac{V}{\beta(1+k_1^2)}}\xi,k\right]. \tag{32}$$

This function takes negative values for negative g.

The excitation described by the function (32), is called a *knoidal* wave. It represents an infinite series of repeating elevations for $g>0$, or hollows for $g<0$. The period Δu, i.e., the distance between neighboring elevations, is equal to the period $2K(k)$ of the elliptic function $dn(u,k)$:

$$dn(u,k) = dn(u+2K(k),k).$$

Here $K(k)$ is the elliptic integral of the first kind. The period of the function (32) can be expressed in terms of ξ,

$$L = 2\sqrt{\beta(1+k_1^2)/V}K(k). \tag{33}$$

Small k correspond to large periods. Given $k_1^2 \ll 1$, one can apply the asymptotic expression for the function $dn(u,k)$ and obtain:

$$dn(u,k) \approx \operatorname{sch} u + \frac{1}{4}k_1^2(\operatorname{sh} u \operatorname{ch} u + u)\operatorname{th} u \operatorname{sch} u.$$

As $k_1 \to 0$, the period L is infinite, and the function (32) transforms into the function (19) of the soliton excitation.

The KdV equation possesses stationary periodic solutions of another type. To obtain these solutions one should choose an integration constant equal to $-B<0$. Then eqn. (13) takes the form

$$\beta\Phi_{\xi\xi} = V\Phi + gB - 3g\Phi^2. \tag{34}$$

Integrating further, one obtains the equation

$$\Phi_\xi^2 = \frac{2}{\beta}\mathcal{P}_B(\Phi), \tag{35}$$

where the polynomial

$$\mathcal{P}_B(\Phi) \equiv \frac{1}{2}V\Phi^2 - g(\Phi^2 - B)\Phi \tag{36}$$

can be written as follows

$$\mathcal{P}_B(\Phi) = g\Phi(\Phi_0 - \Phi)(\Phi + \Phi_1). \tag{37}$$

In eqn. (37) the following notations,

$$\Phi_0 - \Phi_1 = V/2g \quad \text{and} \quad \Phi_0\Phi_1 = B \tag{38}$$

are used. Introducing the ratios,

$$\Phi/\Phi_0 = t^2 \quad \text{and} \quad \Phi_1/\Phi_0 = k_1^2/k^2, \tag{39}$$

one obtains

$$\mathcal{P}_B(\Phi) = g\Phi_0^3 t^2(1-t^2)(k^2t^2 + k_1^2)/k^2. \tag{40}$$

Therefore, the solution of eqn. (35) becomes

$$\int_1^{\sqrt{\Phi/\Phi_0}} \frac{dt}{\sqrt{(1-t^2)(k^2t^2+k_1^2)}} = s\xi\sqrt{g\Phi_0/2\beta k}, \quad s = \pm 1. \tag{41}$$

Taking into account the definition of the Jacobian elliptic function $cn(u,k)$ via inversion of the integral

$$\int_1^{\mathrm{cn}(u,k)} \frac{dt}{\sqrt{(1-t^2)(k^2t^2+k_1^2)}} = u,$$

where k is the modulus, and $k_1 = \sqrt{1-k^2}$ one obtains from eqn. (41) the second knoidal wave

$$\Phi_c(\xi,k) = \Phi_0 \mathrm{cn}^2(\xi\sqrt{g\Phi_0/2\beta k}, k). \tag{42}$$

Using the relations (38) and (39), one finds the expression of the maximum amplitude Φ_0 in terms of the phase velocity and the modulus of the elliptic function

$$\Phi_0 = k^2 V[2g(1-2k_1^2)]^{-1}. \tag{43}$$

As k_1 approaches zero, $k \to 1$, and eqn. (43) is just eqn. (16).

Using the formula (43), one can express the function (42) in terms of the phase velocity of the excitation,

$$\Phi_c(\xi,k) = \frac{k^2 V}{2g(1-2k_1^2)} \mathrm{cn}^2 \left[\xi \sqrt{\frac{kV}{4\beta(1-2k_1^2)}}, k\right]. \tag{44}$$

The function $\mathrm{cn}^2(u,k)$ has the period $2K(k)$, i.e.,

$$\mathrm{cn}^2(u,k) = \mathrm{cn}^2(u+2K(k),k).$$

Therefore, the period of the function (44) defined as follows

$$\Phi_c(\xi+L,k) = \Phi_c(\xi,k),$$

takes the value

$$L = 4K(k)\sqrt{\beta(1-2k_1^2)/kV}. \tag{45}$$

If $k_1^2 \ll 1$, or $k \simeq 1$, the functions $dn(u,k)$ and $\mathrm{cn}(u,k)$, and, therefore, the functions $\Phi_d(\xi,k)$ and $\Phi_c(\xi,k)$ differ slightly from each other, and in the limit $k_1 \to 0$, both solutions transform into soliton-like ones, since

$$\lim_{k\to 1} \mathrm{dn}(u,k) = \lim_{k\to 1} \mathrm{cn}(u,k) = \mathrm{sch}\, u \tag{46}$$

as $L \to \infty$. As the period decreases, their difference becomes more considerable however, and in the limit $k_1 \to 1$, or $k \to 0$, the following asymptotes are valid:

$$\mathrm{dn}(u,k) \approx 1 - \frac{1}{2}k^2 \sin^2 u + \ldots, \tag{47}$$

$$\mathrm{cn}(u,k) \approx \cos u + \frac{1}{4}k^2 \sin u (u - \sin u \cos u).$$

The theory of solitons described by the KdV equation has already become classical. Many original papers, reviews, and books have been devoted to the theory of solitons which occur in diverse classical and quantum phenomena. The review by Scott et al. [248] in 1973 was one of the first reviews which played an important role in the dissemination of the term 'soliton'. In this review, the authors discussed the application of the idea of soliton to different fields of applied sciences.

The relations of the idea of soliton with various problems of mathematics, and theoretical and experimental physics were discussed in the book 'Solitons in Action' edited by Lonngren and Scott [90]. The application of the soliton concept to some phenomena in solids has been considered in the book by Bishop and Schneider [142]. In the review [41], the present author has discussed the application of the soliton concept for describing some biological phenomena.

The second edition of the present book treats the results of theoretical studies of excited states in quasi-one-dimensional periodic structures on the basis of the theory of nonlinear equations which include the interaction of intramolecular excitations, or excess electrons with a local breakdown of translational symmetry conditioned by displacements of the equilibrium positions of molecules. Such structures are particularly realized in α-helix protein molecules.

In Chapters 1-5, we present the theory of molecular solitons arising in quasi-one-dimensional molecular chains and carrying intramolecular excitations within loss. We discuss also the possible manifestations of solitons in a few biological and physical phenomena (particularly, in Chapter 6). The theory of translocation of electrons by solitons along quasi-one-dimensional molecular chains is developed in Chapter 7. Here we also study the process of the pairing of two electrosolitons in a singlet spin state with the formation of a Bose particle, or a bisoliton which carries a double electric charge and vanishing spin.

In Chapter 8, a description of spatially periodic excitations is given. We propose a novel, bisoliton mechanism of high-T_c superconductivity of nonmetallic compounds. Unlike the superconductivity theory of metals - the BCS theory - based on perturbation theory, our novel mechanism is developed on the basis of nonlinear theory beyond the perturbation approach. Such superconductivity is isotopic invariant and characterized by a small coherence length of about 15Å.

In Chapter 9, we develop a quantum theory of motion of a quasiparticle along a molecular chain while taking into account heat motions. The sine-Gordon equation and its applications are discussed in Chapters 10 and 11. In Chapter 12, we study the transfer of protons in hydrogen-bonded molecular chains. The novel version of Pekar's theory of polarons which involves the spatial dispersion of optical phonons describing a polarization of ionic crystals is developed in Chapter 13. In Chapter 14, the problems of the generation of solitons and bisolitons at a boundary of a soft molecular chain are discussed. A new theory of the mechanism of high-T_c superconductivity of nonmetallic oxide ceramics based on the concept of bisolitons is developed in Chapter 15.

Chapter 1

COLLECTIVE EXCITATIONS IN QUASI-PERIODICAL MOLECULAR STRUCTURES

1. Crisis in Bioenergetics

All living organisms are open systems. Their vital activity is inseparably linked with continuous expense of energy and an exchange of substances. Light plays a role as the energy source for plants and photosynthesizing bacteria. Animals cannot directly transform the energy of the sun into other types of energy. Energy demand is increased by the utilization of nutrient substances. Carbohydrates, lipids, and albumens are nutrient substances for highly organized living systems. In the process of respiration, as a result of oxidation, these substances release energy.

The process of fuel oxidation (coal, firewood, petroleum) is a basic source of energy utilized in our daily lives. Generally speaking, the combustion energy of fuel is transformed into heat energy which is then converted into other types of energy. In nuclear power stations, nuclear energy is also transformed at first into heat energy. The employment of an intermediate stage resulting in heat energy considerably lowers the performance coefficient of energy plants.

The process of oxidation of organic substances in living organisms and the use of light energy in plants requires an intermediate stage, too, which involves the formation of chemical energy, rather than heat energy.

In the cells of living organisms, oxidation energy is stored following the synthesis of adenosine triphosphate (ATP) acid molecules from the molecules of adenosine diphosphate (ADP) and phosphoric acid. A molecule of ATP consists of adenine, ribose sugar, and three phosphate groups. The ADP molecule has two phosphate groups.

The process of synthesis of ATP molecules is called *phosphorylation*. This involves replacement of the hydrogen atom of the end phosphate group of the ADP molecule by the third phosphate with the release of a water molecule. The ATP molecules which form under phosphorylation diffuse to sites where energy is utilized. A reverse reaction occurs accompanied by the release of about $0.5\,eV$ of energy in a surrounding of water molecules and complexes of enzymes and co-enzymes. This reaction is recognized as the hydrolysis of ATP molecules.

The synthesis and hydrolysis of ATP molecules play a universally intermediate role in numerous energy transformations in the cells of bacteria, plants, and animals, from the simplest to the most complex cases. Due to the large dimension of organic molecules and enzymes, the release and utilization of hydrolysis energy are separated by distances considerably exceeding interatomic distances. In this context, a problem of great importance is that of elucidating the mechanism of transduction of the hydrolysis energy over such distances.

The special Meeting of the New York Academy of Sciences which was held in 1973 was devoted to a discussion of energy transfer mechanisms in biological systems (see [91]). The following three basic problems were discussed:

(i) Does a crisis in bioenergetics really exist?

(ii) What is its origin?

(iii) How can this crisis be overcome?

As a matter of fact, the origin of these problems emphasized the absence of a complete understanding of the mechanism of the effective transfer of energy released in the hydrolysis of ATP molecules. The explanations proposed encounter insurmountable obstacles, which are briefly the following. The energy which is released in a single occurrence of hydrolysis is rather small (about $0.5\,eV$) and only twenty times greater than the average energy of a thermal fluctuations. It is insufficient to excite the electronic state of molecules. The energy released in the hydrolysis of ATP molecules is only sufficient to excite intramolecular vibrations. Vibrations of the atomic group $C = O$, usually denoted as Amide-I, appear to be suitable for this purpose. This group is incorporated into peptide groups of all protein molecules and its vibrations have been well studied in experiments involving infrared absorption. Excitation of these vibrations requires about $0.21\,eV$. In this connection, some participants of the 1973 Meeting were convinced that the energy of hydrolysis of ATP molecules is transferred along protein molecules in the form of Amide-I vibrations. This suggestion was questioned by other participants, particularly by A. Weber, who asserted that since the lifetime of vibrations of isolated peptide groups is of the order of 1 ps in condensed medium, they could not take part in the transduction of hydrolysis energy over distances considerably exceeding the dimension of peptide groups. The supporters of the first hypothesis assigned a key role to the stabilization and lifetime increase of Amide-I vibrations of proteins to their special structure. However, they were no able to argue completely this point of view. For this reason, the problem of the crisis in bioenergetics was unresolved.

In 1973, Kislukha and the present author [153] proposed an idea for the transfer of vibrational energy of a collective and nonlinear nature in α-helical protein molecules without loss basing this argument on a physical model. Such excitations are called Davydov solitons [202,253-257]. The subsequent development of this approach at the Institute for Theoretical Physics (see [37,40-42,45,46,156-161]), originated numerous investigations carried out in the USSR, the USA, the UK and other countries. An efficiency of transfer of vibrational energy in biological systems by solitons was demonstrated. Such investigations led to considerable progress in solving the problem related to a crisis in bioenergetics, and revealed particularly the essential role of solitons in some biological phenomena. Before presenting the basic results of these studies, we give some necessary information about the structure of protein molecules.

2. Alpha-Helical Structure of Proteins

Protein molecules play the most active role in bioenergetics. They are closely related to all basic manifestations of life. The proteins-enzymes take part in all chemical processes occurring in cells. Proteins transform chemical energy into mechanical energy and are responsible for all inter- and intracellular motions. Accompanied by fatty-like molecules, or lipids, they ensure the transport of ions and molecules inside and outside cells through cellular membranes. Proteins participate in the oxidation of food products to support energetically the process of synthesis of ATP molecules.

All proteins are polymeric molecules having a great molecular weight. It is worth pointing out that the average molecular weight of protein accounts for 10^5, to an order of magnitude, while for a water molecule this is only 18. The origin of

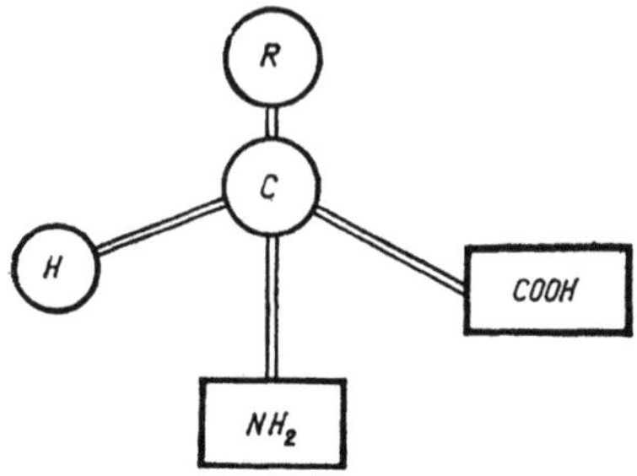

Figure 2. The structure of amino acid.

proteins results from the polymerization of diverse amino acids. Amino acid is a chemical compound whose central carbon atom is bonded to an amino group, NH_2, carboxylic group, $COOH$, hydrogen atom, and a certain group, R, of atoms due to which amino acids are distinguishable (Fig. 2).

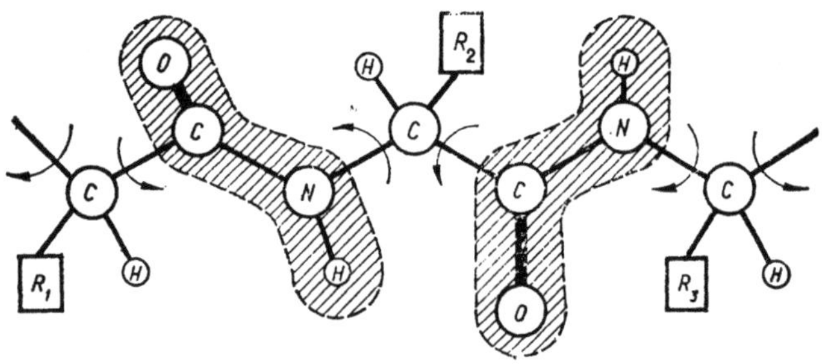

Figure 3. Segment of protein chain with two peptide groups (shaded).

There exists a large number of amino acids in nature, but the proteins of living organisms originate from 20 different amino acids only.

A segment of the protein chain, involving two peptide groups, is shown in Fig. 3. Atoms of each peptide group lie in the same plane. As a whole, a protein molecule possesses various spatial configurations due to possible rotations about single bonds (such rotations are indicated by arrows in Fig. 3). In particular, the protein chain can coil into a helix. The helical structure of the protein molecule was firstly discovered by Pauling and Corey in 1953 who called it the alpha-helix and revealed that this structure is realized as result of the formation of three hydrogen-bonded chains which link the peptide groups of the protein molecule. The arrangement of peptide groups in the α-helical protein molecule is plotted in Fig. 4, where peptide groups are denoted by ellipses and hydrogen bonds by lines connecting these groups.

In the next Section, we study the collective excitations corresponding to the intrapeptide vibrations of Amide-I type, within the simplest model of a protein molecule which involves only a periodic arrangement of peptide groups along a single hydrogen-bonded chain (see Fig. 4).

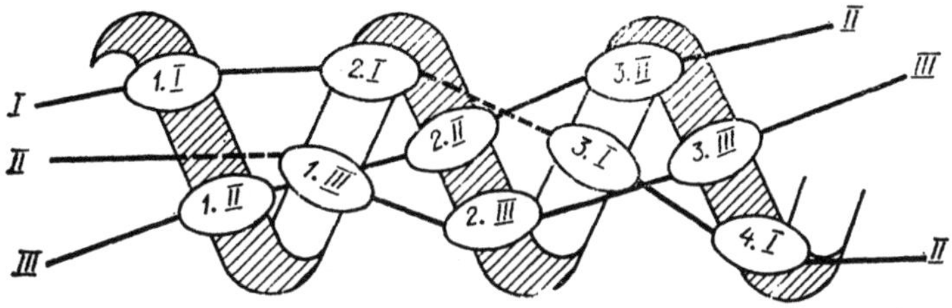

Figure 4. The arrangement of peptide groups in the alpha-helix protein molecules: peptide groups are denoted by ellipses and hydrogen bonds are lines connecting these groups.

3. Collective Excitations in a Single Chain of Peptide Groups

Amide-I vibrations possess an energy, $\mathcal{E}_0$, of about $0.21\,eV$ and a rather large electric dipole moment $d = 0.3$ debye which leads to a resonance interaction between neighboring peptide groups resulting in collectivization, or cooperativity of excitation.

Let us assume that the peptide groups of mass M are located at nodes $na (n = 0, \pm 1, \ldots)$ along the z-axis. Let the electric dipole moment of transition to the excited state with energy $\mathcal{E}_0$ be directed along the z-axis, too. The Hamiltonian of an infinite chain of peptide groups is given by the following sum

$$H = H_{ex} + H_{ph} + H_{int}, \tag{3.1}$$

where

$$H_{ex} = \sum_n B_n^+ [\mathcal{E}_0 B_n - J(B_{n+1} + B_{n-1})], \tag{3.2}$$

$$H_{ph} = \frac{1}{2} \sum_n [M^{-1} \mathbf{p}_n^2 + \kappa (\mathbf{u}_n - \mathbf{u}_{n-1})^2], \tag{3.3}$$

$$H_{int} = \frac{1}{2} \chi \sum_n B_n^+ B_n (\mathbf{u}_{n+1} - \mathbf{u}_{n-1}). \tag{3.4}$$

In eqs. (3.1)-(3.4) the summation is carried out over all integers, $n = 0, \pm 1, \pm 2, ...$; B_n^+ and B_n are the creation and annihilation operators of the peptide group excitation labeled by n, respectively; $\mathbf{u}_n$ is the operator of longitudinal displacement from its equilibrium position na; $\hat{\mathbf{p}}_n$ is the operator conjugated to $\mathbf{u}_n$; κ is the longitudinal elasticity of the chain; χ is a coupling constant of intermolecular excitations with displacements $\mathbf{u}_n$; and finally

$$-J = -2d^2a^{-3}$$

is the resonance interaction energy. In solid state theory, the interaction operator, given by eqn. (3.4), is taken into account usually within the perturbation treatment. In the particular case of soft chains (κ is small) with a large coupling constant χ, all three terms in eqn. (3.1) are of the same order. Using the unitary operator which produces displacements of peptide groups from their equilibrium positions, one can introduce the coherent states, $\beta_n(t)$, of displacements and conjugate momenta, $\pi_n(t)$, at the state $\psi(t)$ as follows:

$$\beta_n(t) = \langle\psi(t)|\mathbf{u}_n|\psi(t)\rangle, \quad \pi_n(t) = \langle\psi(t)|\mathbf{p}_n|\psi(t)\rangle \tag{3.5}$$

(for details see [49]). The product of uncertainties of displacements and conjugated momenta becomes minimal at these coherent states.

Rewriting the Hamiltonian (3.1) in terms of the variables $\beta_n(t)$ and $\pi_n(t)$, and the functions $\varphi_n(t)$ obeying the normalization condition

$$\sum_n |\varphi_n(t)|^2 = 1, \tag{3.6}$$

one obtains the following Hamiltonian of the chain under study

$$\langle\psi(t)|H|\psi(t)\rangle = \sum_n \varphi_n^* \left\{ [\mathcal{E}_0 + W + \frac{1}{2}\chi(\beta_n - \beta_{n-1})]\varphi_n - -J(\varphi_{n+1} + \varphi_{n-1}) \right\}, \tag{3.7}$$

where

$$W \equiv \frac{1}{2}\sum_n [\pi_n^2 M^{-1} + \kappa(\beta_n - \beta_{n-1})^2] \tag{3.8}$$

is the chain deformation energy. The invariance of the Hamiltonian (3.7) with respect to the phase transformation, $\varphi_n \to \varphi_n \exp(i\gamma)$, the time shift, $t \to t+\tau$, and its translational invariance, $n \to n+m$, too, result in the conservation laws of the number of excitations, energy E, and momentum $\mathcal{P}$, respectively.

The following set of equations correspond to the Hamiltonian function (3.7)

$$i\hbar\frac{d\varphi_n}{dt} = [\mathcal{E}_0 + W - \frac{1}{2}\chi(\beta_{n+1} - \beta_{n-1})]\varphi_n - J(\varphi_{n+1} + \varphi_{n-1}), \tag{3.9}$$

$$M\frac{d^2\beta_n}{dt^2} + \kappa[2\beta_n - \beta_{n+1} - \beta_{n-1}] = \frac{1}{2}\chi\{|\varphi_{n+1}|^2 - |\varphi_{n-1}|^2\},$$

$$\pi_n = M\frac{d\beta_n}{dt}.$$

This set defines the following two fields: the first one, described in $\varphi_n(t)$, gives the distribution of intrapeptide vibrational excitations along the chain, and the second one, $\beta_n(t)$, characterizes the displacements of peptide groups into the new equilibrium positions. For this reason, the energy excitations described by the set of eqns. (3.9), can be called the two-component ones.

If the functions φ_n and β_n change negligibly over a single segment of the chain, one can apply the continuum approximation,

$$\varphi_n(t) \Rightarrow \Phi(z,t)\exp[i(kz - Et/\hbar)], \quad \beta_n(t) \Rightarrow \beta(z,t). \tag{3.10}$$

In this case, one has

$$\varphi_{n\pm1}(t) \Rightarrow e^{i[k(z\pm a) - Et/\hbar]}\left\{1 \pm a\frac{\partial}{\partial z} + \frac{a^2}{2}\frac{\partial^2}{\partial z^2} + \dots\right\}\Phi(z,t), \tag{3.11}$$

$$\beta_{n\pm1}(t) = \left\{1 \pm a\frac{\partial}{\partial z} + \frac{a^2}{2}\frac{\partial^2}{\partial z^2} + \dots\right\}\beta(z,t)$$

where $\Phi(z,t)$ and $\beta(z,t)$ are real smooth functions of a variable z. With the help of (3.10) and (3.11), the total energy, eqn. (3.7), of the peptide group displacements is transformed as follows

$$W \equiv \frac{1}{2a}\int\left[M\left(\frac{\partial\beta}{\partial t}\right)^2 + \kappa\rho^2\right]dz, \tag{3.12}$$

where

$$\rho = -a\partial\beta/\partial z \tag{3.13}$$

is the function which characterizes a decrease, $a \to a-\rho$, of the distance between neighboring peptide groups. Within the same approach, the set of (3.9) transforms to the following

$$\hbar\frac{\partial\Phi}{\partial t}+2aJ\sin ka\frac{\partial\Phi}{\partial z} = 0, \tag{3.14}$$

$$\left[\Lambda+a^2J\cos ka\frac{\partial^2}{\partial z^2}+\chi\rho(z,t)\right]\Phi(z,t) = 0, \tag{3.15}$$

$$\left[\frac{\partial^2}{\partial t^2}-V_0^2\frac{\partial^2}{\partial z^2}\right]\rho(z,t)+\frac{\chi}{\kappa}V_0^2\frac{\partial^2\Phi^2}{\partial z^2} = 0. \tag{3.16}$$

In eqn. (3.16), $V_0 = a\sqrt{\kappa/M}$ is the longitudinal sound velocity of the chain. The excitation energy E is expressed in terms of the spectral parameter Λ of eqn. (3.15) through the relation

$$\Lambda = E-\mathcal{E}_0-W+2J\cos ka. \tag{3.17}$$

The normalization condition (3.6) is rewritten now in the following manner

$$\frac{1}{a}\int\Phi^2(z,t)dz = 1. \tag{3.18}$$

For a fixed $|k| \leqslant \pi/2$, eqns. (3.14)-(3.16) determine the energy and the functions $\Phi(z,t)$ and $\rho(z,t)$ of the states of the chain corresponding to a single intrapeptide excitation with energy $\mathcal{E}_0$. In the particular case of excitations propagating with a constant velocity V, one can introduce a moving coordinate frame

$$\zeta = z-z_0-Vt, \tag{3.19}$$

with an arbitrary origin z_0. Assuming that $\Phi(z,t) = \Phi(\zeta)$ and $\rho(z,t) = \rho(\zeta)$, one obtains

$$\frac{\partial\Phi}{\partial t} = -V\frac{d\Phi}{d\zeta}, \quad \frac{\partial\rho}{\partial t} = -V\frac{d\rho}{d\zeta}. \tag{3.20}$$

The set of eqns. (3.14)-(3.16) transforms to the following set of ordinary differential equations

$$[\hbar V-2aJ\sin ka]\frac{d\Phi}{d\zeta} = 0, \tag{3.21}$$

$$\left[\Lambda+a^2J\cos ka\frac{d^2}{d\xi^2}+\chi\rho(\zeta)\right]\Phi(\zeta) = 0, \tag{3.22}$$

$$\frac{d^2}{d\zeta^2}\left[(1-s^2)\rho(\zeta)-\frac{\chi}{\kappa}\Phi^2(\zeta)\right] = 0. \qquad (3.23)$$

Since $\rho(\zeta) = -a\,d\beta(\zeta)/d\zeta$, one has

$$\frac{\partial\beta}{\partial t} = -V\frac{d\beta}{d\zeta} = V\rho/a. \qquad (3.24)$$

Therefore, the total energy of the chain deformation (3.12) becomes

$$W = \frac{MV_0^2}{2a^3}(1+s^2)\int\rho^2(\zeta)d\zeta. \qquad (3.25)$$

For excitations normalized by the condition (3.6), it follows from eqns. (3.21) and (3.23) that

$$\sin ka = \hbar V/2Ja, \qquad (3.26)$$

$$\rho(\zeta) = \chi\Phi^2/\kappa(1-s^2). \qquad (3.27)$$

Equation (3.26) results in a constraint on the permissible velocity of the excitation motion,

$$V \leqslant V_g = 2aJ/\hbar \qquad (3.28)$$

where V_g is the maximum velocity of a resonance transfer of vibrational excitations between peptide groups.

According to the inequality (3.28), the theory developed is valid for such velocities of excitation transfer which are smaller than the maximum velocity of the resonance transfer of vibrational excitations between peptide groups fixed rigidly in the chain. In this sense, the adiabatic approximation is also justified. If the inequality (3.28) is fulfilled, then according to eqn. (3.26), a wavenumber k depends linearly on the velocity of transfer of the collective excitation,

$$ak = \hbar V/2aJ = V/V_g. \qquad (3.29)$$

Therefore, within the long-wave limit,

$$ka \ll 1, \qquad (3.30)$$

and the adiabatic approximation (3.28) also becomes valid.

Substituting $\rho(\zeta)$ into eqn. (3.22), one can transform the latter into the nonlinear Schrödinger equation for stationary collective excitations moving with a constant velocity V and possessing the energy E, given by eqn. (3.17), as follows

$$\left[\Lambda + \frac{\hbar^2}{2m}\frac{d^2}{d\zeta^2} + G\Phi^2(\zeta)\right]\Phi(\zeta) = 0, \tag{3.31}$$

where

$$\hbar^2/2m \equiv a^2 J, \quad G \equiv \chi^2/\kappa(1-s^2). \tag{3.32}$$

Using eqn. (3.32), one sees that the spectral parameter is determined by the relation

$$\Lambda = \frac{1}{a}\int\left[\frac{\hbar^2}{2m}\left(\frac{d\Phi}{d\zeta}\right)^2 - G\Phi^4(\zeta)\right]dx. \tag{3.33}$$

Assuming $ka \ll 1$, and taking eqn. (3.17) into account, one obtains the total energy of excitation

$$E = \mathcal{E}_0 + W - \frac{\hbar^2}{a^2 m} + \frac{\hbar^2/k^2}{2m} + \Lambda. \tag{3.34}$$

This energy characterizes the bound state of two excitations: the intrapeptide vibrations and the local deformation of the chain under study. In the next Section, we study two-component excitations described by the nonlinear Schrödinger equation (3.31) and eqn. (3.27) for the deformation field.

4. Excitons in a Molecular Chain

We start by investigating the solutions of the nonlinear Schrödinger equation (3.31) for the envelope wave function (3.10). In contrast to linear equations possessing unique solutions at given boundary conditions, the nonlinear equation of the form (3.31) admits a branching of solutions in some cases. As is well known, a multiplicity of solutions of nonlinear equations reflects the possible existence of autonomy relative to the boundary conditions. The solution of eqn. (3.31) of plane-wave type represents itself a certain branch. Solutions of this type are unique for velocities exceeding the longitudinal sound velocity of the chain.

If the velocity of the collective excitation is greater that of the longitudinal sound velocity, i.e., $s^2 > 1$, the chain deformation is unable to follow the excitation. Then the parameter of nonlinearity (3.33) becomes negative, and eqn. (3.31) has no stationary solutions normalized by the condition (3.6). For a lengthy chain with length $Na\,(N \gg 1)$, the boundary effects may be neglected, and the normalized stationary solutions (3.31) become independent of ζ, and hence, are

$$\Phi(\zeta) = N^{-1/2}, \quad 0 \leqslant \zeta \leqslant aN. \tag{4.1}$$

The wave function, given by eqn. (3.10), is therefore a plane wave

$$\varphi_k(z,t) = \frac{1}{\sqrt{N}}\exp\{i(kz - Et/\hbar)\}, \tag{4.2}$$

$$z = z_0 + \zeta + Vt.$$

This is characterized by a wavenumber k, which depends on the velocity V through the relation (3.29) if $ka \ll 1$. In that case, in accordance with eqn. (3.27) all intrapeptide distances a change by the same amount,

$$\rho(\zeta) = \chi / N(s^2 - 1) \ll a, \tag{4.3}$$

which tends to zero as $N \to \infty$. The total energy of the chain deformation, given by eqn. (3.25), and the spectral parameter Λ take now the following values:

$$W = \frac{MV_0^2(1+s^2)\chi^2}{2Na^2\kappa^2(s^2-1)^2}, \quad \Lambda = -G/N. \tag{4.4}$$

Both quantities approach zero as $N \to -\infty$.

The excitations corresponding to the dipole intrapeptide vibrations and described by plane waves (4.2), are called *excitatons*. In these stationary states, all peptide groups are excited with the same probability N^{-1}. The stationary excitations of this type depend strongly on the boundary conditions, which determine a discreteness of values of the wavenumber k with an appropriate step $\Delta k = 2\pi / N$.

The total energy of excitations of exciton type is given by the formula (3.34) with

$$W + \Lambda = \frac{\chi^2}{\kappa N}\left[\frac{s^2}{1-s^2} - \frac{1}{2}\right].$$

As $N \gg 1$, the total energy of the exciton really involves the energy of interpeptide vibrations and takes the form

$$E_{ex}(V) = E_{ex}(0) + \frac{1}{2}mV^2 + 0(1/N), \tag{4.5}$$

where

$$E_{ex}(0) \equiv \mathcal{E}_0 - \hbar^2/ma^2. \tag{4.6}$$

The energy levels corresponding to different values of $k = mV/\hbar$ form the *exciton band*. Here, m is the *effective mass of the exciton* in the band, and $E_{ex}(0)$ is the energy of its 'bottom'. By virtue of (3.28), V is upper bounded by

$$V_g = \hbar/ma, \tag{4.7}$$

which corresponds to the maximum group velocity of the exciton in the band.

If $s^2 < 1$, plane waves of the type (4.2) are also the solutions of eqn. (3.31). However, these excitons are metastable since there exist also more stable states having smaller energy (see Chapter 2). Excitons arising in the given chain may be excited by radiation of frequency ω if the conservation laws for energy $E(k) = \hbar\omega$ and quasi-momentum $|k| = \omega/c$ are fulfilled, and if the projection of an electric field into the transition dipole d is nonvanishing. For infrared and visible light, the inequality $a\omega/c \ll 1$ applies. Hence, excitons with small values of ka are excited.

Nevertheless, one should bear in mind that at the moment of transition of the chain considered into its excited state under radiation, the peptide groups are unable to displace into their novel equilibrium positions, $a - \rho(\zeta)$ (the Franck-Condon principle). Therefore, in a rigorous sense, transitions to the states of the total Hamiltonian of the system are forbidden, and, instead, there occur transitions to the states of the Hamiltonian with $\chi = 0$ which describes states of the chain with fixed equilibrium positions na. The eigenfunctions of the latter Hamiltonian are also plane waves of the type (4.2), and their energies are

$$E_{ex}(V) = E_{ex}(0) + \frac{1}{2}mV^2. \tag{4.8}$$

As $N \gg 1$, these coincide with the energies given by eqn. (4.5). However, such excitons refer to nonstationary ones. An inclusion of terms depending on the parameter χ, results in transitions between the states with different k due to the participation of phonons. If $s^2 > 1$, the excitons are damped with the emission of sound phonons (Cherenkov radiation).

Excitons remain unstable even if they have the lowest energetic state of the excitonic band with $k = 0$, since there exists a nonvanishing probability of their transition into more stable, autolocalized states, the so-called spatially inhomogeneous excitations. In order to study this case in detail, let us transform the nonlinear Schrödinger equation (3.31) to the following one,

$$\left[i\hbar\frac{\partial}{\partial t}+\frac{\hbar^2}{2m}\frac{\partial^2}{\partial x^2}+G_0\Phi^2\right]\Phi(x,t) = 0, \quad G_0 = \alpha^2/\kappa, \tag{4.9}$$

and introduce the following quantities,

$$g \equiv \hbar^2/2mG_0, \quad \tau \equiv G_0 t/\hbar. \tag{4.10}$$

Equation (4.9) gives the envelope of the wave function (3.10) of the stationary states with energy $\mathcal{E} = -\Lambda$. One obtains further the equation for the complex function

$$\left[-i\frac{\partial}{\partial \tau}+g\frac{\partial^2}{\partial x^2}+|\Phi(x,\tau)|^2\right]\Phi(x,\tau) = 0. \tag{4.11}$$

The exciton-type solution of this equation,

$$\Phi(x,\tau) = \Phi_0 \exp(-i|\Phi_0|^2\tau) \tag{4.12}$$

describes the regular distribution of excitations in the chain with length aN. The corresponding probability density takes the form

$$|\Phi_0|^2/a = (aN)^{-1}. \tag{4.13}$$

Following Scott et al. [259], we investigate now the stability of the solutions (4.12) relative to negligible inhomogeneous modulations $b(x,\tau)$ of the amplitude Φ_0. For this purpose, let us substitute the solution of the form

$$\Phi(x,\tau) = [1+b(x,\tau)]\Phi_0 \exp(-i|\Phi_0|^2\tau) \tag{4.14}$$

into eqn. (4.11). Taking into account the first-order terms of the perturbation $b(x,\tau)$, eqn. (4.11) transforms to the linear equation,

$$\left[-i\frac{\partial}{\partial \tau}+g\frac{\partial^2}{\partial x^2}\right]b(x,\tau)+|\Phi_0|^2[b^*(x,\tau)+b(x,\tau)] = 0. \tag{4.15}$$

Substituting the function

$$b(x,\tau) = b_1 \exp[i(kx+\Omega\tau]+b_2 \exp[i(kx+\Omega^*\tau)] \tag{4.16}$$

into eqn. (4.15), one can show that this is just the solution of this equation if the following dispersion relation

$$\Omega^2 = gk^2(gk^2 - 2|\Phi_0|^2) \tag{4.17}$$

holds. If the inequality

$$k^2 > k_{\rm rc}^2 \equiv 2|\Phi_0|^2/g \tag{4.18}$$

is valid, the frequency Ω of the function (4.16) becomes real. Hence, the function (4.14) describes a stable modulated wave with $k = 0$.

In the opposite case,

$$k^2 < k_{\rm cr}^2, \tag{4.19}$$

Ω becomes imaginary. Hence, small initial modulations increase in the course of time, and the plane-wave-type solutions of the form (4.12), corresponding to an immobile soliton, are unstable relative to small spatial modulations.

In the next Section we study the other, soliton-type solutions of eqn. (4.9)

$$\Phi(x,t) = \sqrt{\frac{1}{2}}\, aQ\,\mathrm{sech}(Qx)\exp(-i\omega t) \tag{4.20}$$

with

$$Q = maG_0/2\hbar^2, \quad \omega = \hbar Q^2/2m. \tag{4.21}$$

The function $\Phi(x,t)$ represents an excitation which is localized on the segment $\Delta x = 2\pi/Q$. We recall that the critical value $k_{\rm cr}$ determines the condition of instability of the exciton in the chain with length aN, in accordance with (4.18). Hence, it is related with a linear dimension, Δx, of the soliton via the following simple formula

$$k_{\rm cr} = 4(\pi N/a\Delta\chi)^{1/2}. \tag{4.22}$$

Excitonic states with a definite magnitude of energy are uniformly distributed over the entire length of the chain. For this reason, they are unable to transfer energy. Excitations localized on a small segment of the chain are nonstationary, and they are describable by the superposition of plane waves, or in other words, by a wave packet composed of plane waves given by eqn. (4.2)

$$\psi(z,t) = \int_{k_-\Delta k}^{k_0+\Delta k} C(k)\varphi_k(z,t)dk, \quad \Delta k = \pi/2l_0. \tag{4.23}$$

Excitation of the type (4.23) propagates with a group velocity

$$v_{gr} \equiv \hbar k_0 / m \tag{4.24}$$

along a chain. Due to the difference in phase velocity $\hbar k/m$ of individual components φ_k of a given wave packet, the region enveloped by the excitation, expands, or runs. This 'expansion' of the wave packet (4.23) is one of the features peculiar to nonstationarity of wave-packet states.

5. Solitons in a Molecular Chain

It is clear that if $s^2 < 1<$ the parameter of nonlinearity of eqn. (3.31) is positive. This equation then possesses the following normalized particular solution

$$\Phi(\zeta) = \sqrt{aQ/2}\,\mathrm{sch}(Q\zeta), \tag{5.1}$$

where

$$Q = maG/2\hbar^2 = ma\chi^2/2\kappa h^2(1-s^2), \tag{5.2}$$

$$\Lambda = -\hbar^2 Q^2/2m. \tag{5.3}$$

Due to eqn. (3.24), the decrease of intrapeptide separations is determined by the function

$$\rho(\zeta) = a\chi Q/2\kappa(1-s^2)\mathrm{ch}^2(Q\zeta). \tag{5.4}$$

The displacements of peptide groups from equilibrium positions is given by the following formula

$$\beta(\zeta) = \frac{\chi}{\kappa(1-s^2)}[1-\mathrm{th}(Q\zeta)]. \tag{5.5}$$

An excitation described by the amplitude function (5.1) is called a *soliton*. It is mainly distributed in a segment $\Delta\zeta \approx 2\pi/Q$ relative to the coordinate frame ζ which moves together with the excitation, with velocity $V = sV_0 < V_0$. The intrapeptide separations belonging to this segment are reduced. Therefore, the solitonic excitations are substantially two-component, and they displace intrapeptide vibrations as well as producing a local deformation along the chain.

Solitonic-type solutions give a second stable branch of solutions of the

Schrödinger equation. In a certain sense, these solutions are isolated from the boundary conditions due to their localization on a rather small region of the chain.

A soliton is described by a wave whose profile $\Phi(\xi)$ is unchanged under propagation. Such waves refer to stationary ones, for which the following relation holds,

$$\frac{\partial\Phi}{\partial t} = -V\frac{\partial\Phi}{\partial x},$$

where V is the velocity of wave propagation.

Due to the existence of the factor $(1-s^2)$ in the denominators of eqns. (5.2) and (5.4), the range of applicability of the theory is restricted. This constraint is related to larger displacements (5.5) of equilibrium positions which are inconsistent with the harmonic approach applied in eqn. (3.3). We demonstrate in Chapter 5 that a more rapid increase of the interaction energy for peptide groups becoming clear to each other, comparing with a quadratic law, provides finite values of the dynamical properties of solitons moving with an arbitrary velocity, which is less or equal to the longitudinal sound velocity.

Let us now evaluate the energy of a soliton moving with a constant velocity V. Substituting (5.1) into eqn. (3.25), one obtains the energy of chain deformation

$$W = \frac{ma^2\chi^4(1+4s^2)}{12\kappa^2\hbar^2}, \quad s^2 \ll 1. \tag{5.6}$$

Using eqns. (5.3) and (5.6), the total soliton energy is rewritten as follows

$$E_{\mathrm{sol}}(V) = E_{\mathrm{sol}}(0)+\frac{1}{2}M_{\mathrm{sol}}V^2, \quad s^2 \ll 1, \tag{5.7}$$

where

$$E_{\mathrm{sol}}(0) = E_{\mathrm{ex}}(0)-\frac{ma^2\chi^4}{24\kappa^2\hbar^2} \tag{5.8}$$

is the zeroth soliton energy, and

$$M_{\mathrm{sol}} = m\left[1+\frac{M\chi^4}{6\kappa^3\hbar^2}\right] \tag{5.9}$$

its effective mass.

The energy of an immobile soliton (5.8) differs from that of an immobile exciton (4.6) by the quantity

$$\Delta E = E_{ex}(0) - E_{sol}(0) = \frac{ma^2\chi^4}{24\kappa^2\hbar^2}. \tag{5.10}$$

ΔE becomes rather large for a strong coupling χ of intrapeptide vibrations with displacements of peptide groups and small values of elasticity κ of longitudinal deformations. The smallness of κ in proteins is provided by a weakness of hydrogen bonds which link peptide groups with the molecular chain.

The effective soliton mass, given by eqn. (5.9), exceeds considerably that of an exciton, m, in soft chains since the motion of a soliton is accompanied by the local deformation of the chain. Due to the larger soliton mass, a soliton is able to transfer a considerable part of its kinetic energy, even for a small velocity of motion.

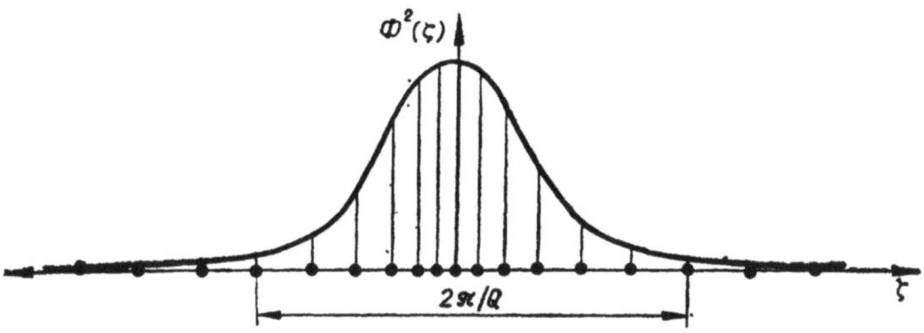

Figure 5. The distribution of probabilities of intramolecular excitations (vertical lines) and distances between peptide groups in the region of the soliton excitation.

A qualitative distribution of the probabilities of intramolecular excitations and interpeptide distances in the region of soliton excitation localization are plotted in Fig. 5 (within the coordinate frame ζ). Figure 6 demonstrates the dependence of soliton and exciton energies on the velocity. One should recall here that the theory based on the harmonic approach for describing the interaction between peptide groups, is justified for small soliton velocities, $V \ll V_0$.

The existence of the energy gap (5.10) in the spectrum of the excited states of the molecular chain is one of the reasons for the great stability of a soliton. The breakdown of a soliton, i.e., its splitting into a free exciton and its homogeneous deformation, requires an expense of the energy (5.10). Solitons do not emit phonons since they always move with a velocity less than that of the longitudinal sound velocity in the chain. In other words, their kinetic energy is not

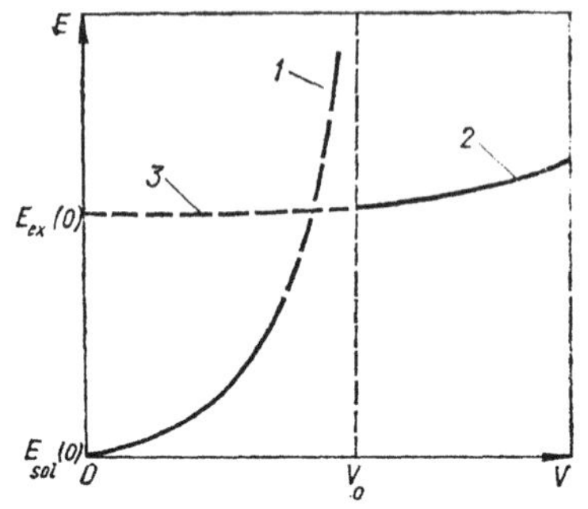

Figure 6. The dependence of the soliton and the exciton energies on their velocities: (1) the soliton energy; (2) the exciton energies; (3) the energy of metastable excitons; and (V_0) the longitudinal sound velocity.

transformed into heat energy. This is a second important peculiarity which results in the high stability of solitons in soft molecular chains. Finally, the third reason for the high stability of solitons under study consists in their *topological stability*. The displacements of equilibrium positions of peptide groups in the chain are given by the function (5.5). Therefore, to the right of the soliton, $\zeta > 0$, all peptide groups are in their nondisplaced positions na, and to the left, $\zeta < 0$, they are displaced by the same distance

$$\beta_0 = 2\chi/\kappa(1-s^2).$$

Therefore, as $n \to \infty$, all peptide groups occupy the positions na, and as $n \to -\infty$, their positions are characterized by $na+\beta_0$. In order to annihilate a soliton, one should return all peptide groups in the left side of chain to their initial positions na. A topological stability can be explained by the analogy of the impossibility of untieying a knot on an infinite rope without cutting it.

The bell-like shape of the soliton amplitude (5.1) is independent of the manner of its generation. It is self-consistent due to the compensation in eqn. (3.31) of dispersion (the operator $\hbar^2(2m)^{-1}d^2/d\zeta^2$) which is responsible for the dispersion of excitation, and nonlinearity (the operator $G\Phi^2(\zeta)$) which localizes it. For $s^2 \ll 1$, the soliton shape remains unchanged even if its kinetic energy decreases. Conversely, the form of the wave packet given by eqn. (4.23) is determined by the initial distribution of amplitudes $C(k)$. A wave packet is 'blurred' in the course of time.

As the energy of resonance interaction, $\hbar^2/2ma^2$, and the elasticity κ of the chain, increase, the difference in the properties of solitons and excitons becomes smaller. The region embraced by the excitation, $\Delta\xi \approx 2\pi/Q$, increases, and the effective mass of the soliton approaches that of the exciton. The internal soliton energy approaches the energy $E_{ex}(0)$ of the bottom of the exciton band.

As is well known, absorption of light by molecular systems is not accompanied by a variation of the coordinates of heavy particles at the moment of quantum transition (Franck-Condon principle). A soliton formation is related to displacements of equilibrium positions of peptide groups. Thus, light is unable to excite solitons. The probability of emission of light by a soliton is negligible for the same reason, too. The theory of resolving this problem has been developed by Eremko and the present author [44].

Solitons may be excited by local external fields, e.g. under chemical reactions. If such local effects act at the end of the molecular chain, the probability of soliton excitation becomes the largest. This statement is based on the soliton topological stability.

6. Solitons in Real α-Helical Molecules

We recall that real α-helical molecules involve three parallel chains of peptide groups linked by hydrogen bonds. So, in order to formulate an adequate theory of describing the excited states of a protein molecule, one needs to include the resonance interaction between peptide groups belonging to different chains. This approach has been developed in the papers [45,61,62,254,255] and presented in detail in the books [38,49,282].

A solution of the equations which include the resonance interaction of peptide groups belonging to a single chain and to different ones, and variations of the pitch and diameter of the helix as well, demonstrate that together with exciton-type solutions being unstable at small velocities, the formation of stable and stationary solutions becomes possible due to the intrapeptide vibration of Amide-I type in the α-helical molecule. The latter solutions describe excitations moving along a molecule with a velocity less than the longitudinal sound velocity. These excitations are divided into three types: *symmetric solitons* describing the excitation moving in phase along three chains, and two types of *asymmetric solitons* for which the excitation propagates along three chains with a rather small difference in phases.

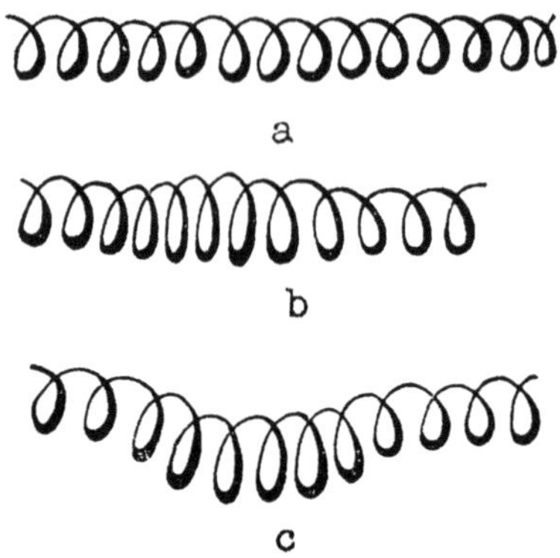

Figure 7. Migration of a soliton in a protein molecule; (a) unexcited helix protein molecule; (b) migration of a symmetric soliton; (c) migration of an asymmetric soliton.

For symmetric solitons, the helical pitch decreases and its diameter increases in the region of excitation moving with constant velocity (see Fig. 7). In the case of asymmetric solitons, the helical diameter increases and a local bending of a molecule occurs there.

In order to obtain the result in analytical form, infinitely long molecules have been studied within the continuum approach. The investigation of restricted and discrete molecules has been carried out numerically using computers. This numerical study of soliton excitations in proteins yield considerable progress to a more complete understanding of these states.

The numerical integration of discrete differential equations given by Davydov et al. [45], was carried out firstly by Hyman et al. [202] in 1979. The authors [202] concluded that 'a numerical study of the Davydov nonlinear dynamical model of α-helix proteins confirm his prediction regarding formation of solitons'. They studied symmetric excited states in the three chains of the helical protein molecule. Each chain involves 200 peptide groups. The following parameters of a protein molecule were chosen:

$$\mathcal{E}_0 = 0.205\, eV \quad J = 1.55 \times 10^{-22} \text{ joule}$$

$$M = 70 m_p L = 2.46 \times 10^{-22} \text{ joule}$$

$$V_0 = 10^4 ms^{-1}.$$

The authors [202] used the following initial conditions ($t = 0$):

$$\beta_{n\alpha} = 0, \quad \varphi_{n\alpha} = \begin{cases} 1 & \text{if } n=1 \\ 0 & \text{if } n \neq 1 \end{cases}. \tag{6.1}$$

The numerical studies were carried out with different values of the coupling parameter of intrapeptide excitations with displacements of the equilibrium positions. It was revealed that under the initial conditions (6.1), the well-distinguished solitons are formal and displaced in a given molecule if $\chi \geqslant 0.3 \times 10^{10}$ newtons. Solitons with values of χ closer to the threshold value move with a velocity of about $1.5 \times 10^5 cm\, s^{-1}$. Hence, neglecting friction and the work produced, one can conclude that these solitons could travel a distance of 1,700Å (which corresponds to the length of the α-helical myosin molecule in muscle fibers) during $1{,}000 ps$ [254]. Thus the paper [202] is of interest in that the study of the process of the formation of solitons from a definite initial state was carried out and elucidated the role of discreteness of that chain.

In 1979, Eilbeck, aided by Scott, produced a computer film [170] which demonstrates the propagation of the internal vibrational excitation of the end group along a chain of peptide groups. This film shows visually that under a superthreshold value of the coupling parameter χ, a soliton-type excitation propagates along a given molecule. This excitation has the form of a local impulse whose shape and width, of the order of a few peptide groups, remain unchanged under motion. Eilbeck's film is of importance for the following two reasons. Firstly, it confirms the earlier numerical studies carried out at the Los Alamos Laboratory [202]. And, secondly, it clearly demonstrates the stability of solitons relative to their interaction with acoustic waves, which were excited simultaneously with the soliton. Moving more rapidly than the soliton, the acoustic wave is reflected a few times by the chain ends and passes through the soliton leaving it unchanged. Eilbeck's film and the numerical study [202] also emphasize that a soliton generates at the very start of a peptide chain. Therefore, solitons can arise inside comparatively short segments of α-helical proteins.

The necessity to overcome the threshold value of the coupling parameter χ for exciting a soliton revealed in numerical studies, is associated with conditions under which a soliton is excited, as was shown by Brizhik and the present author [21,22,147]. According to theory based on the continuum approach, the slow solitons could exist in an infinite chain for any value of χ. As χ decreases, the properties of the soliton approach those of the exciton continuously.

Numerical studies carried out by Lomdahl and coworkers [226] confirmed the particle-like properties of solitons. Given two peptide groups excited at opposite ends of a protein molecule, two solitons arise and move in opposite directions. Colliding, they preserve their shape and velocity. Therefore, the collision of solitons may be likened to the elastic collision of two particles.

To be rigorous, one should emphasize that the condition for the formation of a soliton-type excitation which is not so strongly localized, is determined by the magnitude of the dimensionless nonlinearity parameter,

$$2g_0 = G_0/J = \chi^2/\kappa J \tag{6.2}$$

at small velocities ($s^2 \ll 1$). A larger value of the resonance interaction J leads to a greater tendency of distribution of excitation over a chain. On the other hand, increasing the ratio χ^2/κ facilitates the localization of excitation. For a rather small parameter g_0 of nonlinearity, localization is manifested in a sufficiently weak manner. Conversely, the excitation is localized at a single peptide group if the nonlinearity parameter is rather large. This statement has been confirmed by numerical studies of Scott and his coworkers [202] who demonstrated that the excitation is localized at the end of the chain for rather large values of χ, and fixed κ and J.

Therefore, the smooth localized excitations of soliton-type which become insensitive to a discrete structure of chain correspond to average values of the nonlinearity parameter. If the nonlinearity parameter increases, a discreteness of chain begins to be manifested as much as possible. Kuprievich [216], Kislukha [76], and Vakhnenko and Gaididei [26] demonstrated that a chain discreteness is manifested in a form of effective periodic relief whose depth increases as the nonlinearity parameter does. This effect has been investigated in detail by Vakhnenko and Gaididei [26] on the basis of the set of equations (3.9) given in Section 3. The authors [26] consider the case of rather narrow exciton bands where the maximum velocity of exciton in the band, $V_g = 2aJ/\hbar$, is considerably smaller than that of the longitudinal sound velocity, $V_0 = a(\chi/M)^{1/2}$, i.e., if the inequality

$$V_g/V_0 = (\hbar)^{-1}2J(M/\kappa)^{1/2} \ll 1 \tag{6.3}$$

holds.

The theoretical studies mentioned above are based on a quasiclassical approach within which displacements of molecules of a chain from their equilibrium positions are described by the language of coherent displacements ignoring quantum fluctuations, or zeroth vibrations. A quantum theory of solitons which takes fluctuations and heat motion of molecules in the chain into account is presented in Chapter 9.

Chapter 2

SOME PHENOMENA WITH PARTICIPATION OF MOLECULAR SOLITONS

7. Solitons and Molecular Mechanism for Muscle Contraction

One of the most interesting problems in bioenergetics consists of elucidating the molecular basis of the conversion of chemical energy released in ATP hydrolysis into the energy of motion. The present author applies the concept of soliton motion in helical protein molecules [33,34,38,155,160,162] to explain on a molecular level the mechanism of contraction of transversely striated muscles.

Skeletal muscles of vertebral animals form a tissue which involves millions of muscle fibres. A large part of the muscle fibre volume is occupied by so-called contraction units, or *myofibrils*. Each muscle fibre is surrounded by a membrane which is called the sarcoplasmatic membrane, or *sarcolemma*. Myofibrils consist of consecutively arranged identical units, or *sarcomers*.

Sarcomers are separated from each other by transverse membranes. Thin protein filaments are stretched in a sarcomer, in both directions relative to the membranes. They penetrate partially into a volume between thick protein filaments arranged in the internal region of a sarcomer. Thick filaments are formed by a bundle of myosin molecules, one end of which is thickened and called the head of the molecule. The rest of the molecule, the tail, with a length of about 1,700Å, has a α-helix structure. Myosin molecules are packed in such a way that their heads are directed towards the ends of a thick filament and slightly protrude from the filament surface over its entire length, except for the middle part.

In a muscle tissue, myofibrils are embedded into a liquid, or *sarcoplasma*, which involves mitochondria and a reservoir of glycogene. The latter is utilized by mitochondria to synthesize ATP molecules. Muscle tissue contains also a system of transversal tubes which originate from the protuberance of sarcolemma. A dense network of finest tubes is arranged parallel to myofibrils. These tubes set against the transversal ones by boundary segments called *cisterns*. A region of a transversal tube together with two neigboring cisterns constitutes the so-called *triad structure*. Boundary cisterns comprise calcium ions.

Transversal tubes transfer a nervous impulse in the form of a depolarization wave inside a tissue. In triad structures, a depolarization of transversal tubes results in a sharp increasing of Ca^+ permeability of the membranes of boundary cisterns. Further, calcium ions are transferred into cytoplasma where their concentration is lower. As the concentration of calcium ions in cytoplasma increases, the hydrolysis of ATP molecules begins, for the ATP molecules joined with heads of myosin protein molecules constituting thick filaments. The energy of hydrolysis of ATP molecules induced contraction of sarcomers.

Electron microscope studies have established that a contraction of the sarcomer length is related to the penetration of thin filaments into a volume between the thick ones; the lengths of the thin filaments remaining unchanged. Thus, there arises the problem of explaining on a molecular level the mechanism of the sliding of thin filaments relative to the thick ones. Which forces are responsible for such sliding? What is a pathway of transforming ATP hydrolysis energy into that of sliding?

The current point of view, supported by biologists, is the following: a sliding of thin filaments relative to the thick ones is due to an active motion of the heads of myosin molecules forming the thick filaments. It is suggested that under ATP hydrolysis, the head of a myosin molecule elongates and forms a bond, or 'bridge' with a globular actin molecule being incorporated into a thin filament. A head turns further and displaces a thin filament towards a sarcomer centre. Finally, a head detaches from the action molecule and returns to its initial shape and position within a thick filament.

The notion of cross-bridges between thick and thin filaments which are closed, which pull thin fibres causing their displacements, and become further broken, is unable to explain the molecular nature of such a phenomenon. In particular, the following questions remain unsolved. By what manner is the ATP hydrolysis energy expended on the lengthening of heads, bond formation, pull, and bridge breakdown? What is the molecular mechanism of changes occurring in a myosin molecule head resulting in these phenomena? Finally, why does the head of a huge myosin molecule alone take an active part in the sliding mechanism, since as is well known, the molecular weight of a myosin molecule is 500,000, and contains a lengthly 'tail' formed by α-helix protein, and one can expect that this tail will also take part in sliding?

Based on theoretical studies of solitons in helical protein molecules, the present author suggested the original hypothesis to elucidate the mechanism of contraction of a length of sarcomers. According to this hypothesis by a diffusion process, calcium ions initiate the hydrolysis of ATP molecules attached to an initial series of heads of myosin molecules at the boundaries of thick filaments. The energy released results in the formation of asymmetric solitons in long helical regions of myosin molecules. These solitons further propagate from the heads of myosin molecules to their ends which are situated in a sarcomer centre. Motion of the asymmetric soliton is accompanied by local bending and swelling of a given molecule. The heads of the myosin molecules at rest situated over bent regions of helical molecules attach to thin filaments causing their slight displacements in the direction of the soliton motion, i.e., towards a sarcomer centre.

According to this model, the heads of myosin molecules attach and detach further from thin filaments, as in the model of formation and breakdown of bridges. The origin of this motion is not, however, due to lengthening, rotation, and contraction of the lengths of the heads, but due to the movement of solitons inside a thick fibre which is accompanied by the motion of bent regions of helical molecules. The kinetic energy of solitons thus transforms into the energy of contraction, or results in a tension if a muscle is under load. Within the model proposed, all parts of a myosin molecule, including its head, are active units of the contraction mechanism.

Having expended their kinetic energy on the work necessary for contraction of muscle fibre, the solitons are damped and eventually annihilated, and brought to rest near the centre of thick filaments whereupon their rest energy is converted to heat motion. Therefore, the kinetic component of the soliton energy is solely used for the contraction of muscle fibres.

Vibrometers by Ragulskis and his coworkers from the Research Institute 'Vibrotechnique' of the Kaunas Polytechnical Institute [240] serve a good illustration of the mechanism of the displacement of thin filaments due to the movement of the swollen region of thick filaments. In these motors, the elastic vibrations excited in piezoelectric films and possessing the form of pulses, move along the film surface and, affected by sliding friction, cause rotational or translational motion of the bodies in contact with them.

8. Solitons and the Molecular Mechanism of the Effect of Radiation on Cells

Devyatkov and his coworkers [56,57] revealed the effect of electromagnetic waves with millimeter wavelengths on the growth rate of some bacteria. The resonance behavior of the interaction of millimeter electromagnetic radiation with living organisms has also been observed by others.

The effect of electromagnetic radiation of a very low intensity on living organisms, which does not induce any appreciable heating of tissue, is usually called the nonheating effect. The electric field of radiation with small strengths, of about $10^{-7}\ Vcm^{-1}$, is unable to affect directly biological objects whose intracellular fields are $10^5\ Vcm^{-1}$.

One of the essential peculiarities of the nonheating effect of radiation on living organisms is its sharp resonance nature. Devyatkov emphasizes [56]: 'The explanation of the mechanism of the resonance radiation effect as well as other related peculiarities is of great importance for science. A rigorous scientific explanation of the effect of a millimeter-wavelength electromagnetic field has not yet been given'.

A novel hypothesis for elucidating the mechanism of such a resonance effect was suggested by Eremko [63,174]. This hypothesis is based on the concept of solitons which play an active role in cell vitality. We have already pointed out that solitons are ideal carriers of the hydrolysis energy of ATP molecules along proteins. Eremko showed that under the influence of electromagnetic radiation with definite wavelengths, solitons can transform to metastable excitons. The latter are rapidly decelerated, transmitting their energy to their environment. Therefore, the transformation of solitons into excitons violates the effectiveness of the transmission of energy and information.

The energy of transition from the soliton state with a local deformation of chain (with $s^2 \approx 0$) to the bottom of the exciton band without deformation is given by the quantity

$$\Delta E = \frac{ma^2\chi^4}{24\kappa^2\hbar^2} = \chi^4/48\kappa^2 J \tag{8.1}$$

in accordance with eqn. (5.10). The optical transition for the soliton to the exciton state occurs so rapidly that the local deformation has no time to annihilate. For this reason, together with ΔE, the energy of the optical transition should involve the deformation energy W which is equal to $2\Delta E$ by virtue of eqn. (5.6). Hence, the total energy of transition from the soliton to the exciton state with a wavenumber k is given by the formula

$$\hbar\omega_k = 3\Delta E + \hbar^3 k^2 / 2m. \tag{8.2}$$

Eremko demonstrated [174] that for slow solitons ($s^2 \ll 1$), the probability of absorption of electromagnetic radiation with frequency ω, wavenumber k, and electric field amplitude is determined by the following expression

$$\mathcal{P}(\omega) = \sum_k |\vec{\mathcal{E}}\vec{D}_k|^2 G_k(\omega), \tag{8.3}$$

where

$$\vec{D}_k = \vec{D}\operatorname{sech}(\pi k / 2Q) \tag{8.4}$$

is the dipole electric moment of quantum transition from the soliton state to the exciton state with wavenumber k. In eqn. (8.4) the parameter $1/Q = 4a\kappa J/\chi^2$ characterizes the spatial width of the soliton (see Section 5). The function $G_k(\omega)$ depends on the dimensionless parameter

$$\mu \equiv 2\chi^2 M^{1/2} / \hbar\chi^{1/2}. \tag{8.5}$$

In eqn. (8.3) a summation is carried out for all values of wavenumber belonging to the exciton band. Due to the presence of the factor (8.4) in eqn. (8.3), the excitation of the exciton with wavenumber $k \approx 0$ becomes most probable if $\vec{\mathcal{E}} \| \vec{D}$. Therefore, the resonance transition frequency takes the value

$$\hbar\omega_0 = 3\Delta E. \tag{8.6}$$

Due to eqn. (5.5), the square of the maximum displacement of the equilibrium positions of molecules (peptide groups) is determined by $\beta^2_{\max} = \chi^2/\kappa^2$ under a soliton excitation, and the amplitude of elastic zeroth vibrations of a particle of mass M in the chain with elasticity coefficient κ is equal to $<x^2> = \hbar/2\sqrt{M\kappa}$. Therefore, one can argue that the dimensionless parameter (8.5) is given by the ratio

$$\mu = \beta^2_{\max} / <\chi^2>. \tag{8.7}$$

The function $G_k(\omega)$ involved in eqn. (8.3), and described by simple functions in the two following limiting cases: (i) $\mu \gg 1$, and (ii) $\mu \ll 1$, with k near a resonance ($k \approx 0$). The first case, $\mu \gg 1$, corresponds to that when the square of the maximum displacement from the equilibrium position is considerably larger compared with the square of the amplitude of zeroth vibrations. In this case, the function $G_0(\omega)$ takes a Gaussian form,

$$G_0(\omega) = \hbar^{-1}(2\pi / B)^{1/2} \exp\left[-\frac{(\omega - \omega_0)^2}{2B^2} \right], \tag{8.8}$$

where the quantity B, characterizing the width of the absorption band, is

$$B^2 = 3.6\chi^2 Q^2 V_0 / \kappa a \hbar \pi^3. \tag{8.9}$$

In the second case, $\mu \ll 1$, when the square of the maximum displacement of the equilibrium position of the molecules is considerably less than the square of the amplitude of the zeroth vibrations, the function $G_0(\omega)$ takes the Lorentz form,

$$G_0(\omega) \approx \frac{2\pi^{-1}\gamma_0}{(\omega - \omega_0)^2 + \gamma_0^2} \tag{8.10}$$

with

$$\gamma_0 \approx \chi^6 \sqrt{M} / 48\hbar^2 \kappa^{3/2}. \tag{8.11}$$

Using the values of the parameters inherent for α-helix proteins given by Scott,

$$J = 19.5Nm^{-1}, \quad \chi = 8 \cdot 10^{-11} N, \quad \text{and } M = 70 m_p,$$

one obtains

$$\mu \approx 16 \times 10^{-2}. \tag{8.12}$$

A rather small probability for the transformation of the soliton into an exciton can take place under Raman scattering of light. Vakhnenko et al. [172] have demonstrated that the frequency of scattering of light by solitons shifts into the long-wave region. This frequency shift, $\Delta\omega$, depends on the value of μ given by eqn. (8.5). For this process, a photon expends its energy on the transformation of a soliton into an exciton with wavenumber $k \approx 0$.

If $\mu \gg 1$, the frequency shift is determined by the following formula

$$\Delta\omega_1 = 3\Delta E/\hbar, \tag{8.13}$$

where ΔE is given by eqn. (8.1). In this case, due to the Franck-Gordon principle, there occurs a dissociation of the soliton whilst preserving the local chain deformation. For this reason, the energy of the final state is given by the sum of the energy of soliton formation, ΔE, and the deformation energy, $2\Delta E$. Therefore, the differential cross-section of scattering, $(d^2\sigma/d\omega d\Omega)_{\omega=\omega_1}$, of the photon with frequency $\omega_1 = \omega_0 - \Delta\omega_1$ becomes proportional to $(J/8\pi\hbar B)^{1/2}$ with B given by eqn. (8.9).

In the opposite case, $\mu \ll 1$, the frequency shift is rather weak,

$$\Delta\omega_2 = \Delta E/\hbar \tag{8.14}$$

and the energy of the final state corresponds only to that of soliton formation, without the energy of local deformation.

One can suggest that the second case, with

$$\Delta\omega_2 \approx (0.4 - 0.7) cm^{-1},$$

takes place for vibrational solitons of α-helix protein molecules which correspond to the intrapeptide vibrations of Amide-I type. The differential cross-section of scattering of a photon with frequency $\omega_2 = \omega_0 - \Delta\omega_2$ is given by the formula

$$(d^2\sigma/d\omega d\Omega)_{\omega=\omega_2} \sim (\gamma_0(\kappa M)^{1/2}\alpha^{-2})^{4\mu}(J/2\hbar\gamma_0)^{1/2},$$

where γ_0 is given by (8.11).

9. A Possible Mechanism for General Anesthesia

A large class of molecules, utilized in intravenous anesthetics, form hydrogen bonds with protein molecules situated inside and outside a cytoplasmatic membrane. The binding of anesthetic molecules to proteins inhibits the normal function of a cell. One of the possible mechanisms of such an inhibition by the binding of anesthetic molecules, barbiturates, to proteins was proposed by Layne [227] in 1984 on the basis of a soliton model.

Molecules of barbiturates contain the $H-N-C=O$ groups of atoms which are very similar to the peptide groups in proteins. When a barbiturate molecule attaches to a protein the hydrogen bonds between neighboring peptide groups in a protein molecule are broken, and there new hydrogen bonds are produced with the oxygen and hydrogen atoms of the barbiturate molecule. At a site of dissociating hydrogen bonds of the protein molecule, the interpeptide separation increases, and this results in the violation of the free motion of a soliton, and hence, to the interference of the process of transduction of information and energy along a protein molecule.

Layne suggested that a decrease of efficiency of the transduction of energy and information by solitons under the binding of barbiturates to proteins, is of importance for the two following cell structures:

(a) in the helical proteins of the inner mitochondrial membranes which appear to participate in ATP synthesis;

(b) in the membrane helical proteins which are responsible for chemical reception and transduction of signals inside a cell.

10. Intracellular Dynamics and Solitons

The α-helix structure is often found at those sites of protein molecules where the transduction of energy from one end of a molecule to the other takes place, and where a protein molecule couples a few processes, too. For instance, a haemoglobin molecule of red blood corpuscles transmitting oxygen, involves 32 helix segments. A bacteriorhodospin molecule incorporated into the membranes of Halobacteria which live in salty lakes and basins, intersects the cell membrane seven times in the form of parallel α-helix segments. α-helix protein segments incorporate into cytochrome and proteins intersecting inner membranes of mitochondria.

The modern investigation of the spatial structure of cells demonstrate that a whole cellular interior is spanned by a network of protein microfilaments and microtubes which retain all intracellular organelles in a definite position, determine the shape of cells, any changes, and all motions inside a cell. Such a network of microfilaments and microtubes is called a *cytoskeleton*.

Protein molecules incorporated into a cytoskeleton realize the transduction of energy and all intracellular coupling. All these processes expensed the energy released in the hydrolysis of ATP molecules. The hydrolysis process takes place with the participation of enzymes and is controlled by a variation of the concentration of calcium ions, in a manner similar to muscle fibres.

The existence of contractile proteins, such as actin, myosin and troponin, deenin and tubulin, in nonmuscle cells confirm the hypothesis of a unique pathway for the convertion of the chemical energy of hydrolysis of ATP molecules into that of mechanical motion. Intracellular motion also occurs by the pathway of sliding. Nevertheless, this mechanism differs from that which applies to muscle fibers in some details.

In muscle fibers, the thick filaments are located at the sarcomer centre in a form of parallel rays. The ends of thin actin filaments are coupled with z-th plates separating sarcomers from each other, and are placed in the thick filament interior. However, in nonmuscle cells, myosin filaments have no definite position and float in cytoplasma.

Helix segments constitute a considerable part of the protein structure of a cytoskeleton. Besides motion, they transfer energy and information from one site in a cell to another by means of the propagation of soliton excitations.

Transmembrane glycoproteins play an important role in the vitality of a cell, too. Glycoproteins are formed by the covalent binding of proteins with carbohydrate residues, or polysaccharides. Their long protein fraction has a α-helix structure which spans the whole thickness of the outer cellular membrane. The polysaccharide part is hydrophobic and is located on the outer surface of a cell. The internal parts of glycoproteins are strongly coupled with microfilaments and microtubes of the cellular cytoskeleton. Therefore, glycoproteins realize the coupling of the cellular interior and exterior.

Glycoproteins determine cellular individuality, their adhesion, and intercellular interaction. They transmit signals inside cells which are caused by the cellular environment through the binding of hormones, neurons, immunoglobulins, and other molecules.

One can conclude that the nonlinear dynamics of solitons can undoubtedly provide a key for the understanding of the mechanism of transduction of information from the exterior to the interior of a cell.

Chapter 3

EXCITATION OF SOLITONS IN ONE-DIMENSIONAL SYSTEMS

11. Time-Dependent Evolution of Initial Excitation

Let us refer an excitation energy to the bottom of the exciton band, $E_{ex}(0)$ given by eqn. (4.6). Relative to the laboratory coordinate frame, the nonlinear Schrödinger equation for the wave function $\psi(z,t)$ can be written as follows

$$\left\{i\hbar\frac{\partial}{\partial t}+\frac{\hbar^2}{2m}\frac{\partial^2}{\partial z^2}+G\,|\psi(z,t)|^2\right\}\psi(z,t) = 0, \tag{11.1}$$

where m is the effective exciton mass, and $G = \chi^2/\kappa(1-s^2)$ is a parameter of nonlinearity.

The present treatment is restricted by the study of those solitons whose velocity is considerably smaller than that of the longitudinal sound velocity, i.e., $s^2 \ll 1$. Then, a parameter of nonlinearity, G, of dimension energy, appearing in eqn. (11.1), becomes constant.

Introducing the following dimensionless quantities,

$$x = \frac{2}{a},\quad \tau = \frac{\hbar t}{2ma^2},\quad g = \frac{ma^2 G}{\hbar^2}, \tag{11.2}$$

one can transform eqn. (11.1) into the standard form,

$$\left\{i\frac{\partial}{\partial \tau}+\frac{\partial^2}{\partial x^2}+2g\,|\psi(x,\tau)|^2\right\}\psi(x,\tau) = 0. \tag{11.3}$$

For any nonvanishing and positive g, and, hence, for any value of the coupling parameter χ, eqn. (11.3) possesses the following normalized particular solution,

$$\psi(x,\tau) = \frac{\sqrt{g}\exp[i(2kx-\omega\tau)]}{2\mathrm{ch}[g(x-4k\tau)/2]} \tag{11.4}$$

where

$$\omega = 4[k^2 - g^2/16], \quad 2k = mVa/\hbar. \tag{11.5}$$

This analytical solution (11.4) characterizes a solitary wave, or soliton, which propagates along the x-th axis with a constant velocity V, with the form of its envelope, $|\psi(x,t)|^2$ being time-independent.

It was mentioned in Section 6 that in solving numerically the set of equations corresponding to eqn. (11.3), Scott et al. [202] and Eilbeck [170] have shown that solitons become excited in a chain iff the coupling constant exceeds some threshold value. Brizhik and the present author [147] have demonstrated that the existence of a excitation threshold for solitons (or its absence) and the value of this threshold depend on the nature of the initial spatial distribution of the excitation. The main results of this work are presented below. For convenience, let us study the time evolution of the initial excitation described by the nonlinear Schrödinger equation using the method of inverse scattering. As applied to the nonlinear Schrödinger equation, this method was first developed by Zakharov and Shabat [68,69]. They showed that the nonlinear Schrödinger equation with an initial state, given by an arbitrary function $\psi(x,0)$, rapidly decreasing at infinity, corresponds to a linear scattering problem for the following eigenvectors

$$\mathrm{v} = \begin{pmatrix} \mathrm{v}_1 \\ \mathrm{v}_2 \end{pmatrix} \tag{11.6}$$

satisfying the set of equations,

$$\mathrm{v}_1 + i\zeta \mathrm{v}_1 = q(x)\mathrm{v}_2 \tag{11.7}$$

$$\mathrm{v}_2 - i\zeta \mathrm{v}_2 = -q^*(x)\mathrm{v}_1$$

with eigenvalues

$$\zeta = \zeta + i\eta, \tag{11.8}$$

where the function $q(x)$ is given by the relation

$$q(x) = i\sqrt{g}\,\psi(x,0). \tag{11.9}$$

A prime above functions denotes henceforth differentiation with respect to x.

The set of equations (11.7) can be reduced to the second-order equation relative to the function

$$v_1 - q'v_1'q^{-1} + v_1[\zeta_2 + |q|^2 - i\zeta q'q^{-1}] = 0. \tag{11.10}$$

The function v_2 is expressed in terms of v_1 as follows

$$v_2 = [v_1' + i\zeta v_1]/q. \tag{11.11}$$

According to the method of inverse scattering, as a first step one should find the solutions of the set (11.7) with the following asymptotic behavior

$$v(\xi,x) = \begin{cases} \binom{1}{0}\exp(-i\xi x), & \text{for } x \to -\infty \\ \binom{1}{0}a(\xi)e^{-i\xi x} + \binom{0}{1}b(\xi)e^{i\xi x}, & \text{for } x \to \infty \end{cases} \tag{11.12}$$

for a fixed function, determined by eqn. (11.9), and a real eigenvalue $\zeta = \xi$. The complex coefficients $a(\xi)$ and $b(\xi)$ obey the relation

$$|a(\xi)|^2 - |b(\xi)|^2 = 1. \tag{11.13}$$

A quantity, inversely proportional to $|a(\zeta)|^2$, determines the transmission coefficient for a plane wave incident on the potential $q(x)$ from the region $x \to \infty$, while the quantity $|R(\zeta)|^2$, where

$$R(\xi) = b(\xi)/a(\xi), \tag{11.14}$$

determines its reflection coefficient $R(\xi)$.

The functions (11.12) can be continued analytically into the upper half-plane ($\eta > 0$) of the complex variable ζ. In the inverse scattering method (see [65,66,69]) it is proved that if the functions $a(\xi)$ and $b(\xi)$ are known at time $\tau = 0$, one can obtain their values at $\tau > 0$ using the following relations

$$a(\zeta,\tau) = a(\zeta), \quad b(\zeta,\tau) = b(\zeta)\exp(4i\zeta^2\tau). \tag{11.15}$$

For values $\zeta_j (j = 1,2,...,N)$ belonging to the upper half of the ζ-plane and at which the coefficients $a(\zeta_j)$ vanish, the functions $v(\zeta_j,x)$ have the following asymptotics

$$v(\zeta_j,x) \to \begin{cases} \binom{1}{0}\exp(\eta_j x - i\xi_j x), & x \to -\infty; \\ \binom{0}{1}b(\zeta_j)\exp(-\eta_j x + i\xi_j x), & x \to \infty. \end{cases}$$

Since the function $v(\zeta_j,x)$ decreases asymptotically at $|x| \to \infty$ it describes bound states corresponding to the complex eigenvalues ζ_j of eqn. (11.7).

The following quantities

$$\zeta_j,\ C_j(\tau) = b(\zeta_j,\tau)\left[\frac{\partial a(\zeta)}{\partial\zeta}\right]^{-1}_{\zeta=\zeta_j},\quad j = 1,2,\dots,N, \tag{11.16}$$
$$R(\xi,\tau) = b(\xi,\tau)a^{-1}(\xi)$$

form a set of so-called *scattering data*. Using this set, one can obtain the function $q(x,\tau)$ for an arbitrary time $\tau > 0$.

In the inverse scattering method it is proved that the function $q(x,\tau)$ is given by

$$q(x,\tau) = -2\mathcal{K}(x,y), \tag{11.17}$$

where the function $\mathcal{K}(x,y)$ is a solution of the Gelfand-Levitan-Marchenko integral equation [66]

$$\mathcal{K}(x,y) = F^*(x+y,\tau) - \int_x^\infty ds \int_x^\infty dz F^*(s+y,\tau)\times F(z+s,\tau)\mathcal{K}(x,z). \tag{11.18}$$

The function $F(x,\tau)$, appearing in eqn. (11.18), is expressed in terms of the scattering data, eqn. (11.16), through the equality

$$F(x,\tau) = \frac{1}{2\pi}\int_{-\infty}^{\infty} R(\xi,\tau)e^{i\xi x}d\xi + \sum_{j=1}^{N} C_j(\tau)e^{i\zeta_j x}. \tag{11.19}$$

Determining the function $q(x,\tau)$ by the use of eqn. (11.17), and applying further eqn. (11.9), one finds the time dependence of the wave function

$$\psi(x,\tau) = -iq(x,\tau)/\sqrt{g}, \tag{11.20}$$

which obeys the nonlinear Schrödinger equation and takes the value $\psi(x,0)$ at $\tau = 0$.

In the following Sections, the problem will be solved for some particular cases.

12. Initial Excitation in the Form of a Hyperbolic Secant

In order to illustrate the application of the inverse scattering method we consider now the trivial case when an initial excitation is given in the form of a hyperbolic secant.

We assume that at $\tau = 0$ the excitation wave function is just the soliton solution

(11.4), i.e.,

$$\psi(x,0) = \frac{\sqrt{g}}{2} e^{2ikx} \operatorname{sch}(gx/2), \;\; 2k = maV/\hbar. \tag{12.1}$$

For fixed k, the parameter of nonlinearity, g, is determined by eqn. (11.2), and, consequently, depends on the velocity V which is related to k via eqn. (11.5). To obtain the scattering data, one needs only to find the asymptotic solution of eqn. (11.10) with

$$q(x,0) = i\sqrt{g}\,\psi(x,0) = \frac{ig}{2} e^{2ikx} \operatorname{sch}(gx/2). \tag{12.2}$$

For $|x| \to \infty$, eqn. (11.10) transforms to

$$v_1'' - (2ik \pm g/2)v_1' + (\zeta^2 + 2k\zeta \mp ig\zeta/2)v_1 = 0. \tag{12.3}$$

The upper (lower) sign in eqn. (12.3) corresponds to $x \to -\infty (x \to \infty)$. Equation (12.3) has the solutions

$$v_1 \to \begin{cases} \exp\{[i(k+m_1)+g/4]x\}, & x \to -\infty \\ \exp\{[i(k-m_2)-g/4]x\}, & x \to \infty \end{cases} \tag{12.4}$$

where

$$m_1 = \zeta + k - ig/4, \;\; m_2 = \zeta + k + ig/4. \tag{12.5}$$

For $g = 0$, the functions, given by eqn. (12.4), coincide with the asymptotic functions (11.12) provided

$$\zeta = \zeta_1 \equiv \xi_1 + i\eta_1, \tag{12.6}$$

where

$$\xi_1 = -k, \;\; \eta_1 = g/4. \tag{12.7}$$

Therefore, the imaginary part of ζ is proportional, in units of a, to the inverse width of the soliton (11.4), while its real component is related to the soliton velocity, according to eqn. (11.5).

Taking eqn. (12.4) into account, it follows from eqn. (11.11) that

$$v_2 = 0 \;\text{ for }\; |x| \to \infty. \tag{12.8}$$

Since at $x \to \infty$

$$v_2(\xi,x) = b(\xi)\exp(i\xi x),$$

in accordance with (11.12), from eqn. (12.8) one obtains

$$b(\xi) = 0. \tag{12.9}$$

The potential $q(x)$ which provides the validity of the relation eqn. (12.9), is called a *reflectionless* one.

Taking eqn. (12.6) and (12.9) into account, the function $(F(x,\tau)$ appearing in eqn. (11.18), becomes

$$F(x,\tau) = C_1(\tau)\exp(i\xi_1 x). \tag{12.10}$$

Substituting eqn. (12.10) into (11.18) and assuming that

$$\mathcal{K}(x,y) = f(x)\exp(-i\zeta_1^* y), \tag{12.11}$$

one obtains

$$f(x) = \frac{C_1^*(\tau)\exp(-i\zeta_1^* x)}{1+4|C_1(\tau)|^2 g^{-2}\exp(-gx)} \tag{12.12}$$

where

$$C_1(\tau) = C_0\exp(4i\zeta_1^2\tau), \quad C_0 = ig/2. \tag{12.13}$$

By virtue of eqns. (12.11)-(12.13) and (11.17), one finds

$$q(x,\tau) = -2\mathcal{K}(x,y) = -\frac{2C_0^*\exp[-2i\zeta_1^*(x+2\zeta_1^*\tau)]}{1+4|C_0|^2 g^{-2}\exp[-g(x-4k\tau)]}. \tag{12.14}$$

According to eqn. (11.9) and taking the equality (12.7) into account, the square of the modulus of the wave function is given by the following formula

$$|\psi(x,\tau)|^2 = \frac{|q(x,\tau)|^2}{g} = \frac{g}{4}\mathrm{sch}^2[g(x-x_0-4k\tau)/2], \tag{12.15}$$

where

$$x_0 = \frac{2}{g}\ln(2|C_0|/g). \tag{12.16}$$

It follows from eqn. (12.13) that $x_0 = 0$, i.e., at the initial time moment, the soliton is localized at the origin, which corresponds to the function (12.2).

Therefore, we demonstrate that the method of inverse scattering enables the wave function $\psi(x,\tau)$ to be obtained at any τ, based on its initial value given by eqn. (12.2). This is in complete correspondence with the exact solution, eqn. (11.4), at an arbitrary nonvanishing parameter of nonlinearity g.

13. Initial Excitation in the Form of a Rectangular Impulse

Let us assume that the initial excitation differs from the soliton profile, given by eqn. (12.1), and is described by a normalized wave function in the form of a rectangular impulse,

$$\psi(x,0) = \begin{cases} l^{-1/2}\exp(2ikx), & \text{for} \quad 0 \leqslant x \leqslant l; \\ 0, & \text{for} \quad x < 0 \text{ and } x > l \end{cases} \tag{13.1}$$

This corresponds to a constant velocity which is determined through k by eqn. (11.5) and directed along the positive x-axis. We seek the solutions of the system of equations (11.7) in the following form

$$\begin{aligned} &\mathrm{v} = \binom{1}{0}\exp(-\zeta x), && \text{if} \quad x<0, \\ &\mathrm{v}\binom{1}{0}a(\zeta)\exp(-i\zeta x)+\binom{0}{1}b(\zeta)\exp(i\zeta x), && \text{if} \quad x>l \end{aligned}$$

and

$$\mathrm{v}_1 = [A_1\sin(mx)+A_2\cos(mx)]\exp(ikx), \tag{13.2}$$

$$\begin{aligned} \mathrm{v}_2 = Q_0^{-1}\{&[(\zeta+k)A_1+imA_2]\sin(mx)+[(\zeta+k)A_2- \\ &-imA_1]\cos(mx)\}\exp(ikx). \end{aligned}$$

in the intermediate region, $0 \leqslant x \leqslant l$; and where the notations

$$m^2 \equiv Q_0^2+(\zeta+k)^2, \quad Q_0 \equiv \sqrt{g/l} \tag{13.3}$$

are introduced.

Based on the continuity condition for solution, one obtains the constants of integration,

$$A_1(\zeta) = -i(\zeta+k)/m, \quad A_2(\zeta) = 1, \tag{13.4}$$

$$a(\zeta) = [\cos(ml)-\frac{i}{m}(\zeta+k)\sin(ml)]\exp[i(k+\zeta)l], \tag{13.5}$$

$$b(\zeta) = \frac{iQ_0}{m}\exp[-i(\zeta+k)l]. \tag{13.6}$$

Using eqn. (13.4) and (13.5), one finds the reflection coefficient at $\tau = 0$

$$R(\xi) = \frac{b(\xi)}{a(\xi)} = iQ_0\exp[-2(\xi+k)l]S^{-1}(\xi,k), \tag{13.7}$$

where

$$S(\xi,k) = m\cos(ml) - i(k+\xi)\sin(ml) \tag{13.7a}$$

$$m = \sqrt{Q_0^2+(k+\xi)^2}$$

and at an arbitrary $\tau > 0$

$$R(\xi,\tau) = R(\xi)\exp(4i\xi^2\tau). \tag{13.8}$$

As mentioned in Section 11, the parameters of the bound states of the scattering problem under study are determined by the condition $a(\xi) = 0$ which reduces to the equation

$$S(\zeta,k) = 0 \tag{13.9}$$

in accordance with eqns. (13.4) and (13.7). This equation has the following solution

$$\zeta_0 = -k+i\eta_0, \tag{13.10}$$

where η_0 satisfies the transcendental equation

$$\eta_0 = -\sqrt{Q_0^2-\eta_0^2}\,\mathrm{ctg}\left[\frac{g}{Q_0^2}\sqrt{Q_0^2-\eta_0^2}\right]. \tag{13.11}$$

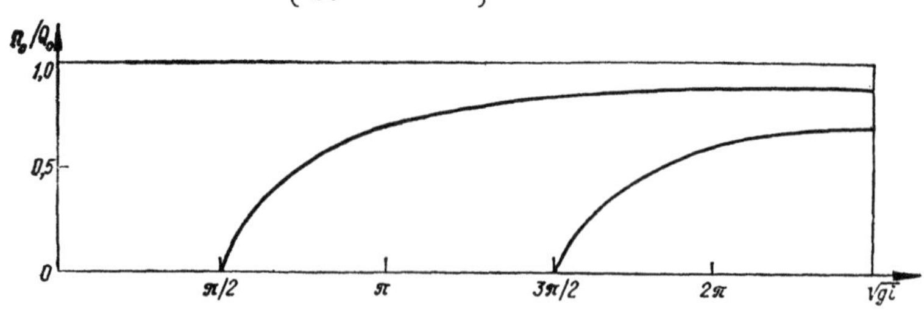

Figure 8. The dependence of a ratio η_0/Q_0, which characterizes the formation of solitons, on a parameter $\sqrt{gl}$ of the excitation pulse.

The dependence of the ratio η_0/Q_0 on the quantity $(gl)^{1/2}$ is plotted in Fig. 8. If

$(gl)^{1/2} < \pi/2$, eqn. (13.11) has no solutions. However, if the following inequality,

$$\pi/2 < \sqrt{gl} < 3\pi/2, \tag{13.12}$$

holds, eqn. (13.11) possesses a single positive $\eta_0 < Q_0$. Assuming $\eta_0 \ll Q_0$, one can obtain approximately

$$\eta_0 \approx -Q_0 \mathrm{ctg}\sqrt{gl} \approx g\sqrt{g/l} = g\left[1 - \frac{\pi}{2\sqrt{gl}}\right], \tag{13.13}$$

with $\mathcal{E} \equiv (gl)^{1/2} - \frac{\pi}{2} \ll 1$.

For gl, obeying the inequalities

$$(2N-1)\frac{\pi}{2} < \sqrt{gl} \ll (2N+1)\frac{\pi}{2}, \tag{13.14}$$

eqn. (13.9) has N solutions. The ith solution corresponds to a localized, or soliton state with the parameter $\eta_i \neq \eta_j < Q_0$. All these states move with the same velocity proportional to $\xi_i = -k$, $i = 1,2,\ldots,N$.

We start by considering the case when gl satisfies the inequality $(gl)^{1/2} < \pi/2$. Then $S(\zeta,k) = 0$, i.e., there are no bound states, and the function $F(x,\tau)$, given by eqn. (11.19), is determined by a single term

$$F(x,\tau) = \frac{i\sqrt{g}e^{-2ikl}}{2\pi\sqrt{l}} \int_{-\infty}^{\infty} S^{-1}(\xi,k)\exp[i\xi(x-2l+4\xi\tau)]d\xi. \tag{13.15}$$

At small distances and at fixed time, or at large times and for fixed x, when the condition

$$|x-2l| \ll \tau, \tag{13.16}$$

holds, the function (13.15) is approximated by the expression

$$F(x,\tau) = \frac{1}{\sqrt{\tau}}F_1(x,\tau) = \frac{A_0\exp[-i(x-2l)^2/16\tau]}{\sqrt{\tau}S(0,k)} \tag{13.17}$$

where

$$A_0 = \frac{i\sqrt{g}}{4\sqrt{\pi l}}\exp[-2i(kl-\pi/8)], \tag{13.18}$$

with

$$S(0,k) \equiv m_0 \cos(m_0 l) - ik \sin(m_0 l), \quad m_0 \equiv \sqrt{2k^2 + g/t}. \tag{13.19}$$

Substituting the function (13.17) into the Gelfand-Levitan-Marchenko equation (11.18), one can obtain

$$\mathcal{K}(x,y) = \frac{A_0^*}{\sqrt{\tau}\, S^*(0,k)} \exp[i(x+y-2l)^2/16\tau] -$$

$$- \frac{|A_0|^2}{\tau |S(0,k)|^2} \int_x^\infty ds \int_x^\infty dz\, \mathcal{K}(x,z) \exp\{i[(s+y-2l)^2 - (z+s-2l)^2]/16\tau\}.$$

This equation has the following solution:

$$\mathcal{K}(x,y) = \frac{A_0^*(0,k) \exp[i(x+y-2l)^2/16\tau]}{\sqrt{\tau}(|S(0,k|^2 + 8\pi |A_0|^2)}. \tag{13.20}$$

Therefore, the wave function of the system under study is

$$\psi(x,\tau) = \frac{2iA_0^* S(0,k) \exp\{i[2k(x-l) + \frac{(x-L)^2}{4\tau} - 4k^2\tau]\}}{\sqrt{g\tau}\{|S(0,k)|^2 + 8\pi |A_0|^2\}}, \tag{13.21}$$

if the condition (13.16) takes place. In eqn. (13.21) $L \equiv 1 + 4k\tau$. This function vanishes asymptotically with increasing time.

If, however, the following inequality

$$x - 2l \gg \tau \tag{13.22}$$

holds, the function (13.15) is approximated as follows

$$F(x,\tau) \approx F_l(x,\tau)/\sqrt{\tau} = \frac{A_0 \exp[-i(x-2l)^2/16\tau]}{\sqrt{\tau}\, S_1(x)}, \tag{13.23}$$

where

$$S_1(x) \equiv \tilde{m} \cos(\tilde{m} l) - i \left\{ k - \frac{x-2l}{4\tau} \right\} \sin(\tilde{m} l),$$

$$\tilde{m} = \{g/l + [k - (x-2l)/4\tau]^2\}^{1/2}$$

with A_0 is determined by eqn. (13.18).

Within the approximation (13.22), the Gelfand-Levitan-Marchenko eqn. (11.18) has the solution

$$\mathcal{K}(x,y) = \frac{A_0^* \exp[i(x+y-2l)^2/16\tau]}{\sqrt{\tau}S^*(x+y)[1+8\pi|A_0|^2/|S(2x)|^2]}.$$

Hence, if the inequality (13.22) is valid, the wave function of the system becomes

$$\psi(x,\tau) = \frac{2iA_0^* \exp[i(x-l)^2/4\tau]}{\sqrt{g\tau}S^*(2x)|1+8\pi|A_0|^2|S(2x)|^{-2}]}.$$

Therefore, for gl obeying the inequality, $(gl)^{1/2} < \pi/2$, in the chain there arise excitations described by functions whose amplitude decreases asymptotically with time.

Let us consider now the case when gl satisfies the following inequality

$$\pi/2 \leqslant \sqrt{gl} < 3\pi/2.$$

At these values of gl, the function (11.19) involves two terms,

$$F(x,\tau) = F_0(x,\tau) + \frac{1}{\sqrt{\tau}}F_1(x,\tau), \tag{13.24}$$

where $\tau^{-1/2}F_1(x,\tau)$ is given by eqn. (13.17) with the condition (13.16), or by eqn. (13.23) with (13.22), and the bound-state function becomes

$$F_0(x,\tau) = c(\tau)\exp(i\zeta_0 x). \tag{13.25}$$

The parameter $c(\tau)$, characterizing a bound state, is given by

$$c(\tau) = \alpha\exp(4i\zeta_0^2\tau), \tag{13.26}$$

where

$$\alpha = \frac{m_1^2\exp(2l\eta_0)}{Q_0[m_1 l\cos(m_1 l) - \sin(m_1 l)]}, \quad m_1 = \sqrt{g/l - \eta_0^2}. \tag{13.27}$$

Due to eqn. (13.24), the solution of eqn. (13.18) can be represented in the form

$$\mathcal{K} = \mathcal{K}_0 + \mathcal{K}_1/\sqrt{\tau}, \tag{13.28}$$

where the functions $\mathcal{K}_0$ and $\mathcal{K}_1$ are defined by equations

$$\mathcal{K}_0(x,y) = F_0^*(x+y) - \int_x^\infty ds \int_x^\infty dz F_0^*(s+y)F_0(s+z)\mathcal{K}_0(x,z), \tag{13.29}$$

$$\mathcal{K}_1(x,y) = F_1^*(x+y) - \int_x^\infty ds \int_x^\infty dz \{F_0^*(s+y)F_0(s+z)\mathcal{K}_1(x,z) + \tag{13.30}$$

$$+[F_0^*(x+y)F_1(s+z)+F_0(z+s)F_1^*(s+y)]\mathcal{K}_0(x,z)\}.$$

The solution of eqn. (13.29) is obtained in a way similar to that shown in Section 12,

$$\mathcal{K}_0(x,y) = \frac{c^*(\tau)}{u(x,\tau)}\exp[-i\zeta_0^*(x+y)], \tag{13.31}$$

where,

$$u(x,\tau) \equiv \frac{1}{2}e^{-z}\operatorname{ch} z, \quad z \equiv 2\eta_0(x-x_0-4k\tau), \tag{13.32}$$

$$x_0 = \frac{1}{2\eta_0}\ln(|\alpha|/2\eta_0), \quad \eta_0 > 0.$$

Taking this formula into account, and assuming that the inequality (13.12) is valid, eqn. (13.30) has the solution

$$\mathcal{K}_1(x,y) = f_2(x,\tau)+\left[f_1(x,\tau)+\frac{|c(\tau)|^2}{2\eta_0\zeta_0}f_2(x,\tau)\times\right. \tag{13.33}$$

$$\left.\times\exp(-3\eta_0 x-ikx)\right]\frac{e^{-i\zeta_0^* y}}{u(x,\tau)},$$

where the functions f_1 and f_2 are given by the expressions

$$f_1(x,\tau) = \frac{A_0 c^{*2}(\tau)\exp[3x(ik-\eta_0)]}{i\zeta_0^* S(0,k)u(x,\tau)},$$

$$f_2(x,\tau) = \frac{A_0^*}{S^*(0,k)}\left[1+\frac{|c(\tau)|^2\exp(-4\eta_0 x)}{2\eta_0\zeta_0 u(x,\tau)}\right].$$

Therefore, by virtue of the relations (11.17), (11.20), and (13.28), the wave function of the system now takes the form

$$\psi(x,\tau) = \psi_s(x,\tau)+\psi_1(x,\tau). \tag{13.34}$$

Due to eqn. (13.31), the first term,

$$\psi_s(x,\tau) = \frac{2i\eta_0\exp\{2i[kx-2(k^2-\eta_0^2)\tau]\}}{\sqrt{g}\operatorname{ch} z} \tag{13.35}$$

describes a soliton whose envelope is

$$|\psi_s(x,\tau)|^2 = 4\eta_0^2 g^{-1}\operatorname{sch}^2 z,$$

and has amplitude $2\eta_0/\sqrt{g}$ and width $\sim(2\eta_0)^{-1}$. The maximum of the soliton amplitude is attained at the point $x+4k\tau$ where the velocity is $V = 2k\hbar/ma$. Here, x_0 is given by eqn. (13.29). The second term in eqn. (13.34) describes the wave packet moving with the velocity inherent for the basic soliton and damping asymptotically in time as $\tau^{-1/2}$.

Hence, if the nonlinearity parameter is less than the critical value, $g_{cr} = \pi^2/4l$, the initial rectangular impulse transforms into a rapidly oscillating wave packet damping in time. If however, the nonlinearity parameter g exceeds this critical value, there is formed in the system a soliton together with a rapidly oscillating and damping wave packet, or 'tail'. As g increases, so does the soliton amplitude, but its width decreases. If g approaches $9\pi^2/4l$, the major contribution to the normalization of the wave function is provided by the soliton, since in this limit

$$\int |\psi_s(x,\tau)|^2 dx \to 0.85.$$

The role of the tail appears to be negligible. In other words, the normalized step, whose width is close to $9\pi^2/4l$, well approximates the ideal case, eqn. (12.2), corresponding to the excitation of a single soliton in the chain.

N solitons are generated in the system if the nonlinearity parameter g obeys the inequality

$$(2N-1)^2\pi^2(4l)^{-1} < g < (2N+1)^2\pi^2(4l)^{-1}.$$

The major contribution to the normalization of the wave function, however, is still provided by the soliton with the parameter $\eta_0 \approx Q_0$.

Let us now consider the energy which is carried by the soliton. Within the inertia-free approach, i.e., when the parameter of nonlinearity g is independent of the soliton velocity V, the excitation, described by eqn. (11.3), transfers the energy

$$E_0(V) = \int\left\{\left|\frac{\partial\psi}{\partial x}\right|^2 - g_0|\psi|^4\right\}dx, \tag{13.36}$$

in units of $\hbar^2/2ma^2$.

In the nonlinearity parameter $g = g_0/(1-s^2)$ depends on the velocity, an inertia of displacement of peptide groups is manifested. Then the energy integral involves also that of chain deformation

$$E(V) = \int\left\{\left|\frac{\partial\psi}{\partial x}\right|^2 - g(1-2s^2)|\psi|^4\right\}dx. \tag{13.37}$$

The energy (13.37) is an integral of motion. That is, it may be evaluated using the function of the initial excitation given by eqn. (13.1). Hence, one obtains

$$E(V) = E(0)+E_k, \tag{13.38}$$

where

$$E(0) = -g_0 l^{-1}, \; E_k = 4k^2(1+\beta g/l), \; \beta = \hbar^2/m^2a^2V_0^2. \tag{13.39}$$

Therefore, for vanishing k, the energy is less than the energy at the bottom of the excitation band by the bound g_0/l. This energy is carried partially by the tail, and the remainder,

$$E_s(V) = E_s(0)+E_{s,k}, \tag{13.40}$$

where

$$E_s(0) = -\frac{16\eta_0^3}{3g_0}, \; E_{s,k} = 16\eta_0 k^2 g_0^{-1}(1+20\eta_0^2\beta/3), \tag{13.41}$$

by the soliton.

Thus, as the nonlinearity parameter increases within the range given by (11.4), the soliton energy grows by increasing its degree of localization. In the limit, $g \to 9\pi^2/4l$, when, in particular, $\eta_0 \to Q_0$, the ratio of the soliton kinetic energy, $E_{s,k}$, to that of the initial excitation, E_k, becomes

$$E_{s,k}/E_k = 0.85\frac{1+20\beta g/3l}{1+\beta g/l}.$$

Therefore, the threshold nature of the soliton formation in molecular chains is provided by a way it is excited, i.e., by a choice of the original wave function form, since the nonlinear Schrödinger equation (11.3), does possess a soliton solution (11.4). The manner of exciting the system in question by impulse in the form of a rectangular step, as considered above, explains adequately the results of the numerical study of the corresponding discrete problem [202]. Namely, for a given initial condition, the solitons become excited in the chain, starting with a certain critical value of the coupling constant of exciton-phonon interaction and accompanied by damping an oscillating tail. The contribution of the latter to the wave function normalization decreases as the parameter of nonlinearity increases in the range determined by (13.12). A further increase of the nonlinearity parameter results in the formation of additional solitons in the system. These solitons move with the same velocity as the basic one which is proportional to the wave vector of the initial condition. However, the additional solitons are more diffuse and possess smaller amplitudes. As in the preceding case, these solitons are accompanied by a rapidly oscillating tail damping in time as $\tau^{-1/2}$.

Chapter 4

DYNAMICAL PROPERTIES OF SOLITONS

14. Free Solitons

The comprehensive study of a large number of nonlinear equations describing free solitons is nowadays being carried out. However, the possibility of using the 'soliton' to explain many phenomena in physics and applied sciences is inevitably related to a study of the physical conditions in a given system which result in nonlinearity and dispersion, and to an investigation of the dynamical properties of solitons, too.

The dynamical properties of solitons are manifested in their interaction with external fields and other degrees of freedom of a system. Various approaches to developing a perturbation theory for solitons in one-dimensional systems have been developed by Karpman and Maslov [75,26], Bishop, Krumhansl and co-workers [142,141,214], Keener and McLaughlin [209], and others [30,211,212]. In this Section we outline the results obtained by Eremko and the author [46] from the study of external effects on solitons generated in quasi-one-dimensional molecular chains by local excitations.

Within the continuum approach, $ak \ll 1$, the properties of free solitons in a molecular chain are described by the following set of equations

$$\left\{ i\hbar\frac{\partial}{\partial t} + \frac{\hbar^2}{2m}\frac{\partial^2}{\partial z^2} - [E_{\text{ex}}(0) + W] - \chi\rho \right\} \varphi(z,t) = 0, \tag{14.1}$$

$$\left[\frac{\partial^2}{\partial t^2} - V_0^2\frac{\partial^2}{\partial z^2} \right] \beta(z,t) - \frac{a\chi}{M}\frac{\partial}{\partial z}|\varphi|^2 = 0, \tag{14.2}$$

$$\frac{1}{a}\int |\varphi|^2 dz = 1, \tag{14.3}$$

where $V_0 = a(\kappa/M)^{1/2}$ is the velocity of longitudinal sound in the chain, κ is the longitudinal elasticity coefficient, M is the mass of the molecules, m is an effective exciton mass, function $\beta(z,t)$ characterizes coherent displacements of molecules from their equilibria, $E_{ex}(0)$ is the energy of the bottom of the exciton band, and W is the deformation energy of the chain in question determined by eqn. (3.12).

Defining a function

$$\rho(z,t) = -\partial\beta(z,t)/\partial z, \tag{14.3}$$

which describes the concentration of intermolecular separations, one can transform eqn. (14.2) to the following

$$\left[\frac{\partial^2}{\partial t^2} - V_0^2\frac{\partial^2}{\partial z^2}\right]\rho(z,t) + \frac{a\chi V_0^2}{x}\frac{\partial^2}{\partial z^2}|\varphi|^2 = 0. \tag{14.4}$$

For convenience, one represents the function $\varphi(z,t)$ as

$$\varphi(z,t) = \sqrt{a}\,B(z,t)\exp\{-i[E_{ex}(0)+W]t/\hbar\}, \tag{14.5}$$

$$\int |B(z,t)|^2 dz = 1. \tag{14.6}$$

Then eqn. (14.4) takes the form

$$\left[\frac{\partial^2}{\partial t^2} - V_0^2\frac{\partial^2}{\partial z^2}\right]\rho(z,t) + \frac{a\chi V_0^2}{\kappa}\frac{\partial^2}{\partial z^2}|B(z,t)|^2 = 0. \tag{14.7}$$

Now we consider the particular case of excitations moving with a constant velocity V which is less than that of longitudinal sound. It follows from eqn. (14.4) that

$$\rho(z,t) = \frac{a\chi}{\kappa}|B(z,t)|^2, \quad s^2 \ll 1. \tag{14.8}$$

Substituting eqns. (14.5) and (14.8) into eqn. (14.1), one finds the nonlinear Schrödinger equation

$$\left[i\hbar\frac{\partial}{\partial t} + \frac{\hbar^2}{2m}\frac{\partial^2}{\partial z^2} + G\,|(z,t)|^2\right]B(z,t) = 0, \tag{14.9}$$

where

$$G = a\chi^2/\kappa \tag{14.10}$$

is a parameter of nonlinearity which arises due to the interaction of an exciton with a local deformation of the chain generated by an exciton.

The solution of eqn. (14.9), normalized via condition (14.6), takes the following form

$$B_0(z,t) = \frac{\sqrt{q}\exp\{i[k(z-\zeta)-\Theta]\}}{\sqrt{2}\,\mathrm{ch}[q(z-\zeta)]}, \tag{14.11}$$

where

$$\zeta \equiv z_0 + \hbar kt/m, \quad \Theta = \frac{\hbar^2(k^2+q^2)t}{2m}, \quad q = \frac{am\chi}{2\kappa\hbar^2}. \tag{14.12}$$

The function (14.11) describes the solitary wave, or soliton, moving with a constant velocity $V = \hbar k/m$ along the z-axis, and characterized by the phase Θ and the maximum of envelope attained at $z = \zeta$.

15. Soliton Motion with Friction

If a molecular chain is embedded into a medium, a frictional force acts on the moving soliton. As a result, this soliton is decelerated. Such braking may be taken into account formally by introducing dissipative terms on the right-hand side of eqn. (14.2) which determines displacements of molecules from their equilibrium positions. Assuming that frictional forces to be proportional to velocity, eqn. (14.2) is replaced by the following equation,

$$\left[\frac{\partial^2}{\partial t^2} - V_0^2\frac{\partial^2}{\partial z^2}\right]\beta - \frac{\chi V_0^2}{\kappa}\frac{\partial}{\partial z}|B|^2 = -\epsilon\gamma\frac{\partial\beta}{\partial t}. \tag{15.1}$$

For convenience, we include a small positive parameter $\epsilon \ll 1$ in the damping coefficient $\epsilon\gamma$. One can rewrite eqn. (15.1) in terms of the variable $\rho(z,t)$ defined by (14.3)

$$\left[\frac{\partial^2}{\partial t^2} - V_0^2\frac{\partial^2}{\partial z^2}\right]\rho + \frac{\chi V_0^2 a}{\kappa}\frac{\partial^2}{\partial z^2}|B|^2 = -\epsilon\gamma\frac{\partial\rho}{\partial t}. \tag{15.2}$$

Therefore, soliton motion with friction is determined by eqn. (15.2), combined with eqn. (14.9). In order to solve this set of equations, Eremko and the present author [46] applied the 'method of slowly varying coefficients' proposed by van der Pol and developed in detail by Bogoliubov and Mitropolskii [19].

Let us assume that, at $t = 0$, a soliton is described via the function (14.11) chosen at $k = k_0$. To determine the change in this function due to friction, we search for a solution of the set of eqns. (14.9) and (15.2) in the form

$$B(z,t) = B_0(z,t) + \epsilon B_1(z,t), \tag{15.3}$$

$$\rho(z,t) = \rho_0(z,t) + \epsilon\rho_1(z,t), \tag{15.4}$$

where

$$\rho_0(z,t) = \frac{a\chi}{\kappa}\,|B_0(z,t)|^2. \tag{15.5}$$

The function $B_0(z,t)$ is given by eqn. (14.11). The time dependence of its parameters k, ζ, and Θ is defined by the equations

$$dk/dt = \epsilon f(k), \tag{15.6}$$

$$d\zeta/dt = \hbar k/m, \quad d^2\zeta/dt^2 = \epsilon\hbar f(k)/m, \tag{15.7}$$

$$\frac{d\Theta}{dt} = \frac{\hbar(k^2+q^2)}{2m} + \epsilon\varphi(k). \tag{15.8}$$

The problem is thus equivalent to obtaining the unknown functions $f(k)$, $\varphi(k)$, and $B_1(z,t)$. In addition, we require that the latter function does not contain terms involving the derivatives $\partial B_0/\partial\zeta$ and $\partial B_0/\partial\Theta$, since these terms increase in time and are called *secular terms* [19]. Bearing in mind eqn. (12.5) and (14.11) and substituting eqn. (15.4) into eqn. (15.2), one obtains within the terms of first order in ϵ,

$$\rho_1(z,t) = -\frac{a\hbar\chi\,\mathrm{sh}[q(z-\zeta)]}{2\kappa V_0^2 m\,\mathrm{ch}^2[q(z-\zeta)]}(\gamma k + f(k)). \tag{15.9}$$

Here we take into account a weak time-dependence of $\rho_1(z,t)$. We substitute further the expressions (15.3) and (15.4) into eqn. (14.9). Taking eqns. (15.9) and (15.6) into account, one derives the equation for determining $B_1(z,t)$:

$$\left\{i\hbar\frac{\partial}{\partial t} - \frac{\hbar^2}{2m}\frac{\partial^2}{\partial z^2} + \frac{2a\chi}{\kappa}|B_0(z,t)|^2\right\}B_1(z,t) + \frac{a\chi}{\kappa}B_0^2B_1^* = F(B_0), \tag{15.10}$$

where

$$F(B_0) \equiv [f(k)(z-\zeta) + \varphi(k) - \chi\rho_1(z,t)]B_0(z,t). \tag{15.11}$$

The condition imposed on $B_1(z,t)$, which requires that $B_1(z,t)$ does not involve the so-called secular terms, is equivalent to that of orthogonality of the function $F(B_0)$ to $\partial B_0/\partial\zeta$ and $\partial B_0/\partial\Theta$ [19]. Its fulfilment results in the following equations for the unknown functions

$$f(k) = \frac{\chi}{\hbar}\int |B_0(z,t)|^2\rho_1(z,t)\text{th}[q(z-\zeta)]dz, \tag{15.12}$$

$$\varphi(k) = \frac{\chi}{\hbar}\int |B_0(z,t)|^2\rho_1(z,t)dz_0. \tag{15.13}$$

Evaluating the integrals, one obtains

$$\varphi(k) = 0,\ \ f(k) = k\gamma D/(1+D), \tag{15.14}$$

where

$$D \equiv \frac{m_{\text{sol}}-m}{m}. \tag{15.15}$$

Substituting eqn. (15.14) into (15.6) and integrating, one determines the law governing the decrease of soliton velocity due to frictional forces to velocity

$$V(t) = V(0)\exp(-\alpha t), \tag{15.16}$$

where

$$\alpha \equiv \epsilon\gamma(m_{\text{sol}}-m)/m_{\text{sol}}.$$

If a frictional force is proportional to the square of velocity, i.e., it is given by $-\epsilon\lambda(\partial\beta/\partial t)^2$, then eqn. (15.2) is replaced by

$$\left[\frac{\partial^2}{\partial t^2} - V_0^2\frac{\partial^2}{\partial z^2}\right]\rho(z,t) + \frac{a\chi V_0^2}{\kappa}\frac{\partial^2}{\partial z^2}|B(z,t)|^2 = -\epsilon\lambda\frac{V}{a}\frac{\partial}{\partial t}\rho^2(z,t).$$

In a similar way, one can find explicitly the law relating to the decrease of the soliton velocity,

$$V(t) = \frac{V(0)}{1+\beta V(0)t} \approx V(0)[1-\beta V(0)t], \tag{15.17}$$

where

$$\beta \equiv 2\epsilon\lambda\chi^3 aq^2/15m_{\text{sol}}\kappa^2V_0^2.$$

Therefore, the character of velocity decrease is close to that obtained in the linear case with the proportionality coefficient $\beta[V(0)]^2$.

16. Soliton Motion Affected by External Inhomogeneous Fields

The presence of external inhomogeneous fields or local inhomogeneities in the molecular arrangement results in the z-dependence of the energy of the bottom of the exciton band $E_{ex}(0)$. Such an energy, appearing in eqn. (14.1), takes into account the nonresonance interaction with neighboring molecules and external inhomogeneous fields. Carrying out the following substitution

$$E_{ex}(0) \to E_{ex}(0) - \epsilon U(z), \tag{16.1}$$

in eqn. (14.1), one obtains

$$\left[i\hbar\frac{\partial}{\partial t} + \frac{\hbar^2}{2m}\frac{\partial^2}{\partial z^2} + \chi\rho\right]B(z,t) = \epsilon U(z)B(z,t), \tag{16.2}$$

which, together with eqn. (14.7), forms a set of equations which determines the soliton motion in a chain. One can seek its solution in the same form as eqns. (15.13) - (15.15). Within the first-order perturbation treatment one obtains from eqns. (14.7) and (16.2) that

$$\left[i\hbar\frac{\partial}{\partial t} + \frac{\hbar^2}{2m}\frac{\partial^2}{\partial z^2} + \frac{a\chi^2}{\kappa}|B(z,t)|^2\right]B(z,t) = \epsilon[U(z) + \rho_1(z,t)]B(z,t), \tag{16.3}$$

$$\left[\frac{\partial^2}{\partial t^2} - V_0^2\frac{\partial^2}{\partial z^2}\right]\rho_1(z,t) = -\frac{1}{\epsilon}\frac{\partial^2\zeta}{\partial t^2}\frac{\partial\rho_0}{\partial\zeta}. \tag{16.4}$$

If excitation velocities obey the inequality $V \ll V_0$, the function ρ_1 depends weakly on time. Hence, taking eqns. (14.11) and (15.7) into account and further integrating, one has

$$\rho_1(z,t) = -\frac{a\hbar\chi}{2\kappa m V_0^2}f(k)\mathrm{th}[q(z-\zeta)]. \tag{16.5}$$

Using eqns. (15.3) and (16.5), one can rewrite eqn. (16.3) as follows

$$\left[i\hbar\frac{\partial}{\partial t} + \frac{\hbar^2}{2m}\frac{\partial^2}{\partial z^2} + \frac{2a\chi^2}{\kappa}|B_0|^2\right]B_1 + \frac{a\chi^2}{\kappa}B_0^2B_1^* = \Phi(B_0), \tag{16.6}$$

where

$$\Phi(B_0) = [U(z) - \chi\rho_1 + (z-\zeta)f(k) + \varphi(k)]B_0(z,t). \tag{16.7}$$

Assuming the absence of terms $B_1(z,t)$ increasing in a time, one obtains the following two equations

$$\varphi(k) = -\frac{1}{\hbar}\int |B_0(z,t)|^2 U(z)dz, \tag{16.8}$$

$$f(k) = -\frac{2q}{\hbar(1+D)}\int |B_0(z,t)|^2 U(z)\text{th}[q(z-\zeta)]dz, \tag{16.9}$$

where D is given by eqn. (15.15).

Substituting eqn. (16.9) into (15.6) and taking eqn. (15.15) into account, one finds the equation which determines the motion of the soliton centre-of-mass; that is, the motion of maximum of the solitary wave envelope,

$$m_{\text{sol}}\frac{d^2\zeta}{dt^2} = F(\zeta), \tag{16.10}$$

where

$$F(\zeta) = -3\epsilon q\int |B_0(z,t)|^2 U(z)\text{th}[q(z-\zeta)]dz. \tag{16.11}$$

Equation (16.10) is just the equation of Newton's type for a particle with a soliton mass. A soliton is affected by the potential field $U(z)$ at values of z where $|B_0|^2 \neq 0$. For $z>\zeta$ ($z<\zeta$) the potential $U(z)$ provides a negative (positive) contribution.

One should note that eqn. (16.10) involves the effective soliton mass rather than that of the exciton which enters directly into eqn. (16.3). This is due to the two-component nature of the soliton in the chain of peptide groups. Then, nonlinearity of the collective excitation is determined by the interaction of an exciton with a local deformation of the chain whose inertial properties change the effective mass of the free exciton.

Some authors study the effect of an external perturbation $U(z)$ upon the nonlinear Schrödinger equation by solving the following equation

$$\left\{i\hbar\frac{\partial}{\partial t} + \frac{\hbar^2}{2m}\frac{\partial^2}{\partial z^2} + G|B|^2\right\}B(z,t) = \epsilon U(z)B(z,t). \tag{16.12}$$

This equation corresponds to a physical system with a nonlinearity which arises due to the self-interaction of $B(z,t)$ or its interaction with other inertia-free fields, e.g. in a plasma when the inertia of heavy ions is neglected. In this case, the excitation becomes one-component, and the function $f(k)$, appearing in eqn. (15.7), takes the form

$$f(k) = -\frac{2q}{\hbar}\int |B_0(z,t)|^2 U(z)\text{th}[q(z-\zeta)]dz.$$

The equation of motion of the maximum of the envelope becomes

$$m\frac{d^2\zeta}{dt^2} = F(\zeta), \tag{16.13}$$

where $F(\zeta)$ is given by eqn. (16.11), and the mass m is just that which appears directly in eqn. (16.12).

Chapter 5

SOLITONS IN MOLECULAR SYSTEMS WITH NONLINEAR INTERMOLECULAR INTERACTIONS

17. The Equation of Soliton Motion for General Intermolecular Interaction

In the preceding Sections, the harmonic approach has been applied to describe the interaction between molecules in a one-dimensional chain where solitons may exist. The solutions obtained in this way are valid if the inequality, $V \ll V_0$, holds, since if V approaches V_0, the energy and mass of the solitons become infinite, and neighboring molecules approach each other up to the point of contact. Nevertheless, the harmonic approximation cannot be justified if molecules are considerably close to each other. A rigorous theory should involve anharmonic terms in the equations describing molecular motions. A particular case where cubic anharmonicity in intermolecular interactions is included has been considered by Zolotariuk and the present author [147]. A more general case of an arbitrary and rather general nonlinear interaction between molecules has been further studied by the same authors [48,166,167,168]. We give below the basic results of these papers.

Within the continuum approach, the Hamiltonian of an exciton (electron) interacting with molecules, has the following form

$$H = \frac{1}{a}\int\left\{\frac{\hbar^2}{2m}|\psi_x|^2+\frac{1}{2}Mu_t^2+MV_0^2U(\rho)-\Theta\rho|\psi|^2\right\}dx, \tag{17.1}$$

where m is an effective mass of the exciton (electron), M denotes the mass of a molecule, $u(x,t)$ is its displacement from the equilibrium position, $x = na$, V_0 is the velocity of longitudinal sound in the linear approximation, Θ is a parameter characterizing the coupling of an exciton (electron) with displacement and having the dimension of energy, and

$$\rho = -u_x \tag{17.2}$$

is the relative decrease of the equilibrium separation between neighboring molecules. Notice that the zeroth energy refers to the bottom of the exciton band. We use, henceforth, subscripts for designating the appropriate derivatives with respect to the independent variables. An exciton (electron) wave function satisfies the normalization condition,

$$\frac{1}{a}\int|\psi(z,t)|^2dx = 1. \tag{17.3}$$

In eqn. (17.1), $U(\rho)$ is the dimensionless intermolecular potential which attains its minimum, being equal to zero, at $\rho = 0$ and increases as molecules approach each other, $\rho > 0$. We assume that for all positive ρ the following conditions,

$$U_{\rho\rho}(\rho) > 0, \quad U_{\rho\rho}(0) = 1, \tag{17.4}$$

are valid.

From the Hamiltonian (17.1), it follows that

$$i\hbar\psi_t + \frac{\hbar^2}{2m}\psi_{xx} + \Theta\rho\psi = 0, \tag{17.5}$$

$$u_{tt} - V_0^2 U_{\rho\rho} u_{xx} = \frac{\Theta}{M}(|\psi|^2)_x.$$

This set of equations describes the interaction of the complex field $\psi(x,t)$ with the nonlinear real field $u(x,t)$ possessing an inertia characterized by a finite velocity V_0 of longitudinal sound. Due to the translational invariance of the set of equations, one can seek its solutions in the form of excitations moving with a constant velocity. They are described by the following wave function

$$\psi(x,t) = \Phi(\xi)\exp\left\{\frac{t}{\hbar}\left[mxV - \left[\mathcal{E} + \frac{1}{2}mV^2\right]\right] + i\eta\right\} \tag{17.6}$$

and the function $\rho(x,t) = \rho(\xi)$, too. The latter characterizes the relative displacements of molecules. Here

$$\xi \equiv (x - x_0 - Vt)/a, \tag{17.7}$$

and x_0 and η are arbitrary constants.

Using eqn. (17.6), one can transform the set of equations, given by (17.5), to the following set

$$D\Phi_{\xi\xi}+\Theta\rho\Phi = -\mathcal{E}\Phi,\quad D\equiv \hbar^2/2ma^2, \tag{17.7a}$$

$$MV_0^2\left[U_{\rho\rho}(\rho)-\left(\frac{V}{V_0}\right)^2\right]\rho_\xi = \Theta(\Phi^2)_\xi.$$

We restrict ourselves by searching for the solutions satisfying the boundary conditions

$$\Phi(\pm\infty) = \Phi_\xi(\pm\infty) = \rho(\pm\infty) = 0 \tag{17.8}$$

as $|\xi|\to\infty$, and the normalization condition

$$\int\Phi^2(\xi)d\xi = 1. \tag{17.9}$$

Taking these boundary conditions into account and introducing the dimensionless parameters,

$$\lambda = \frac{-\mathcal{E}}{D},\quad \sigma = \frac{\Theta}{D},\quad \tau = \frac{\Theta}{MV_0^2},\quad s = V/V_0, \tag{17.10}$$

one transforms the set (17.7a) of equations to its dimensionless form

$$\Phi_{\xi\xi}+\sigma\rho\Phi = \lambda\Phi, \tag{17.11}$$

$$W_\rho(\rho) = \tau\Phi^2, \tag{17.12}$$

where

$$W(\rho) = U(\rho)-\frac{1}{2}s^2\rho^2. \tag{17.13}$$

It follows from the conditions (17.4) and (17.8), and eqn. (17.12) that the allowed value of soliton velocity are in the interval

$$0\leqslant s\leqslant 1.$$

A single excitation, or autolocalization, is broken if the excitation velocity V is greater than V_0, i.e., $s^2>1$, since the inertial $u(x,t)$ is not able to follow the field $\Phi(\xi)$. Such excitations are described by blurred wave packets.

As follows from eqn. (17.12) and the boundary conditions (17.7), the function $W\rho$ must be positive for all $\rho>0$ belonging to the domain of the definition of $U(\rho)$ for soliton solutions to exist.

Multiplying eqn. (17.1) by Φ_ξ and further integrating from $-\infty$ to ξ, and taking the boundary conditions (17.8) into account, one obtains the equation

$$(\Phi_\xi)^2 = \lambda\Phi^2(\xi) - \sigma A(\xi), \tag{17.14}$$

where

$$A(\xi) = \int_{-\infty}^{\xi} \rho(\xi')d\Phi^2(\xi'). \tag{17.15}$$

Let us assume that the functions $\Phi(\xi)$ and $\rho(\xi)$ take their maximum values, say Φ_0 and $\rho_0 <$ respectively, at $\xi = 0$. Hence, the derivative $\Phi_\xi(0)$ vanishes, and from eqn. (17.1) the following relation holds:

$$\lambda = \sigma A(0)\Phi_0^{-2}. \tag{17.16}$$

As ξ' varies from $-\infty$ to ξ, the function $\rho(\xi')$ changes from 0 to $\rho(\xi)$. Hence, using the relation

$$d\Phi^2(\xi) = \tau^{-1}dW_\rho,$$

which follows from eqn. (17.12), one can introduce the variable ρ in eqn. (15.2) and carry out an integration by parts. The result is

$$A(\xi) = \frac{1}{\tau}\int_0^{\rho(\xi)} \rho' dW_{\rho'} = \frac{1}{\tau}W_\rho(\rho)G(\rho), \quad \rho = \rho(\xi), \tag{17.17}$$

where

$$G(\rho) = \rho - W(\rho)/W_\rho(\rho). \tag{17.18}$$

Applying eqn. (17.12) and the relation

$$(\Phi_\xi)^2 = W_{\rho\rho}^2\rho_\xi^2/4\tau W_\rho(\rho), \tag{17.19}$$

which follows from it, one can transform eqn. (17.1) into

$$\rho_\xi = \pm\frac{2W_\rho}{W_{\rho\rho}}\sqrt{\lambda - \sigma G(\rho)}. \tag{17.20}$$

According to eqns. (17.3) and (17.4), the spectral parameter λ is given by the expression

$$\lambda = \sigma G(\rho_0) \tag{17.21}$$

and substituting this into (17.20) yields finally

$$d\xi(\rho) = \frac{\pm W_{\rho\rho}d\rho}{2\sqrt{\sigma}\,W_\rho\sqrt{G(\rho_0)-G(\rho)}}. \tag{17.22}$$

The implicit formula for $\rho(\xi)$ is then derived by integrating eqn. (17.22),

$$\xi(\rho) = \pm\frac{1}{2\sqrt{\sigma}}\int_{\rho(\xi)}^{\rho_0}\frac{W_{\rho\rho}d\rho}{W_\rho\sqrt{G(\rho_0)-G(\rho)}}. \tag{17.23}$$

Using this expression for $\rho(\xi)$ one obtains from eqn. (17.12)

$$\Phi(\xi) = \sqrt{W_\rho/\tau}. \tag{17.24}$$

This equality enables us to transform the normalization condition (17.24) into the following form

$$2\int_{-\infty}^{0}\Phi^2(\xi)d\xi = \frac{2}{\tau}\int_{0}^{\rho_0}W_\rho(\rho)\,|\,d\xi(\rho)\,| = 1. \tag{17.25}$$

Substituting the differential (17.20) into eqn. (17.25), one obtains the equation

$$\int_{0}^{\rho_0}\frac{W_{\rho\rho}d\rho}{\sqrt{G(\rho_0)-G(\rho)}} = \tau\sqrt{\sigma}, \tag{17.26}$$

for the value ρ_0 with the known $W(\rho)$. If ρ_0 is given, with the help of eqn. (17.24) one can evaluate the maximum amplitude of the soliton,

$$\Phi_0 = [\tau^{-1}W_\rho(\rho_0)]^{1/2}. \tag{17.27}$$

The energy and momentum of a soliton moving with velocity V are expressed in terms of the function $W(\rho)$ and its derivatives. In particular, the total energy $E(V)$ is represented by the sum of the following three terms,

$$E(V) = \frac{1}{2}mV^2+\mathcal{E}(V)+Q(V), \tag{17.28}$$

where

$$\mathcal{E}(V) = -\lambda D = -\tau G(\rho_0)MV_0^2 \tag{17.29}$$

is the energy of an exciton (electron) which interacts with a deformed chain, and

$$Q(V) = 2MV_0^2\int_{-\infty}^{0}[W(\rho)+s^2\rho^2]d\xi \tag{17.30}$$

is the deformation energy of the chain in question. Using eqn. (17.22), one can express (17.30) in terms of ρ,

$$Q(V) = \frac{MV_0^2}{\sqrt{\sigma}} \int_0^{\rho_0} \frac{W_{\rho\rho}[W(\rho)+s^2\rho^2]d\rho}{W_\rho \sqrt{G(\rho_0)-G(\rho)}}. \tag{17.30a}$$

The momentum is determined similarly,

$$P(V) = \left[m + M\int_{-\infty}^{\infty} \rho^2 d\xi\right] V = V\left[m + \frac{M}{\sqrt{\sigma}} \int_0^{\rho_0} \frac{\rho^2 W_{\rho\rho} d\rho}{W_\rho \sqrt{G(\rho_0)-G(\rho)}}\right]. \tag{17.31}$$

Now we investigate the solitons which arise in systems with different types of potential energy of intermolecular interaction. In the particular case of the harmonic-type interaction, $U(\rho) = (1/2)\rho^2$, the function $W(\rho)$ and its derivatives are determined by the following expressions

$$W(\rho) = \frac{1}{2}(1-s^2)\rho^2, \quad W_\rho = (1-s^2)\rho, \tag{17.32}$$

$$W_{\rho\rho} = 1-s^2, \quad 0 \leqslant s^2 < 1.$$

According to eqn. (17.18), one has

$$G(\rho) = \frac{1}{2}\rho. \tag{17.33}$$

Equation (17.16) then takes the form

$$\int_0^{\rho_0} \frac{d\rho}{\sqrt{\rho_0 - \rho}} = \tau\sqrt{\sigma/2}(1-s^2)^{-1}. \tag{17.34}$$

Evaluating the integral in (17.34) and assuming that $\Theta = 2\chi$, one obtains

$$\rho_0 = \frac{\sigma\tau^2}{8(1-s^2)^2} = \frac{\chi^3}{D[(1-s^2)MV_0^2]^2}. \tag{17.35}$$

Therefore, the maximum amplitude of the soliton becomes

$$\Phi_0 = \left[\frac{(1-s^2)\rho_0}{\tau}\right]^{1/2} = \frac{1}{2}\left[\frac{\sigma\tau}{2(1-s)^2}\right]^{1/2} = \frac{\chi}{\sqrt{2D(1-s^2)MV_0^2}}. \tag{17.36}$$

The values ρ_0 and Φ_0 given by eqns. (17.35) and (17.36), respectively, are just these obtained in Chapter 1 for the appropriate quantities.

Equation (17.23), determining the function $\rho(\xi)$, takes now the form

$$\xi = \pm\frac{1}{\sqrt{2\sigma}}\int_{\rho(\xi)}^{\rho_0}\frac{d\rho}{\rho\sqrt{\rho_0-\rho}}. \tag{17.37}$$

Inverting this, one obtains

$$\rho(\xi) = \rho_0\,\mathrm{sch}^2\mu\xi, \tag{17.38}$$

where

$$\mu = \sqrt{\sigma\rho_0/2}. \tag{17.39}$$

With the help of eqns. (17.25) and (17.38), one can find the explicit form of the envelope function (17.19), namely

$$\Phi(\xi) = \Phi_0\,\mathrm{sch}(\mu\xi). \tag{17.40}$$

According to eqn. (17.29), the spectral parameter $\mathcal{E}(V)$ takes the value

$$\mathcal{E}(V) = -\frac{\Theta\rho_0}{2} = -\frac{1}{2}\tau\rho_0 MV_0^2 = -\frac{\chi^4}{D[(1-s^2)MV_0^2]^{2'}}. \tag{17.41}$$

The deformation energy of the chain, eqn. (17.30a), is expressed in terms of $\mathcal{E}(V)$, too,

$$Q(V) = -\frac{2(1+s^2)}{3(1-s^2)}\mathcal{E}(v). \tag{17.42}$$

Hence, the total soliton energy, including the energy of deformation, is given by

$$E(V) = \frac{1}{2}mV^2+\frac{1-5s^2}{3(1-s^2)}\mathcal{E}(V). \tag{17.43}$$

in agreement with eqn. (17.28). Using (17.31), one can find the explicit formula for the soliton momentum

$$P(V) = V\left[m+\frac{\sigma\tau^3 M}{12(1-s^2)^3}\right] = V\left[\frac{4M\chi^4}{3D[(1-s^2)Mv_0^2]^2}\right] \tag{17.44}$$

At small velocities, when the inequality $s^2 \ll 1$ is satisfied, eqns. (17.43) and (17.44) are simplified,

$$E(V) = E_0+\frac{1}{2}M_sV^2, \quad \rho(V) = M_sV, \tag{17.45}$$

where

$$E_0 = \frac{1}{3}\mathcal{E}(0) = -\frac{\sigma\tau^3}{48}MV_0^2 = -\frac{\chi^4}{3D(MV_0^2)^2}, \tag{17.46}$$

is the soliton rest energy, and

$$M_s = m + \frac{\sigma\tau^3}{12}M = m + \frac{4\chi^4 M}{3D(MV_0^2)^3}, \tag{17.47}$$

is its effective mass, which naturally exceeds that of an exciton.

So, we obtain the expressions which determine the magnitudes of displacements from the equilibrium positions (17.35), the maximum soliton amplitude (17.36), its energy, eqn. (17.43), and its momentum, eqn. (17.44). Both latter quantities approach infinity as the soliton velocity becomes close to its limit, V_0. Therefore, in this range of velocities, the given theory becomes invalid. The harmonic approach cannot be justified for large displacements of equilibrium distances between molecules. We demonstrate now that if power anharmonic terms are taken into account, all the quantities which characterize a soliton become finite, even at the limiting soliton velocity.

We begin by assuming that the potential energy of the intermolecular interaction has the form

$$U(\rho) = \frac{1}{2}\rho^2 + \frac{\gamma}{\nu+1}\rho^{\nu+1}. \tag{17.48}$$

If $s^2 = 1$, the function (17.26) and its derivatives take the following values

$$W(\rho) = \frac{\nu}{\nu+1}\rho^{\nu+1}, \quad W_\rho = \gamma\rho^\nu, \quad W_{\rho\rho} = \nu\gamma\rho^{\nu-1}. \tag{17.49}$$

Hence, the subsidiary function (17.18) is proportional to ρ; that is,

$$G(\rho) = \frac{\nu}{\nu+1}\rho. \tag{17.50}$$

Equation (17.26) which determines the value ρ_0, can be transformed to

$$I_\nu \rho_0^{(2\nu-1)/2} = \frac{\tau}{\gamma}\sqrt{\sigma/\nu(\nu+1)}, \tag{17.51}$$

where

$$I_\nu \equiv \int_0^1 \frac{y^{\nu-1}dy}{\sqrt{1-y}}. \tag{17.52}$$

The integral (17.52) is evaluated for any ν by means of the recurrence relation

$$(2\nu+1)I_{\nu+1} = 2\nu I_\nu, \quad I_1 = 2, \quad I_2 = \frac{4}{3}, \ldots. \tag{17.53}$$

Therefore, the arbitrary $\nu > 1$ and all $\gamma \neq 0$, there occurs a contraction of the intermolecular distance by the finite value

$$\rho_0 = \left[\frac{\tau}{\gamma/\nu}\sqrt{\sigma/\nu(\nu+1)}\right]^{2/(2\nu-1)}. \tag{17.54}$$

However, the condition $\rho_0 < 1$ imposes a restriction on the minimum value of γ. Evaluating ρ_0, with the help of eqn. (17.27) one can determine the amplitude of the wave function envelope

$$\Phi_0 = \sqrt{\gamma\rho_0^\nu/\tau}. \tag{17.55}$$

According to (17.29), the spectral parameter $\mathcal{E}(V_0)$ has the following value for any $\nu > 1$

$$\mathcal{E}(V_0) = -\frac{\nu\Theta\rho_0}{\nu+1} = -MV_0^2\frac{\nu\tau\rho_0}{\nu+1}. \tag{17.56}$$

Equation (17.23), which determines the function $\rho(\xi)$, transforms in this case to

$$\xi = \pm\frac{1}{2}\sqrt{\nu(\nu+1)/\sigma}\int_{\rho(\xi)}^{\rho_0}\frac{d\rho}{\rho\sqrt{\rho_\nu-\rho}}. \tag{17.57}$$

Resolving this, one finds

$$\rho(\xi) = \rho_0 \operatorname{sch}^2(\mu\xi), \tag{17.58}$$

where

$$\mu = \sqrt{\sigma\rho_0/\gamma(\nu+1)}. \tag{17.59}$$

In accordance with (17.24), the envelope of the soliton wave function takes the simple form

$$\Phi(\xi) = \sqrt{\nu/\tau}\,\rho^{\nu/2}(\xi) = \Phi_0 \operatorname{sch}^\nu(\mu\xi), \tag{17.60}$$

where Φ_0 is given by eqn. (17.55).

Bearing in mind eqn. (17.30a), one obtains the energy of chain deformation

$$Q(V_0) = MV_0^2\sqrt{\nu(\nu+1)/\sigma}\left[\frac{4}{2}\rho_0^{3/2}+\frac{\gamma I_{\nu+1}}{\nu+1}\rho_0^{\nu+1/2}\right]. \tag{17.61}$$

Substituting eqn. (17.54) for ρ_0 into (17.56) and (17.61) and taking into account the recurrence relation (17.53), we find the total energy of the soliton moving with velocity V_0

$$E(V_0) = \frac{1}{2}mV_0^2+\mathcal{E}(V_0)+Q(V_0), \tag{17.62}$$

where

$$\mathcal{E}(V_0) = -MV_0^2\left[\left[\frac{\nu\tau}{\nu+1}\right]^{\nu}\frac{\sqrt{\sigma\tau}}{\nu\gamma I_\nu}\right]^{2/(2\nu-1)}, \tag{17.63}$$

$$Q(V_0) = -\frac{2\mathcal{E}(V_0)}{2\nu+1}+\frac{4}{3}MV_0^2\left\{\left[\frac{\nu(\nu+1)}{\sigma}\right]^{\nu-2}\left[\frac{\tau}{\gamma I_\nu}\right]^3\right\}^{1/(2\nu-1)}. \tag{17.64}$$

The soliton momentum is derived in a similar way,

$$P(V_0) = \left\{m+\frac{4}{3}\left[\left[\frac{\nu(\nu+1)}{\sigma}\right]^{\nu-2}\left[\frac{\tau}{\nu I_\nu}\right]^3\right]^{1(2\nu-1)}\right\}V_0. \tag{17.65}$$

Therefore, using the expressions derived, one can evaluate all the quantities characterizing the soliton which moves with the limiting velocity V_0 for an arbitrary $\nu > 1$.

In the particular case of cubic anharmonicity, $\nu = 2$, one obtains

$$\rho_0 = \frac{1}{2}\left[\frac{3\sigma}{4}\right]^{1/3}\left[\frac{\tau}{\gamma}\right]^{2/3} \tag{17.66}$$

$$\Phi(\xi) = \Phi_0 \mathrm{sch}^2(\mu\xi),\ \ \Phi_0 = \frac{1}{2}\left[\frac{3\sigma}{4}\right]^{1/3}\left[\frac{\tau}{\gamma}\right]^{1/3} \tag{17.67}$$

$$\mu = \frac{1}{2}\left[\frac{\tau\sigma^2}{6\gamma}\right]^{1/3}. \tag{17.68}$$

In this case, the soliton energy and momentum, and the spectral parameter $\mathcal{E}(V_0)$, are determined as follows

$$E(V_0) = \frac{1}{2}mV_0^2+\frac{\tau}{\gamma}\left[1-\frac{3}{5}(\gamma\sigma)^{1/2}\left[\frac{\tau}{6}\right]^{2/3}\right]MV_0^2, \tag{17.69}$$

$$P(V_0) = V_0\left[m + \frac{\tau M}{\gamma}\right], \tag{17.70}$$

and

$$\mathcal{E}(V_0) = -MV_0^2(\sigma\tau^5/(6\gamma^2))^{1/3}. \tag{17.71}$$

These formulas are just the same as those obtained in [166].

18. Cubic Anharmonicity

In the preceding Section, we consider only the case of the limiting soliton velocity, i.e., $s^2 = 1$. Speaking generally, s^2 ranges over the following interval,

$$0 \leqslant s^2 \leqslant 1. \tag{18.1}$$

Assuming that $\nu = 2$, one obtains

$$W(\rho) = \frac{1}{2}\gamma\rho^2\left[\delta + \frac{2}{3}\rho\right], \quad \delta \equiv (1-s^2)/\gamma, \tag{18.2}$$

and

$$W_\rho = \gamma\rho(\delta+\rho), \quad W_{\rho\rho} = \gamma(\delta+2\rho). \tag{18.2a}$$

Therefore, the subsidary function (17.18) takes now the form

$$G(\rho) = \rho\left[\delta + \frac{4}{3}\rho\right]/2(\delta+\rho). \tag{18.3}$$

A difference of functions (18.3), appearing in the integrals, can be written as follows

$$G(\rho_0) - G(\rho) = (\rho_0-\rho)\Theta(\rho_0,\rho), \tag{18.4}$$

where

$$\Theta(\rho_0,\rho) = \frac{\delta^2 + (4/3)[\delta(\rho_0+\rho)+\rho_0\rho]}{2(\delta+\rho_0)(\delta+\rho)}. \tag{18.5}$$

As may be easily shown, the function (18.5) varies for all values of δ and ρ_0 in the limits,

$$\frac{1}{2} \leqslant \Theta(\rho_0,\rho) \leqslant \frac{2}{3}.$$

In particular, one can use the following approximate expression

$$\Theta(\rho_0,\rho) \approx \Theta(\rho) \equiv \frac{\delta^2+(4/3)\rho_0(2\delta+\rho_0)}{2(\delta+\rho_0)^2} \tag{18.6}$$

which, substituted into eqn. (17.26), yields

$$\int_0^{\rho_0} \frac{(\delta+2\rho)d\rho}{\sqrt{\rho_0-\rho}} = \frac{\tau}{\gamma}\sqrt{\sigma\Theta(\rho_0)}. \tag{18.7}$$

Evaluating this integral, one obtains

$$1-s^2+\frac{4}{3}\gamma\rho_0 = \frac{1}{2}\sqrt{\sigma\Theta(\rho_0)/\rho_0}. \tag{18.8}$$

Assuming $\Theta(\rho_0)$ is a constant, eqn. (18.8) transforms into the cubic equation for ρ_0. Since its left-hand side is a linearly increasing function of ρ_0, and the right-hand one is inversely proportional to ρ_0, this equation possesses a unique real (positive) solution. In addition, it follows from the form of this equation that its solution, i.e., the maximum soliton amplitude ρ_0 increases as the soliton velocity does too, and decreases if the parameters γ of anharmonicity increases. According to (17.27), the amplitude of the soliton wave function envelope takes the value

$$\Phi_0 = \sqrt{(1-s^2+\gamma\rho_0)\rho_0/\tau}. \tag{18.9}$$

The displacement function $\rho(\xi)$ is determined by inverting the integral

$$\xi = \pm\frac{1}{2\sqrt{\sigma}}\int_{\rho(\xi)}^{\rho_0} \frac{f(\rho)d\rho}{\rho\sqrt{(\rho_0-\rho)\,\Theta(\rho_0,\rho)}}, \tag{18.10}$$

where

$$f(\rho) = (\delta+2\rho)/(\delta+\rho). \tag{18.11}$$

The function (18.11), as well as $\Theta(\rho_0,\rho)$, weakly depends on ρ. One can easily prove that the following inequality holds,

$$\sqrt{3/2} \leqslant f(\rho)[\Theta(\rho_0,\rho)]^{-1/2} \leqslant \sqrt{6}. \tag{18.12}$$

Hence, one can approximate the integrand as follows, $f(\rho)\times[\Theta(\rho_0,\rho)]^{-1/2} \approx 2$. Then, inverting the integral (18.10) one obtains

$$\rho(\xi) = \rho_0 \operatorname{sch}^2(\mu\xi), \quad \mu = \frac{1}{2}\sqrt{\sigma\rho_0}. \tag{18.13}$$

In fact, μ slowly varies with ξ, and at the soliton centre we have

$$\mu \approx \frac{\sqrt{\sigma\rho_0 \Theta(\rho_0,\rho_0)}}{f(\rho_0)} = \frac{\sqrt{\sigma\rho_0/2}\sqrt{\delta^2+(4/3)\rho_0(2\delta+\rho_0)}}{\delta+2\rho_0}, \tag{18.14}$$

while at its ends we have

$$\mu \approx \sqrt{\sigma\rho_0\Theta(\rho_0,0)} = \left[\frac{\sigma\rho_0(\delta+(4/3)\rho_0)}{2(\delta+\rho_0)}\right]^{1/2}. \tag{18.15}$$

The envelope of the soliton wave function is given by the expression

$$\Phi(\xi) \approx \sqrt{\rho_0/\nu}\operatorname{sch}(\mu\xi)\sqrt{1-s^2+\gamma\rho_0\operatorname{sch}^2(\mu\xi)}. \tag{18.16}$$

The spectral parameter and the deformation energy of the chain are given in terms of the amplitude ρ as follows,

$$\mathcal{E}(V) = -MV_0^2\frac{\tau\rho_0(1-s^2+(4/3)\gamma\rho_0)}{2[1-s^2+\gamma\rho_0]} \tag{18.17}$$

and

$$Q(V) = \frac{MV_0^2}{2\sqrt{\sigma}}\int_0^{\rho_0}\frac{f(\rho)(1+s^2+2\gamma\rho/3)\rho}{\sqrt{(\rho_0-\rho)\Theta(\rho_0,\rho)}}d\rho, \tag{18.18}$$

where the function $f(\rho)$ is determined by eqn. (18.11). Assuming that $f(\rho)[\Theta(\rho_0,\rho)]^{-1/2} \approx f(\rho_0)[\Theta(\rho_0,\rho_0)]^{-1/2}$, one obtains

$$Q(V) \approx \frac{2}{3}\rho_0^{3/2}\frac{f(\rho_0)MV_0^2}{\sqrt{\sigma\Theta(\rho_0,\rho_0)}}\left[1+s^2+\frac{8\gamma\rho_0}{15}\right] \tag{18.19}$$

which can be simplified by putting $f(\rho_0)[\Theta(\rho_0,\rho_0)]^{-1/2} \approx 2$.

The total soliton energy is given by eqn. (17.28) where the spectral parameter $\mathcal{E}(V)$ and the deformation energy $Q(V)$ are given by (18.17) and (18.19), respectively. In a similar way one can evaluate soliton momentum,

$$P(V) \approx V\left\{m+\frac{4M\rho_0^{3/2}f(\rho_0)}{3\sqrt{\sigma\Theta(\rho_0,\rho_0)}}\right\}. \tag{18.20}$$

Note that the expressions derived in this Section are obtained in [47,148].

19. Quartic Anharmonicity

In this case, the intermolecular interaction is given by

$$U(\rho) = \frac{1}{2}\rho^2\left[1+\frac{1}{2}\gamma\rho^2\right], \tag{19.1}$$

and the function $W(\rho)$ and its derivatives take the form

$$W(\rho) = \frac{1}{2}\gamma\rho^2\left[\delta+\frac{1}{2}\rho^2\right], \quad \delta \equiv (1-s^2)/\gamma, \tag{19.2}$$

and

$$W_\rho(\rho) = \gamma\rho(\delta+\rho^2), \quad W_{\rho\rho} = \gamma(\delta+3\rho^2). \tag{19.2a}$$

Hence,

$$G = \rho\left[\delta+\frac{3}{2}\rho^2\right]/2(\delta+\rho^2) \tag{19.3}$$

and

$$G(\rho_0)-G(\rho) = (\rho_0-\rho)\Theta(\rho_0,\rho), \tag{19.4}$$

where

$$\Theta(\rho_0,\rho) = \frac{2\delta^2+\delta[3(\rho_0^2+\rho^2)+\rho_0\rho]+3\rho_0^2\rho^2}{4(\delta+\rho_0^2)(\delta+\rho^2)} \tag{19.5}$$

varies slowly with ρ_0,ρ, and δ in the following limits

$$\frac{1}{2} \leqslant \Theta(\rho_0,\rho) \leqslant \frac{7}{8}. \tag{19.6}$$

Therefore, as in the preceding Section, one can approximate $\Theta(\rho_0,\rho)$ as follows

$$\Theta(\rho_0) = (2\delta^2+7\delta\rho_0^2+3\rho_0^4)/4(\delta+\rho_0^2)^2. \tag{19.7}$$

The equation for the amplitude ρ_0 now takes the form

$$\int_0^{\rho_0}\frac{[\delta+3\rho^2]d\rho}{\sqrt{\rho_0-\rho}} = \frac{\tau}{\gamma}\sqrt{\sigma\Theta(\rho_0)}. \tag{19.8}$$

Evaluating this integral, one can transform this equation to

$$1-s^2+\frac{8}{3}\gamma\rho_0^2 = \frac{\tau}{2}\sqrt{\sigma\Theta(\rho_0)/\rho_0}. \tag{19.9}$$

Equation (19.9) possesses a single real positive root in all cases. This root determines the amplitude of the function of the relative displacements $\rho(\xi)$. Further, all the quantities characterizing the soliton are expressed in terms of this root. As in the case of cubic anharmonicity, it follows immediately from eqn. (19.9) that as the soliton velocity increases, or χ decreases, the soliton amplitude grows, but its growth is slower than in the case of cubic anharmonicity.

According to eqn. (18.10), the displacement function $\rho(\xi)$ is determined by inverting the same integral (18.10) where the functions $\Theta(\rho_0,\rho)$ and $f(\rho)$ are now given by the expression (19.5) and

$$f(\rho) = (\delta+3\rho^2)/(\delta+\rho^2), \tag{19.10}$$

the latter being weakly dependent on ρ. One can easily prove that

$$2\sqrt{2/7} \leqslant \frac{f(\rho)}{\sqrt{\Theta(\rho_0,\rho)}} \leqslant 3\sqrt{2}. \tag{19.11}$$

As in the preceding Section, the displacement function can be represented as follows

$$\rho(\xi) = \rho_0\,\mathrm{sch}^2(\mu\xi), \tag{19.12}$$

where μ is a slowly varying function of ξ, such that $\mu \approx \sqrt{\sigma\rho_0\Theta(\rho_0)}/f(\rho_0)$ and

$$\mu \approx \sqrt{\sigma\rho_0\Theta(\rho_0,0)} = \frac{1}{2}\sqrt{\frac{\sigma\rho_0(2\delta+3\rho_0^2)}{\delta+\rho_0^2}} \tag{19.13}$$

at the centre of the soliton and $\xi\to\pm\infty$, respectively.

In this case, the envelope of the soliton wave function becomes

$$\Theta(\xi) \approx \sqrt{\rho_0/\tau}\,\mathrm{sch}(\mu\xi)\sqrt{1-s^2+\gamma\rho_0^2\mathrm{sch}^4(\mu\xi)}. \tag{19.14}$$

The spectral parameter $\xi(V)$ and the chain deformation energy $Q(V)$ which appear in eqn. (17.28) for the energy of soliton take the form

$$\mathcal{E}(V) = -\frac{\tau}{2}\rho_0 MV_0^2\left[1-s^2+\frac{3}{2}\gamma\rho_0^2\right]/(1-s^2+\gamma\rho_0^2), \tag{19.15}$$

$$Q(V) \approx \frac{2}{3}\rho_0^{3/2}\frac{f(\rho_0)MV_0^2}{\sqrt{\sigma\Theta(\rho_0)}}\left[1+s^2+\frac{4}{35}\gamma\rho_0^2\right], \tag{19.16}$$

where the functions $\Theta(\rho_0,\rho_0)$ and $f(\rho_0)$ are given by (19.7) and (19.20), respectively. Therefore, as in the case of cubic anharmonicity, within the treatment presented, the soliton momentum is determined by eqn. (18.20) with the appropriate magnitudes of $\Theta(\rho_0,\rho_0)$ and $f(\rho_0)$.

20. Repulsive Core Potential

The formation of a soliton in a linear molecular chain is accompanied by a contraction of the intermolecular separation a. Within the coordinate frame ξ moving with the soliton, this naturally the following inequality

$$\rho(\xi) \leqslant 1, \quad l = 1-d/a, \tag{20.1}$$

where d is the molecular diameter. In this relation, the intermolecular interactions satisfying the inequality (20.1) are of a particular interest.

As an example, let us consider the interaction given by the function

$$U(\rho) = \frac{1}{2}\rho^2\left[1+\frac{2\gamma/\rho}{3(l-\rho)}\right], \tag{20.2}$$

with $0 \leqslant \rho < 1$. If $\rho \ll l$, this potential can be approximated as follows

$$U(\rho) \approx \frac{1}{2}\rho^2+\frac{\gamma}{3}\rho^3+\frac{\gamma}{3l}\rho^4+\dots. \tag{20.2a}$$

By analogy with the two preceding Sections, one can obtain the soliton solutions in the general case of the potential (20.2) and evaluate further the dynamical properties of soliton for $0 \leqslant s^2 \leqslant 1$. However, the formulas derived in this way are rather complicated compared with the foregoing ones. For this reason and to simplify the calculations for the purpose of clarity, one should consider the case $s^2 = 1$, for which

$$W(\rho) = \gamma l\rho^3/3(l-\rho), \tag{20.3}$$

and

$$W_\rho = \gamma l\rho^2(3l-2\rho)/3(l-\rho)^2, \tag{20.3a}$$

$$W_{\rho\rho} = 2\gamma l\rho[3l^2 - 3l\rho + \rho^2]/(l-\rho)^3,$$

$$G(\rho) = \rho[2l-\rho]/(3l-2\rho). \tag{20.4}$$

Therefore, one obtains

$$G(\rho_0) - G(\rho) = (\rho_0 - \rho)\Theta(\rho_0, \rho), \tag{20.5}$$

where

$$\Theta(\rho_0, \rho) = \frac{6l^2 - 3l(\rho_0 + \rho) + 2\rho_0\rho}{(3l-2\rho_0)(3l-2\rho)}. \tag{20.6}$$

It is easily shown that $\Theta(\rho_0,\rho)$ varies slowly in the range, $0 \leqslant \rho_0 < l$, $0 \leqslant \rho < l$, and

$$\frac{3}{2} < \Theta(\rho_0, \rho) < 2. \tag{20.7}$$

One can use further the approximate relation, $\Theta(\rho_0,\rho) \approx \Theta(\rho_0,\rho_0)$.

The equation for determining the amplitude ρ_0 now takes the form

$$\int_0^{\rho_0} \frac{[3l^2 - 3l\rho + \rho^2]\rho d\rho}{(l-\rho)^3\sqrt{\rho_0 - \rho}} = 3\tau\sqrt{\sigma\Theta(\rho_0,\rho_0)}\,/\,2\gamma l. \tag{20.8}$$

Evaluating the integral on the left-hand side of eqn. (20.8), one has

$$\frac{l}{4(l-\rho_0)^2}\left\{5l - 2\rho_0 + \frac{3l^2}{\sqrt{\rho_0(l-\rho_0)}}\operatorname{arctg}\sqrt{\frac{\rho_0}{l-\rho_0}}\right\} = \tag{20.9}$$

$$= 2 + 3\tau\sqrt{\sigma\Theta(\rho_0,\rho_0)/\rho_0}\,/\,2\gamma l.$$

One can prove easily that this equation possesses a single real (positive) solution which is just the amplitude of the displacement field $\rho(\xi)$. According to eqn. (17.24), the envelope of the wave function becomes

$$\Phi_0 = \frac{\rho_0}{l-\rho_0}\sqrt{\gamma l(3l-2\rho_0)/3\tau}. \tag{20.10}$$

The displacement function $\rho(\xi)$ is obtained from the equation

$$\xi = \pm\frac{3}{\sqrt{\sigma}}\int_{\rho(\xi)}^{\rho_0} \frac{f(\rho)d\rho}{\rho\sqrt{(\rho_0-\rho)\Theta(\rho_0,\rho)}}, \tag{20.11}$$

where

$$f(\rho) = \frac{2l^2 - 3l\rho + \rho^2}{(l-\rho)(3l-2\rho)}. \tag{20.12}$$

Inverting the integral in (20.11), one finds the explicit solution

$$\rho(\xi) = \rho_0 \operatorname{sch}^2(\mu\xi), \tag{20.13}$$

where μ is a slowly varying function which takes the following particular values: at $\xi \to 0$,

$$\mu \approx \sqrt{\sigma\rho_0/2}\frac{l-\rho_0}{3\sqrt{3l^2 - 3l\rho_0 + \rho_0^2}}, \tag{20.14}$$

and at $\xi \to \pm\infty$,

$$\mu \approx \frac{1}{2}\sqrt{\frac{\sigma\rho_0(2l-\rho_0)}{3l-2\rho_0}}. \tag{20.15}$$

Then the envelope of the wave function is

$$\Phi(\xi) \approx \frac{\sqrt{\gamma/3\tau}}{\rho_0^{-1}\operatorname{ch}^2(\mu\xi) - l^{-1}}\sqrt{3 - 2\rho_0 l^{-1}\operatorname{sch}^2(\mu\xi)}. \tag{20.16}$$

The spectral parameter $\mathcal{E}(V)$, the deformation energy $Q(V)$, and the momentum of the soliton are determined by the following expressions

$$\mathcal{E}(V_0) = -\tau\rho_0 M V_0^2 \frac{2l-\rho_0}{3l-2\rho_0}, \tag{20.17}$$

$$Q(V_0) = \frac{2MV_0^2}{\sqrt{\sigma}}\int_0^{\rho_0}\frac{f(\rho)[3(l-\rho)+\gamma l\rho]\rho d\rho}{(l-\rho)\sqrt{(\rho_0-\rho)\Theta(\rho_0,\rho)}}, \tag{20.18}$$

$$P(V_0) \approx V_0\left[m + \frac{6M}{\sqrt{\sigma}}\int_0^{\rho_0}\frac{f(\rho)\rho d\rho}{\sqrt{(\rho_0-\rho)\Theta(\rho_0,\rho)}}\right]. \tag{20.19}$$

Assuming that $f(\rho)[\Theta(\rho_0,\rho)]^{-1/2} \approx f(\rho_0)[\Theta(\rho_0,\rho_0)]^{-1/2}$ in the integrands appearing in (20.18) and (20.19), and evaluating further these integrals, one obtains finally

$$Q(V_0) \approx \frac{4MV_0^2 f(\rho_0)}{\sqrt{\sigma\Theta(\rho_0,\rho_0)}}\left\{2\left[1-\frac{\gamma l}{3}\right] - \gamma l^2\rho_0^{1/2} + \right. \tag{20.20}$$

$$\left. + \gamma l^3(l-\rho_0)^{-1/2}\operatorname{arctg}\sqrt{\frac{\rho_0}{l-\rho_0}}\right\}$$

and

$$P(V_0) \approx V_0 \left[m + \frac{8Mf(\rho_0)\rho_0^{3/2}}{\sqrt{\sigma\Theta(\rho_0, \rho_0)}} \right]. \tag{20.21}$$

The total soliton energy is determined by eqn. (17.28) where the spectral parameter $\mathcal{E}(V)$ and the deformation energy $Q(V)$ are given by eqns. (20.17) and (20.20), respectively.

Chapter 6

SUPER-SOUND ACOUSTIC SOLITONS

To explain phenomena such as conductivity, heat expansion, and related ones, one should take into account the nonlinearity, or anharmonicity of interactions between the molecules of solids. Due to anharmonicity, phonon interaction results in their scattering and heat conduction.

Anharmonic interactions are usually negligible, and their effect is considered as a small perturbation correction. However, for some crystals, such as solid helium, solid hydrogen, ferroelectrics, and others in which structural phase transitions take place, anharmonicity effects become considerable, and the perturbation treatment is invalid.

A strong anharmonicity can be observed in some one-dimensional molecular systems. In such systems, nonlinearity and dispersion effects lead to soliton-type excitations arising. Excitations of this type are viewed as local displacements of the equilibrium positions of molecules which migrate along molecular chains. We call these acoustic solitons. In this Chapter the results of some theoretical studies of acoustic solitons are presented.

21. Solitons in Toda Nonlinear Lattices

One of the first soliton solutions describing the local displacements of the equilibrium positions of molecules in one-dimensional lattices exponentially decaying at infinity, was investigated by Toda [265-270]. He studied a chain of molecules or atoms separated by a distance a and assumed that the displacement of equilibrium positions of molecules occupying the positions na cause additional interactions between neighboring molecules. The energy of this interaction is given by the following function

$$U = \sum_n \Phi(\rho_n), \tag{21.1}$$

where

$$\Phi(\rho_n) = \frac{\kappa}{b}\left\{-\rho_n + \frac{1}{b}[\exp(b\rho_n) - 1]\right\} \tag{21.2}$$

and

$$\rho_n \equiv \beta_{n-1} - \beta_n \tag{21.3}$$

is the contraction of separations between neighboring molecules ($a \to a - \rho_n$). Lattices with an exponential interaction of the type (21.2) are called Toda lattices.

Assuming $b\rho_n \ll 1$, one can represent the function (21.2) by the series expansion

$$\Phi(\rho_n) = \frac{1}{2}\kappa\rho_n^2 + b\frac{\kappa}{6}\rho_n^3 + \dots .$$

Hence, the parameter κ whch appears in eqn. (21.2) and has the dimension erg/cm^2, characterizes the coefficient of longitudinal elasticity of the harmonic lattice. The parameter b determines the anharmonicity.

The equations of motion of molecules in the potential field (21.1) take the following form

$$M\frac{d^2\beta_n}{dt^2} = -\frac{\partial U}{\partial \beta_n} = \frac{\kappa}{b}[e^{b\rho_n} - e^{b\rho_{n+1}}]$$

where M is the mass of a molecule. Using eqn. (21.3), one can rewrite

$$M\frac{d^2\rho_n}{dt^2} = \frac{\kappa}{b}[e^{b\rho_{n+1}} + e^{b\rho_{n-1}} - 2e^{b\rho_n}]. \tag{21.4}$$

The particular solutions of eqn. (21.4) which vanish exponentially at $n = \pm\infty$, are of the soliton type,

$$\exp(b\rho_n) - 1 = \mathrm{sh}^2(qa)\mathrm{sch}^2[q(na - Vt)], \tag{21.5}$$

where

$$V = \frac{V_0}{qa}\mathrm{sh}(qa) \tag{21.6}$$

is the velocity of the soliton motion, and $V_0 = a(\kappa/M)^{1/2}$ is the longitudinal sound velocity in the harmonic chain, i.e., for $b = 0$. As follows from eqn. (21.6), the velocity of acoustic solitons always exceeds the longitudinal sound velocity V_0. As

$V \to V_0$, the amplitude of such supersound solitons approaches zero due to $q \to 0$.

It follows from eqn. (21.5) that the contraction of intermolecular separations is given by

$$\rho_n = \frac{1}{b}\ln\left\{1+\frac{\mathrm{sh}^2(qa)}{\mathrm{ch}^2[q(na-Vt)]}\right\} \approx \frac{1}{b}\left[\frac{\mathrm{sh}(qa)}{\mathrm{ch}[q(na-Vt)]}\right]^2. \tag{21.7}$$

Equation (21.6) characterizes a 'dispersion' of solitons. It also determines implicitly the soliton parameter q describing the width of the excitation region, $\Delta n \approx 2\pi/q$, and the displacement amplitude in terms of velocity V. If the inequality $qa \ll 2\pi$ applies, one can use the continuum approach and represent the function given by eqn. (21.7), as follows

$$\rho(z,t) = \frac{1}{b}\left[\frac{\mathrm{sh}(qa)}{\mathrm{ch}(q\zeta)}\right]^2, \tag{21.8}$$

where $\xi = z-z_0-Vt$, and z_0 corresponds to the soliton centre at $t = 0$. Due to the translational symmetry of the problem under study, z_0 can be chosen at an arbitrary point of the system. Within the continuum approach, eqn. (21.3) becomes

$$\rho(z,t) = -a\frac{\partial}{\partial z}\beta(z,t). \tag{21.9}$$

Hence, the function $\rho(z,t)$ describes the local contraction of the molecular chain. Using the relation (21.9), one finds an explicit formula for the displacement of molecules from their equilibrium positions (with $\beta(\infty)=0$)

$$\beta(\zeta) = \frac{2[1-\mathrm{th}(q\zeta)]\mathrm{sh}^2(qa)}{3a^2bq[1+2\mathrm{ch}^2(q\zeta)]}. \tag{21.10}$$

A mobile supersound acoustic soliton carries the energy

$$E(V) = K+W,$$

where

$$W = \frac{1}{a}\int\Phi[\rho(\zeta)]d\zeta = 2\kappa\mathrm{sh}^2(qa)\left[1+\frac{4}{15}\mathrm{sh}^2(qa)\right]/3aqb^2, \tag{21.11}$$

is the potential energy of soliton, and

$$K = \frac{M}{2a}\int\left[\frac{\partial\beta(\zeta)}{\partial z}\right]^2 d\zeta = \frac{2MV^2\mathrm{sh}^4(qa)}{3qa^3b^2} \tag{21.12}$$

its kinetic energy. Soliton motion is accompanied by the transfer of momentum,

$$P = -\frac{M}{a}\int\frac{\partial B}{\partial t}\frac{\partial \beta}{\partial \zeta}d\zeta = MV\mathrm{sh}^4(qa)/3qb^2a^3. \tag{21.13}$$

An important role of acoustic solitons in heat transfer in nonmetalic crystals with isotopic admixtures has been pointed out by Toda [236]. He demonstrated that the inclusion of soliton-type excitations enables an explanation of the increase of heat conductivity of crystals containing admixtures whose mass slightly differs from that of the atoms of the basic substance. Unlike phonons, solitons do not really lose energy as a result of collisions with atoms of admixtures.

If a molecular chain attaches an excess electron (see Section 24), its local contraction, generated by a soliton, becomes the potential well for that electron, which accompanies the soliton under its motion.

22. Acoustic Solitons in a Molecular Chain with Cubic Anharmonicity

Let us assume that the potential energy of displacement of molecules in a chain from their equilibirum positions, na, is given by

$$U = \frac{1}{2}\sum_n(\kappa\rho_n^2+2\gamma\rho_n^3/3), \tag{22.1}$$

where γ is a parameter of cubic anharmonicity, $\rho_n = \beta_{n-1}-\beta_n$, and β_n describe the displacements of equilibrium positions of molecules.

A set of equations of motion of the molecules with mass M in the chain under study has the form

$$M\frac{d^2\beta_n}{dt^2}+\kappa[2\beta_n-\beta_{n+1}-\beta_{n-1}]+\gamma(\rho_{n+1}^2-\rho_n^2) = 0. \tag{22.2}$$

Equation (22.2) involves a small parameter γ. For this reason, within the continuum approach we take into account the derivatives on β relative to z up to the fourth degree only; that is,

$$\beta_{n\pm1}(t) = \left[1\pm a\frac{\partial}{\partial z^2}+\frac{a^2}{2}\frac{\partial^2}{\partial z^2}\pm\frac{a^3}{6}\frac{\partial^3}{\partial z^3}+\frac{a^4}{24}\frac{\partial^4}{\partial z^4}\right]\beta(z,t).$$

Equation (22.2) transforms into the following differential equation

$$\left[M\frac{\partial^2}{\partial t^2}-\kappa a^2\frac{\partial^2}{\partial z^2}-\frac{\kappa a^4}{12}\frac{\partial^4}{\partial z^4}\right]\beta(z,t)+a\gamma\frac{\partial}{\partial z}\rho^2(z,t) = 0.$$

Multiplying this by the operator $-a\partial/\partial z$ and using eqn. (21.9), one obtains

$$\left\{\frac{\partial^2}{\partial t^2}-V_0^2\left[\frac{\partial^2}{\partial z^2}+\frac{a^2}{12}\frac{\partial^4}{\partial z^4}\right]\right\}\rho(z,t)-\frac{\gamma}{\kappa}\frac{\partial^2}{\partial z^2}\rho^2(z,t) = 0. \qquad (22.3)$$

Neglecting anharmonicity, $\gamma = 0$, the solution of this equation is represented in plane-wave from (the case of long-wave phonons)

$$\rho(z,t) = \rho_0\exp[i(kz-\omega f)]$$

with

$$\omega^2 = k^2V_0^2[1-k^2a^2/12], \quad ka \ll 1, \qquad (22.4)$$

corresponding to the following dispersion relation

$$\omega^2 = \frac{4\kappa}{M}\sin^2(ka/2), \quad V_0^2 = a^2\kappa/M.$$

According to eqn. (22.4), long-wave phonons possess a negative dispersion with the maximum phase velocity (at $k\approx 0$) equal to V_0. The group velocity, $d\omega/dk = V_0\cos(ka/2)$ decreases from V_0 to zero (at $k=\pi/a$) as k increases.

Due to translational symmetry, we seek the solutions of eqn. (22.4) in the form of excitations moving with a constant velocity V. Initially, we put

$$\rho(z,t) = \rho(\zeta), \quad \zeta \equiv (z-z_0-Vt). \qquad (22.5)$$

Carrying out a double integration over ξ under the condition $\rho(-\infty) = 0$, one obtains the nonlinear equation

$$\left[\frac{a^2}{12}\frac{d^2}{d\zeta^2}-(s^2-1)\right]\rho(\zeta)+\frac{\gamma}{\kappa}\rho^2(\zeta) = 0, \quad s = \frac{V}{V_0}, \qquad (22.6)$$

which is usually called the *Boussinesq equation*. Its solutions were studied by Toda and Vaddatti [270]. By means of direct substitution, one can show that the lowest energy of the excitation which moves with velocity $V > V_0$ corresponds to the solution of a solitary-wave type

$$\rho(\zeta) = a^2q^2\kappa/2\gamma\mathrm{ch}^2(q\zeta), \qquad (22.7)$$

where

$$q^2 = 3(s^2-1)/a^2,$$

or

$$V = V_0\sqrt{1+a^2q^2/3}. \tag{22.8}$$

The localized excitation, represented via the bell-shaped function (22.7) which describes a contraction of intermolecular separations in the excitation region, is called a *supersound acoustic solution*.

A soliton 'width', i.e., the region enveloped by the excitation, is determined by the following expression

$$\Delta\zeta \approx 2\pi/q = 2\pi[3(s^2-1)]^{-1/2}a,$$

and its maximum amplitude is

$$\rho(0) = 3\kappa(s^2-1)/2\gamma.$$

Hence, with a decrease of velocity V, the soliton amplitude decreases also, although its width increases. In the limit $V \to V_0$, it vanishes.

Using eqn. (21.9) and imposing the condition $\beta(\infty) = 0$, one finds the function described the displacements of the molecular equilibrium positions,

$$\beta(\zeta) = \frac{a\kappa}{2\gamma}[1-\mathrm{th}(q\zeta)] = \frac{a\kappa}{\gamma[1+\exp(2q\zeta)]}.$$

This determines a step-type transition from $a\kappa/\gamma$ at $\zeta = -\infty$ to zero at $\zeta = \infty$.

The continuum approach is justified if the following inequality holds

$$qa < 2\pi. \tag{22.9}$$

According to eqn. (22.8), this inequality determines the boundaries of the permissible velocities of soliton motion,

$$1 < s^2 < 1+4\pi^2/3.$$

The energy and momentum of the acoustic soliton are given by the following formulas

$$E(s) = \frac{1}{2a}\int\left[\kappa\rho^2 + M\left[\frac{\partial\beta}{\partial t}\right]^2 + 2\gamma\rho^2/3\right]d\zeta,$$

$$P(s) = -\frac{M}{a}\int \frac{\partial\beta}{\partial t}\frac{\partial\beta}{\partial\zeta}d\zeta.$$

Using eqns. (22.7), (22.8, and (21.9), and the following equation

$$\frac{\partial\beta(zt)}{\partial t} = -V\frac{d\beta(\zeta)}{d\zeta}, \tag{22.10}$$

one obtains finally

$$E(s) = \sqrt{3}\kappa^3(s^2-1)^{3/2}[1+9s^2]\{30\gamma^2\}^{-1}, \tag{22.11}$$

$$P(s) = \sqrt{3}\kappa^2(s^2-1)^{3/2}MV[a^2\gamma^2]^{-1}. \tag{22.12}$$

The solution, given by eqn. (22.7), does not take into account the conditions at the boundaries of the molecular chain. Therefore, it is correct only for those chains whose length is considerably greater that of the region enveloped by an excitation. In order to elucidate the possibilities of acoustic soliton formation, one should investigate the role of boundary effects, since such solitons may be formed only at chain ends.

23. Acoustic Solitons in a Molecular Chain with Quartic Anharmonicity

Let us now assume that the potential energy of displacements of molecular equilibrium positions has the form

$$U = \frac{1}{2}\sum_n\left\{\kappa\rho_n^2+\frac{1}{2}\delta\rho_n^4\right\}. \tag{23.1}$$

Within the continuum approach, and by including the dispersion of acoustic waves, one obtains the equation which determines the change of intermolecular equilibrium separations in an excited state moving with velocity V along a chain,

$$\left[\frac{a^2}{12}\frac{d^2}{d\zeta^2}-(s^2-1)\right]\rho(\zeta)+\frac{\delta}{\kappa}\rho^3(\zeta) = 0. \tag{23.2}$$

For a fixed value of $s = V/V_0 > 1$, a particular solution of eqn. (23.2) becomes doubly-degenerated,

$$\rho(\zeta) = \pm\sqrt{2(s^2-1)\kappa/\delta}\,\mathrm{sch}(q\zeta), \tag{23.3}$$

where

$$q^2 = 12(s^2-1)/a^2. \tag{23.4}$$

Excitations described by eqn. (23.3) with a positive sign, are called *supersound acoustic solitons*. Their formation is accompanied by a contraction of that part of the chain which belongs to the excitation region $[a \to a - \rho(\zeta)]$. Excitations corresponding to a negative sign in eqn. (23.3), are called *supersound acoustic antisolitons*. An extension of the chain accompanies their formation.

Both types of supersound solitons have the same energy and momentum for a given $s^2 > 1$,

$$E(s) = 2\kappa\sqrt{s^2-1}\,(2+s^2)/2\sqrt{3}\,\delta, \tag{23.5}$$

$$P(s) = 2\sqrt{s^2-1}\,\kappa MV/\sqrt{3}\,a^2\delta, \tag{23.6}$$

which approach zero as $V \to V_0$.

Supersound acoustic solitons and antisolitons, moving in opposite directions, can arise in an infinite chain with quartic anharmonicity under the local excitation of its arbitrary region. As mentioned before, the formation of these solitons in a chain with cubic anharmonicity becomes possible only at its ends.

24. The Capture of an Excess Electron by Supersound Acoustic Solitons

Molecular, or ionic one-dimensional periodic chains sometimes can capture an electron, which reaches a chain after detaching from an external donor molecule, reaches a chain (see Section 25). The state of such an excess electron in a strongly periodic chain is described by the following Schrödinger equation (within the effective electron mass approximation)

$$\left\{ i\hbar\frac{\partial}{\partial t} - \mathcal{E}_0 + \frac{\hbar^2}{2m^*}\frac{\partial^2}{\partial z^2} \right\} \psi(z,t) = 0 \tag{24.1}$$

where m^* is the effective electron mass, and $\mathcal{E}_0$ is the energy of the bottom of the conduction band.

Let us suggest that in such a chain there is generated an acoustic soliton accompanied by a local deformation of the chain. This formation results in an additional potential well for the electron, called traditionally the *deformation potential* (see, for instance, [35], Section 34). Neglecting the reverse effect of the electron soliton, one can represent the operator of the deformation potential created by the acoustic soliton in the following form

$$W(z,t) = -\sigma\rho(z,t), \tag{24.2}$$

where $\rho(z,t)$ is a function characterizing a decrease of intermolecular equilibrium distances, and σ is a parameter of the electron-deformation interaction (in $\mathrm{erg}\,cm^{-1}$). Under favourable conditions, an electron may be captured by the potential well (24.2) and then move inside it. This mechanism of electron capture was discussed by Zmuidzinas [281] and the present author [39].

Given an acoustic soliton in a chain, the electron motion is determined by the following Schrödinger equation

$$\left[i\hbar\frac{\partial}{\partial z} - \mathcal{E}_0 + \frac{\hbar^2}{2m^*}\frac{\partial^2}{\partial z^2} - W(z,t)\right]\psi(z,t) = 0. \tag{24.3}$$

Let us define the real amplitude function $\varphi(\zeta)$ making use of the relation

$$\psi(z,t) = \varphi(\zeta)\exp[i(kz - \omega t)], \tag{24.4}$$

where

$$k = m^* V/\hbar; \quad \zeta \equiv z - z_0 - Vt, \tag{24.5}$$

and $\hbar\omega$ is the electron energy relative to the laboratory reference frame. Substituting the relation (24.4) into eqn. (24.3), one obtains

$$\left\{\frac{d^2}{d\zeta^2} + \frac{2m^*}{\hbar^2}[E - W(\zeta)]\right\}\varphi(\zeta) = 0. \tag{24.6}$$

The function $\varphi(\zeta)$, obeying eqn. (24.6), describes the electron motion relative to the laboratory reference frame which moves with velocity V together with the soliton. In eqn. (24.6),

$$E = \hbar\omega - \mathcal{E}_0 - \frac{1}{2}m^* V^2$$

is the electron energy inside the potential well $W(\zeta)$.

For chains having a cubic anharmonicity, an acoustic soliton is characterized by the function (22.7). Hence the deformation potential given by eqn. (24.2) takes the form

$$W(\zeta) = -A\,\mathrm{sch}^2(q\zeta) \tag{24.7}$$

with

$$A \equiv \sigma^2 a^2 q^2 \kappa/2\gamma, \quad q \equiv \sqrt{3(s^2-1)/a}. \tag{24.8}$$

The solution of eqn. (24.6) with the potential well in the form (24.7) has been studied by Landau and Lifshitz ([87], §23) who showed that the electron possesses a discrete spectrum of negative energies,

$$E_n = -\hbar^2 q^2 (\nu - n)^2/2m^*, \tag{24.9}$$

where $n = 0, 1, 2, ..., n_{\max}$, and

$$n_{\max} \leqslant \nu \equiv \frac{1}{2}(\sqrt{1 + 8m^* A/\hbar^2 q^2} - 1). \tag{24.10}$$

The lowest energy level,

$$E_0 = -\frac{\hbar^2 q^2 \nu^2}{2m^*} = -\frac{3(s^2-1)\hbar^2 \nu^2}{2m^* a^2} \tag{24.11}$$

corresponds to the following wave function

$$\varphi_0(\zeta) = aq\Gamma(\nu + 1/2)\sqrt{\pi}\,\Gamma(\nu)\mathrm{ch}^\nu(q\zeta), \tag{24.12}$$

where $\Gamma(\nu)$ is the gamma function. For $a \approx 5 \times 10^{-8} cm$, the energy of the lowest electron state inside the well is

$$E_0 \approx -1.5\frac{m}{m^*}(s^2-1)\nu^2 \; eV.$$

Here, m is the free electron mass. This energy approaches zero under a decrease of soliton velocity up to that of sound velocity. An electron capture by an acoustic soliton take place iff E_0 is larger than the average energy of heat vibrations. These formulas are justified only for rather strong anharmonicity when the interaction of an electron with the soliton potential well does not substantially distort the latter.

For chains having a quartic anharmonicity, an acoustic soliton only creates a well for an electron with the following potential energy

$$W(\zeta) = -B\,\mathrm{sch}(q\zeta), \tag{24.13}$$

where

$$B \equiv \sigma\sqrt{2(s^2-1)\kappa/\delta}, \quad q^2 = 12(s^2-1)/a^2. \tag{24.14}$$

To valuate the lowest energy E_0 for an electron inside the given well (24.13), one can apply the variational principle by minimizing the functional

$$J\{\varphi\} = \frac{1}{a}\int\varphi(\zeta)\left[-\frac{\hbar^2}{2m^*}\frac{d^2}{d\zeta^2}-\frac{B}{\mathrm{ch}(q\zeta)}\right]\varphi(\zeta)d\zeta.$$

Using the simplest normalized one-parameter trial function

$$\varphi(\xi) = \sqrt{a\alpha}\,/\,\sqrt{2}\,\mathrm{ch}(\alpha\zeta), \tag{24.15}$$

one obtains

$$J\{\alpha\} = \frac{\hbar^2\alpha^2}{6m^*}-\frac{\pi B\alpha}{2q\hbar^2}. \tag{24.16}$$

A minimum of $J\{\alpha\}$ is attained at

$$\alpha_0 = -3\pi^2 m^* B^2/4q^2\hbar^2, \tag{24.17}$$

which corresponds to the following electron energy

$$E_0 = -3\pi^2 m^* B^2/4q^2\hbar^2.$$

Of course, one should take into account that electron transfer by supersound solitons requires the energy to be expended on their formation. The problem of realizing the conditions of electron transfer by supersound solitons in real systems demands special study. In particular one should investigate the reverse effect of an electron on a soliton, too. Apparently, this effect cannot be negligible. The estimations obtained in [39] demonstrate that the parameters of a molecular chain, obey rather strong conditions which seem to be unrealistic.

25. Supersound Solitons in Nonlinear Quasi-One-Dimensional Molecular Structures

It was shown in Chapter 1 that nonlinear autolocalized intermolecular excitations (solitons) can exist in a molecular chain with harmonic interaction between neighboring molecules in the presence of exciton-phonon interactions. These excitations propagate with a velocity less than the longitudinal sound velocity. As the excitation velocity approaches that of sound waves, the energy of excitation increases infinitely. Such a difficulty is removed by including anharmonic corrections to the interaction between molecules (see Chapter 5). Nevertheless, the longitudinal sound velocity remains the limiting velocity for soliton velocities even in that case, since, if the excitation velocity exceeds that of sound, the local deformation of the chain is unable to follow the soliton motion.

On the other hand, as shown in some works reviewed in preceding Sections, a local thickening of density, or acoustic solitons, can propagate in nonlinear chains taking the dispersion of sound waves into account. Their velocity exceeds that of sound in all cases. Therefore, including nonlinearity and dispersion of displacements of molecular equilibarium positions, one can expect the propagation of bound states of a quasiparticle and local deformation with a velocity exceeding that of the longitudinal sound wave, under the appropriate values of the parameters of a molecular chain. We now present the results of the investigation by Zolotariuk and the author [48,166,168] who studied the solutions of a set of coupled equations describing the interaction between a quasiparticle and the displacements of the equilibrium positions of molecules in a chain taking into account cubic anharmonicity and sound dispersion.

Consider such a chain with a single quasiparticle. Its excited states are described by the following Hamiltonian

$$H = \sum_n \Big[D\psi_n^*(2\psi_n - \psi_{n+1} - \psi_{n-1}) + \frac{1}{2} Ma^2 (du_n/dt)^2 + \tag{25.1}$$

$$+ MV_0^2\, U(u_n - u_{n-1}) + \frac{1}{2}\chi |\psi_n|^2 (u_{n+1} - u_{n-1}) \Big].$$

The square of the absolute value of the function $\psi_n(t)$ determines the probability of finding a quasiparticle at the molecule of mass M occupying the nth node. The functions $u_n(t)$ also satisfy the normalization condition

$$\sum_n |\psi_n(t)|^2 = 1. \tag{25.2}$$

In eqn. (25.1), the function $u_n(t)$ is a displacement of the equilibrium position of the n-th molecule; this displacement being normalized to the lattice constant a. The zeroth energy refers to the bottom of the energy and of the quasiparticle with an effective mass m which is determined by the resonance (or exchange for an electron, in particular) interaction D through the relation

$$m = \hbar^2/2a^2D,$$

U is a dimensionless intermolecular potential, and V_0 is the longitudinal sound wave velocity.

The following set of finite-difference equations corresponds to the Hamiltonian (25.1),

$$i\hbar\frac{d\psi_n}{dt} = \left[2D+\frac{1}{2}\chi(u_{n+1}-u_{n-1})\psi_n - D(\psi_{n+1}+\psi_{n-1})\right],$$

$$a^2\frac{d^2u_n}{dt^2} = V_0^2[U'(u_{n-1}-u_n)-U'(u_n-u_{n+1})]-\chi(|\psi_{n+1}|^2-|\psi_{n-1}|^2)/2M$$

where the prime designates henceforth the differentiation of U relative to tis argument. Substituting the expression

$$\psi_n(t) = \varphi_n(t)\exp\left\{\frac{l}{\hbar}\left[nmaV-\left(\epsilon+\frac{1}{2}mV^2\right)t\right]\right\}, \tag{25.3}$$

where ϵ is the bound stage energy of the quasiparticle in the field of the local deformation moving with velocity V along the chain, into these equations, one obtains the following equations for the real function $\varphi_n(t)$

$$i\hbar\frac{d\varphi_n}{dt} = D(\varphi_{n-1}-\varphi_{n+1})\sin(\hbar V/2aD), \tag{25.4}$$

$$\left[2D+\frac{1}{2}\chi(u_{n+1}-u_{n-1})\right]\varphi_n - D(\varphi_{n+1}+\varphi_{n-1})\cos(\hbar V/2aD) = \tag{25.5}$$

$$= \left(\epsilon+\frac{1}{2}mv^2\right)\varphi_n, \quad \sum_n \varphi_n^2 = 1,$$

$$a^2\frac{d^2u_n}{dt^2} = V_0^2[U'(u_{n-1}-u_n)-U'(u_n-u_{n+1})]+\chi(2M)^{-1}(\varphi_{n+1}^2-\varphi_{n-1}^2). \tag{25.6}$$

Assuming that the velocity V of excitation motion obeys the inequality,

$$\hbar V \ll 2aD, \tag{25.7}$$

one has, within the continuum approach, the set of equations in partial derivatives,

$$\varphi_t = -V\varphi_x, \tag{25.8}$$

$$-Da^2\varphi_{xx}+\chi au_x\varphi = \epsilon\varphi, \tag{25.9}$$

where the subscript denotes the corresponding partial derivative. Taking into account cubic anharmonicity, the function $U(\rho)$ becomes

$$U(\rho) = \frac{1}{2}\rho^2+\frac{1}{3}\gamma\rho^3, \tag{25.10}$$

where γ is a dimensionless parameter characterizing cubic anharmonicity, and $\rho(x,t)$ is the field of relative displacements of neighboring molecules,

$$\rho = -au_x.$$

In the context of the continuum approach, eqn. (25.6) can be rewritten within the terms involving fourth-order derivatives with respect to the coordinate x in the following form

$$\nu a^3 u_{xxxx}+u_{xx}-V_0^{-2}u_{tt} = 2\gamma a u_x u_{xx}+\chi(aMV_0^2)^{-1}(\varphi^2)_x = 0. \tag{25.11}$$

To make easier the study of systems with an arbitrary dispersion, we replace the parameter ν by its value $1/12$. We seek further the stationary solutions belonging to the class of rapidly decreasing functions,

$$\varphi(x,t) = \varphi(\xi),\;\; u(x,t) = u(\xi),\;\; \rho(x,t) = \rho(\xi),$$

where a dimensionless variable $\xi = (x-Vt)/a$ is introduced. Equation (25.8) transforms into an identity. Equations (25.9) and (25.11) take now the form

$$D\frac{d^2\rho}{d\xi^2}+\xi\rho\varphi+\epsilon\varphi = 0, \tag{25.12}$$

$$\mu\frac{d^2\rho}{d\xi^2}+(1-s^2)\rho+\gamma\rho^2 = \chi(MV_0^2)^{-1}\varphi^2, \tag{25.13}$$

where

$$s = V/V_0,\;\; \rho = -du/d\xi,\;\; \int\varphi^2(\xi)d\xi = 1. \tag{25.14}$$

It is convenient to represent these equations in integral form. For this purpose, one should multiply eqn. (25.12) by $d\varphi/d\xi$, and eqn. (25.13) by $d\rho/d\xi$. Integrating and including the vanishing boundary conditions, one obtains the following integral equations,

$$D\left[\frac{d\varphi}{d\xi}\right]^2+2\chi\int_0^{\psi}\rho(\tau)\tau d\tau+\epsilon\varphi^2 = 0, \tag{25.15}$$

and

$$\nu\left[\frac{d\rho}{d\xi}\right]^2 = (s^2-1)\rho^2 - \frac{2}{3}\nu\rho^2 + \frac{2\chi}{MV_0^2}\rho\varphi^3 - \frac{4\chi}{MV_0^2}\int_0^{\varphi}\rho(\tau)\tau d\tau \tag{25.16}$$

where the substitution $\xi \to \varphi(\xi)$ is carried out in the integrands. The bell-like solitons centered at $\xi = 0$ obey the relation

$$(d\varphi/d\xi)_{\xi=0} = 0. \tag{25.17}$$

Using eqn. (25.17), one obtains from eqn. (25.15) the spectral parameter

$$\epsilon = -2\chi\varphi_0^{-2}\int_0^{\varphi_0}\rho(\tau)\tau d\tau, \quad \varphi_0 \equiv \varphi(0). \tag{25.18}$$

Substituting this quantity into eqn. (25.15), the latter transforms into

$$|d\xi(\varphi)| = \left[\frac{D}{2\chi}\right]^{1/2}\left[\left[\frac{\varphi}{\varphi_0}\right]^2\int_0^{\varphi_0}\rho(\tau)\tau d\tau - \int_0^{\varphi}\rho(\tau)\tau d\tau\right]^{-1/2} d\varphi. \tag{25.19}$$

The unknown amplitude φ_0 is defined by the normalization condition for $\varphi(\xi)$. Replacing ξ by $\varphi(\xi)$, one rewrites this condition as follows

$$\int_0^{\varphi_0}\varphi^2\,|d\xi(\varphi)| = \frac{1}{2}.$$

Taking eqn. (25.19) into account, one obtains finally

$$\int_0^{\varphi_0}\tau^2\left[\left[\frac{\tau}{\varphi_0}\right]^2\int_0^{\varphi_0}\rho(y)y dy - \int_0^{\tau}\rho(y)y dy\right]^{-1/2} d\tau = \frac{\chi}{D}. \tag{25.20}$$

Equation (25.20) enables the evaluation of the maximum value of the function $\varphi(\xi)$ for a given $\rho(\varphi)$, which is the solution of the following integral equation

$$\rho(\varphi) = \nu^{-1/2}\int_0^{\varphi}\left\{(s^2-1)\rho^2(\tau) - \frac{2}{3}\nu\rho^3(\tau) + \right. \tag{25.21}$$

$$\left. + 2\chi(MV_0^2)^{-1}\left[\tau^2\rho(\tau) - 2\int_0^{\tau}y\rho(y)dy\right]\right\}|d\xi(\tau)|.$$

Equation (25.21) is derived via the integration of eqn. (25.16).

A solution of the integral equation (25.21) is equivalent to that of the set of eqn. (25.15) and (25.16), and can be derived via the method of successive approximation. Substituting a given function $\rho(\varphi)$ into eqn. (25.19) and integrating further, one obtains implicitly the function $\varphi(|\xi|)$

$$|\xi| = \left[\frac{D}{2\chi}\right]^{1/2} \int_{\varphi(\xi)}^{\varphi_0} \left[\left[\frac{\tau}{\varphi_0}\right]^2 \int_0^{\varphi_0} \rho(y)y dy - \int_0^{\tau} \rho(y)y dy\right]^{-1/2} d\tau. \tag{25.22}$$

Given $\varphi(\xi)$, one has the displacement field

$$\rho(\xi) = \rho[\varphi(\xi)]. \tag{25.23}$$

Using the latter, one evaluates the energy and momentum of the soliton moving with constant velocity V,

$$E(V) = \frac{1}{2}mV^2 + \epsilon + \frac{1}{2}MV_0^2 \int d\xi \left[(1+s^2)\rho^2 + \nu\left[\frac{d\rho}{d\xi}\right]^2 + \frac{2}{3}\gamma\rho^2\right], \tag{25.24}$$

$$P(V) = \left[m + M\int \rho^2 d\xi\right] V. \tag{25.25}$$

In order to solve the integral equation (25.22), one can choose the zeroth-order approximate function $\rho^{(0)}(\varphi)$ in the following form,

$$\rho^{(0)}(\varphi) = \alpha\varphi^2. \tag{25.26}$$

Here α is an unknown factor. Substituting this function into eqn. (25.20), one obtains the zeroth-order approximation for the amplitude function φ_0,

$$\varphi_0^0 = (\chi\alpha/8D)^{1/2}. \tag{25.27}$$

Substituting further φ_0^0 and $\rho^{(0)}(\varphi)$ into eqn. (25.20) and the right-hand side of eqn. (25.21), within the first-order approximation treatment one has

$$d\xi^{(1)}(\varphi) = \pm(2D)^{1/2}\left[\frac{\alpha\xi}{8D} - \varphi^2\right]^{-1/2} d\varphi, \tag{25.28}$$

$$\rho^{(1)}(\varphi) = \left[\frac{2D}{\nu\chi}\right]^{1/2} \int_0^{\varphi} \left\{\frac{\chi(MV_0^2)^{-1} + \alpha(s^2-1) - 2/3\gamma\alpha^2\tau^2}{\alpha\chi/8D - \tau^2}\right\}^{1/2} \tau d\tau. \tag{25.29}$$

Unfortunately, th evaluation of integral (25.29) gives a rather complicated expression. However, if α a positive root of the cubic equation

$$\gamma\chi z^3 + 12D(1-s^2)z - 12\chi D(MV_0^2)^{-1} = 0, \tag{25.30}$$

the integral is easily evaluated. Within the first-order approximation, one obtains

$$\rho^{(1)}(\varphi) = \alpha g \varphi^2, \tag{25.31}$$

where a dimensionless parameter

$$g \equiv (D\gamma/3\nu\chi)^{1/2}, \tag{25.32}$$

is introduced. Substituting the function $\rho^{(1)}(\varphi)$ into eqn. (25.20), one finds the appropriate function $\varphi(\xi)$,

$$\varphi_0^{(1)} = (\alpha g \chi/8D)^{1/2}. \tag{25.33}$$

To derive explicitly the solutions of eqns. (25.15) and (25.16), one should substitute (25.31) and (25.33) into (25.22) and (25.18). The result is the following:

$$\varphi^{(1)}(\xi) = \left[\frac{1}{2}\mu\right]^{1/2} \operatorname{sech}(\mu\xi), \quad \epsilon^{(1)} = -\mu^2 D. \tag{25.34}$$

From the relation (25.31) it follows that

$$\rho^{(1)}(\xi) = (2D\mu^2/\chi)\operatorname{sech}^2(\mu\xi). \tag{25.35}$$

The dimensionless parameter

$$\tilde{\mu} \equiv \alpha g \chi/4D, \tag{25.36}$$

introduced above, characterizes a region of soliton localization.

Within the second-order treatment, one can find the solution by substituting (25.31) and (25.33) into eqn. (25.19). Proceeding further, one obtains the solution within the required approximation. One should note that if the relation

$$\gamma D = 3\nu\chi, \quad \nu \neq 0, \tag{25.37}$$

holds in particular, the first-order approximation coincides with the zeroth order one with $g = 1$. In this case, the solutions are

$$\varphi(\xi) = \left[\frac{1}{2}\mu\right]^{1/2} \operatorname{sech}(\mu\xi), \quad \epsilon = -\mu^2 D, \tag{25.38}$$

$$\rho(\xi) = (2D\mu^2/\chi)\operatorname{sech}^2(\mu\xi) \tag{25.39}$$

and become the exact ones if μ is given by the expression

$$\mu = \chi\alpha/4D, \tag{25.40}$$

with α a positive root of the cubic equation (25.30). Using (25.40), one can transform this equation into

$$4\nu\mu^3 + (1-s^2)\mu - \chi^2/4DMV_0^2 = 0, \tag{25.41}$$

which involves directly μ entering into the exact solutions.

For a fixed value of s, eqn. (25.41) possesses a single positive root $\mu = \mu(s)$ regardless of $s < 1$ (subsonic soliton) or $s > 1$ (supersonic soliton). This proposal is easily proved via a graphical method, according to which the roots of eqn. (24.41) are determined by the intersection of a straight line,

$$z = (s^2-1)\mu + \frac{\chi^2}{4DMV_0^2}$$

with the curve $z = 4\nu\mu^3$. As the velocity increases, μ does too, with the value $s = 1$ not being a special one. The restriction on velocity is determined by the conditions of validity of the continuum approach, the satisfaction of the inequality (25.7), and the requirement that the soliton energy is lower than the bottom of the energy band of the quasiparticle. The simultaneous fulfilment of all these conditions is rather difficult to realize.

If, however, all these conditions are satisfied, the integral equation (25.21) possesses a second type of solutions corresponding only to supersonic solitons. One can obtain these solutions by choosing the zeroth-order function in the following form

$$\rho^{(0)}(\varphi) = \beta\varphi. \tag{25.42}$$

Here, β is a positive number which must be determined. Substituting this function into eqns. (25.19), (25.20), (25.21), and (25.22), within the first-order treatment one obtains

$$d\xi^{(1)}(\varphi) = (3D)^{1/2}\varphi^{-1}[2\chi\beta(\varphi_0^{(0)}-\varphi)]^{-1/2}d\varphi, \tag{25.43}$$

$$4\varphi_0^{(0)} = (6\chi\beta/D)^{1/2}, \tag{25.44}$$

$$\rho^{(1)}(\varphi) = \left[\frac{3D}{2\nu\chi}\right]^{1/2}\int_0^{\varphi}\left[\frac{(s^2-1)\beta - 2\tau F(\beta)/3}{\varphi_0^{(0)}-\tau}\right]^{1/2}d\tau, \tag{25.45}$$

where

$$F(\beta) \equiv \gamma\beta^2 - \chi/MV_0^2. \tag{25.46}$$

If β is, in particular, a root of the equation

$$(s^2-1)\beta = 2F(\beta)\varphi_0^{(1)}/3, \tag{25.47}$$

the integral (25.45) is easily evaluated and results in

$$\rho^{(1)} = (DF(\beta)/\nu\chi)^{1/2}\varphi. \tag{25.48}$$

Substituting (25.44) into (25.47), one finds the equation for β,

$$(s^2-1)(6\beta)^{2/3} = (\chi/3D)F(\beta). \tag{25.49}$$

This is a cubic equation with respect to $\beta^{2/3}$. If $s^2>1$, this has a positive solution (supersonic soliton) which increases as the velocity increases.

Substituting the first-order approximation into eqns. (25.19), (25.20), (25.21), and (25.22), one can determine the second-order approximation. If the relation

$$DF(\beta) = \nu\chi\beta^2 \tag{25.50}$$

holds, one can easily prove that the first approximation is just the zeroth one. Therefore, the solution (25.42) becomes exact if β obeys eqns. (25.49) and (25.50) simultaneously. Evidently, this is possible for a definite value of velocity only. To evaluate this velocity, let us substitute the relation

$$\beta^2 = \chi D/(\gamma D - \nu\chi)MV_0^2, \tag{25.51}$$

which follows from eqn. (25.20), into eqn. (25.49). We then obtain

$$s_a^2 = 1 + 4\nu\mu_a^2, \tag{25.52}$$

where

$$\mu_a = \frac{1}{2}\chi[6MV_0^2D(D\gamma - \nu\chi)]^{-1/2}. \tag{25.53}$$

Under these conditions, by virtue of eqns. (25.19) and (25.22), one obtains the exact solutions

$$\varphi(\xi) = \frac{1}{2}(3\mu_a)^{1/2}\text{sech}^2(\mu_a\xi), \quad \epsilon = -4D\mu_a^2, \tag{25.54}$$

$$\rho(\xi) = (6D\mu_a^2/\chi)\text{sech}^2(\mu_a\xi). \tag{25.55}$$

These characterize a soliton with supersonic velocity which is given by the relation (25.52). In the limiting case, $\nu \rightarrow 0$, this velocity approaches that of the longitudinal sound velocity. In that case, the formulas derived are just those obtained in Section 18 for a medium with cubic anharmonicity.

If the relation (25.37) is valid, and $\nu = 1/12$, the parameter (25.53) is equal to $\mu_a = d/2$, and the velocity (25.52) is given by the expression

$$s_a = 1+d^2/12, \quad d \equiv [\chi^2/DMV_0^2]^{1/2}. \tag{25.56}$$

In this case, the exact solution for supersonic solitons, which move with the velocity (25.56), is as follows

$$\varphi(\xi) = \frac{1}{2}(3d/2)\text{sech}^2(\xi d/2), \quad \epsilon = -Dd^2,$$

$$\rho(\xi) = \frac{3Dd^2}{2\chi}\text{sech}^2(d\xi/2).$$

Chapter 7

THE THEORY OF ELECTRON TRANSFER BY SOLITONS

26. The Donor-Acceptor Model of Electron Transfer by Protein Molecules

Many biological phenomena, such as photosynthesis, cell respiration, the activity of enzymes, and others, related to the electron transfer from donor molecules to acceptor molecules, through molecular structures. The latter are often called electron transport chains.

Unlike energy transfer by neutral excitations such as solitons, excitons, and phonons, the transport of an electric charge becomes possible in the presence of electric fields, concentration gradients, or oxidation-reduction potentials which are sometimes called redox potentials.

As has been commonly suggested, the transport of electrons along electron transport chains of biological objects is provided by quantum mechanical tunnelling. This idea was first proposed by De Voult et al. [275]. A complete and detailed theoretical study of a possible tunnelling mechanism for electron transfer between components of the electron transport chain in chloroplasts and chromophores was carried out by Grigorov and Chernavskii [31], and Blumenfeld and Chernavskii [144].

In some experiments (see, for example, [10]) it was revealed that a transferred electron covers a distance of the order of 30 − 70Å. The centres between which an electron migrates, are separated by a lot of atoms and atomic groups linked by covalent, hydrogen, and van der Waals forces. It is unlikely that an electron transfer over such large distances is realized via a simple tunnelling mechanism.

One possible explanation of electron transfer over such distances requires the assumption that this process is facilitated by the participation of protein molecules sited between a donor and an acceptor of electrons. The possibility of transferring electrons through protein molecules has been debated for a long time. In 1941, Szent-Gyorgyi suggested that proteins possess some properties of semiconductors and, hence, conduction bands in which electrons could transfer. This suggestion was not, however, confirmed.

Petrov, Ukrainskii and Kharkyanen from the Institute for Theoretical Physics, Kiev, pointed out that electron transport from a donor to an acceptor molecule should involve an excess electron which is transferred to a protein molecule from a donor, rather than involve protein electrons [239]. Such an electron could reach a conduction band without having to overcome a forbidden energy band. The authors [239] study also the case of the transport of an electron whose energy does not exceed that of the bottom of the conduction band. With a similar approach and by neglecting the displacements of the equilibrium positions of atoms, Davydov [159] has studied the sub-barrier process of the transfer of electrons along a chain of monomers. It has been demonstrated that if the donor molecule is in contact with a monomeric chain, its outer electron wave function is distributed over this chain with an exponential decay. For this reason, there is an increase in the probability of transfer of an electron to an acceptor molecule placed at the opposite end of the chain under study. The numerical studies carried out in the papers [239,159], are valid only for rigid structures with chemical bonds between monomers. However, the electrons in α-helix protein molecules are transferred over chains of peptide groups linked via weak hydrogen bonds. In this case, the electron-chain interaction results in local chain deformation which should be taken into account even within the zeroth-order treatment.

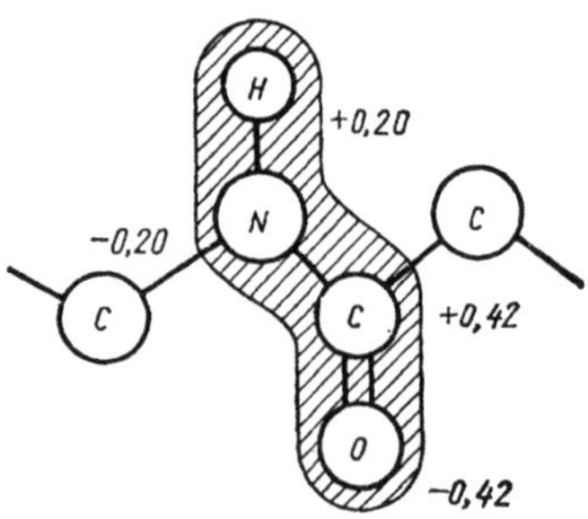

Figure 9. The distribution of electric charges on atoms of the peptide group.

Ukrainskii and Mironov [121] advanced the development of the donor-acceptor model of electron transport in α-helix proteins by demonstrating that the charge distribution of atoms constituting peptide groups (see Fig. 9) results in a permanent dipole moment of about 3.5 debye pair on each group [200]. Turner et al. [274] showed that in the field of such a dipole an electron is bounded with a binding energy $\epsilon_0 = 0.9\,eV$. In this state, the electron distribution is rather extended, and the average distances between the electron and positive and negative dipole centres account for approximately 4 and 5 bohrs, respectively.

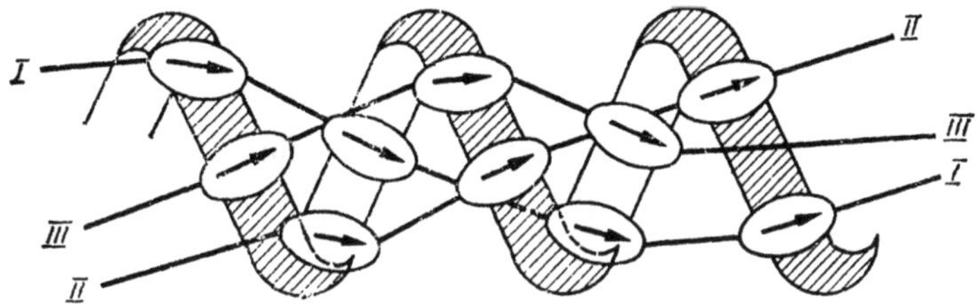

Figure 10. The arrangement of the permanent electric dipole moments in the peptide group of the α-helix protein molecule.

The arrangement of permanent electric dipole moments] of a peptide group of an α-helix molecule is depicted in Fig. 10. One can conclude that for each chain of hydrogen bonds, the peptide group form periodically repeating potential wells for outer electron coming into a chain from a donor. The overlap of wave functions of the ground states of the electron in neighboring potential wells provides a collective origin of such states in a chain, i.e., the conduction band arises. Overlapping of the wave functions of electrons of the peptide groups situated in different chains is of less importance. For this reason, one can consider independently the motion of a single electron which falls into a chain from a donor.

In order to study theoretically the energy states of an electron fallen into a given chain of peptide groups from a donor, Ukrainskii and Mironov [121] have considered a one-dimensional rigid chain whose peptide groups are separated from each other by 4.5 Å. An overlap of the wave functions of an excess electron in the chain of periodically arranged electric dipoles which serves as a model for protein molecule, initiates the exchange interaction, D, and the collectivization of electron states. The electron energy level with $\epsilon_0 = 0.9\,eV$ in the field of a single electric dipole transforms into the energy band of states of a quasiparticle with an effective mass

$$m^* = \hbar^2 / 2a^2 D. \tag{26.1}$$

Neglecting the displacements of the equilibrium positions of peptide groups, the motion of a quasiparticle inside this energy band (or the so-called *conduction band of a quasiparticle*) is described by a plane wave having a definite magnitude of wavenumber $k = 2\pi/\lambda$. Within the long-wave approach, this state corresponds to the energy

$$E_e(k) = E_e(0) + \hbar^2 k^2/2m^*. \tag{26.2}$$

A phase velocity, $\pi\hbar/m^*\lambda$, of the quasiparticle in this state depends on the wavelength λ. The energy of the bottom of the conduction band, $E_e(0)$, is less than the energy level in the isolated potential well as a result of an exchange interaction between neighboring peptide groups,

$$E_e(0) = \epsilon_0 - 2D.$$

A quasiparticle inside the energy band (26.2) possesses an electron charge; however, its effective mass m^* can differ considerably from that of a free electron. Nevertheless, the term 'electron' is reserved for designating this particle.

Generally speaking, the model of rigidly fixed peptide groups is inconsistent to describe adequately the α-helix proteins due to the weakness of the intrapeptide hydrogen bonding. Due to the strong interaction of a quasiparticle with a chain which is described in terms of the deformation potential σ, there arises a tendency of varying intrapeptide distances in the region of a quasiparticle localization. For this reason, the states of a quasiparticle in the chain under study are complicated in nature due to its interaction with a local deformation. Therefore, the motion of an excess electron which has fallen into the chain from a donor should be considered by including the displacements of the equilibrium positions of the molecules of the chain without an electron.

As has been shown by the present author [39] within the continuum approach, the electron motion in an infinite chain of peptide groups is described by the following set of equations,

$$\left[i\hbar\frac{\partial}{\partial t} - \mathcal{E}_0 + \frac{\hbar^2}{2m^*}\frac{\partial^2}{\partial z^2} + \sigma\rho(z,t)\right]\psi(z,t) = 0, \tag{26.3}$$

$$\left\{\frac{\partial^2}{\partial t^2} - V_0^2\frac{\partial^2}{\partial z^2}\right\}\rho(z,t) + \frac{a^2\sigma}{M}\frac{\partial^2}{\partial z^2}|\psi(z,t)|^2 = 0, \tag{26.4}$$

where $\rho(z,t)$ is the function characterizing a decrease of distances between neighboring peptide groups, V_0 is the longitudinal sound velocity parameter (in $ergcm^{-1}$). This set of equations is derived under the assumption that the permanent electric dipole moments of peptide groups are rather rigid and, when affected by an electron, only change their equilibrium positions. It is not noteworthy that the coherent long-wave displacements of the equilibrium positions of peptide groups are considered only.

Introducing the variable where V is the velocity of the motion of the excitation along the chain ($V<V_0$), one can transform the set of equations (26.3) and (26.4) as follows

$$\rho(\zeta) = \sigma|\psi(z,t)|^2/\kappa(1-s^2), \quad s = V/V_0, \tag{26.6}$$

$$\left[i\hbar\frac{\partial}{\partial t} - \mathcal{E}_0 + \frac{\hbar^2}{2m^*}\frac{\partial^2}{\partial z^2} + G(s)|\psi(z,t)|^2\right]\psi(z,t) = 0. \tag{26.7}$$

In eqn. (26.7) of the nonlinear Schrödinger equation type, a nonlinearity parameter G is expressed in terms of the parameter σ of the deformation potential and longitudinal elasticity κ via the relation

$$G(s) = \sigma^2/\kappa(1-s^2). \tag{26.7a}$$

The wave function of an electron whose movement with velocity $V = sV_0$ is accompanied by a local chain deformation, takes the form

$$\psi(z,t) = \Phi(\zeta)\exp[i(kz-\omega t)], \tag{26.8}$$

where in accordance with eqn. (26.1), the real amplitude function $\Phi(\zeta)$ obeys the normalization condition

$$\frac{1}{a}\int\Phi^2(\zeta)d\zeta = 1. \tag{26.8a}$$

Substituting (26.8) into eqns. (26.6) and (26.7), one obtains the following set of equations

$$\rho(\zeta) = \sigma\Phi^2(\zeta)/\kappa(1-s^2), \tag{26.9}$$

$$\left[\Lambda + \frac{\hbar^2}{2m^*}\frac{d^2}{d\zeta^2} + G(s)\Phi^2(\zeta)\right]\Phi(\zeta) = 0. \tag{26.10}$$

where

$$\Lambda \equiv -\frac{1}{a}\int\left[\frac{\hbar^2}{2m^*}\Phi\frac{d^2\Phi}{d\zeta^2}+G\Phi^4\right]d\zeta, \tag{26.11}$$

$$\hbar\omega = \Lambda+\mathcal{E}_h+\hbar^2k^2/2m^*, \tag{26.11a}$$

is the energy of an electron interacting with the local chain deformation induced by it. In order to derive the formula for the total energy, one needs to add the energy of the chain deformation displaced by an electron,

$$W = \frac{1}{2a}\int\left[M\left[\frac{\partial\beta}{\partial t}\right]^2+\kappa\rho^2\right]d\zeta. \tag{26.12}$$

Since

$$\rho(\zeta) = -a\frac{\partial\beta}{\partial\zeta}, \tag{26.13}$$

one has

$$\frac{\partial\beta(\zeta)}{\partial t} = V\rho(\zeta)/a \tag{26.14}$$

which enables eqn. (26.12) to be transformed as follows

$$W = \frac{\kappa(1+s^2)}{2a}\int\rho^2(\zeta)d\zeta. \tag{26.14}$$

The solution of eqn. (26.10), normalized by the condition (26.8a), has the bell-like form

$$\Phi(\zeta) = \frac{\sqrt{aq}}{\sqrt{2}\,\mathrm{ch}(q\zeta a)}, \tag{26.15}$$

under the following values of the parameters,

$$k = m^*V/\hbar,\ \ q = am^*G/2\hbar^2,\ \ \Lambda = -\frac{\hbar^2q^2}{2m^*}. \tag{26.16}$$

The system of the electron and the local deformation which surrounds the electron and moves with a constant velocity V is hence described by the function (26.8) with the envelope given by eqn. (26.15). We call this system an *autolocalized electron*, or *electrosoliton*.

According to eqns. (26.9) and (26.5), a decrease of distances between the peptide groups in the region of an electrosoliton is given by the function

$$\rho(\zeta) = aq\sigma/2(1-s^2)\kappa \mathrm{ch}^2(q\zeta). \tag{26.17}$$

Equation (26.13) is valid if the displacements of the equilibrium positions of the peptide groups are determined by the function

$$\beta(\zeta) = \sigma[1-\mathrm{th}(q\zeta)]/2\kappa(1-s^2). \tag{26.17a}$$

The electrosoliton carries the energy

$$E(V) = W+\hbar\omega,$$

where $\hbar\omega$ and W are given by eqns. (26.11) and (26.24), respectively. Substituting (26.15) and (26.2) in the latter equations, one finds

$$\hbar\omega = \mathcal{E}_0+\frac{\hbar^2}{2m^*}(k^2-q^2),$$

$$W = m^*a^2G^2(1+s^2)/12\hbar^2(1-s^2).$$

Therefore,

$$E(V) = \mathcal{E}_0+\frac{1}{2}m^*V^2+\frac{m^*a^2G^2[5s^2-1]}{24(1-s^2)\hbar^2}. \tag{26.18}$$

At small velocities, i.e., $s^2 \ll 1$, it follows from eqn. (26.18) that

$$E(V) = E(0)+\frac{1}{2}m_{\text{sol}}V^2,\ \ s^2 \ll 1, \tag{26.19}$$

where $E(0)$ is the soliton rest energy,

$$E(0) = \mathcal{E}_0-m^*a^2G^2(0)/24\hbar^2,\ \ G(0) = \frac{\sigma^2}{\kappa} \tag{26.19a}$$

and

$$m_{\text{sol}} = m^*(1+MG^2/6\kappa\hbar^2) \tag{26.19b}$$

is the effective mass of the electrosoliton.

The total momentum of the electrosoliton is given by

$$P(V) = \frac{\hbar}{2ia}\int\left[\psi^*\frac{\partial\psi}{\partial z}-\psi\frac{\partial\psi^*}{\partial z}\right]dz-\frac{M}{a}\int\frac{\partial\beta}{\partial t}\frac{\partial\beta}{\partial\zeta}d\zeta.$$

Using eqns. (26.8) and (26.13), one can write

$$P(V) = m^*V\left[1+\frac{M}{m^*a^3}\int\rho^2(\zeta)d\zeta\right].$$

Evaluating the integral, one finally obtains

$$P(V) = m^*V[1+MG^2/16\kappa\hbar^2(1-s^2)]. \tag{26.20}$$

In particular, at small velocities one has

$$P(V) = m_{\text{sol}}V. \tag{26.20a}$$

So, when the electrosoliton is generated, the total energy of the electron and the displacement field is decreased, due to their interaction, by the quantity (with $V=0$)

$$\Delta E = m^*a^2G^2/24\hbar^2 = m^*a^2\sigma^4/24\kappa^2\hbar^2 \tag{26.21}$$

while the effective electron mass m^* increases, in agreement with eqn. (26.19b). The electrosoliton, moving always with a velocity V which is less than that of the longitudinal sound velocity, does not emit phonons. The interaction of the electron with the displacement field is transformed into a quasi-conducting state. Such a stabilization becomes substantial at small κ and large σ.

Together with the normalized electrosoliton-type solution (26.9), eqn. (26.7) possesses also plane-wave solutions with wavenumber k. Such solutions are metastable for $k<m^*V_0/\hbar$ since their energy exceeds that of electrosolitons by the value of the energy gap (26.21). For $k>m^*V_0/\hbar$ the electrosoliton solutions are absent, and the unique solutions are of plane-wave type. Wave packets, formed by these plane waves, described the motion of an electron in the conduction band. This motion is dampened rapidly due to the radiation of phonons.

Therefore, due to autolocalization, the motion of an electron along an α-helix protein molecule is stabilized and takes a quasiconducting feature. This specific type of 'superconductivity' without an electron pairing is inherent only for one-dimensional systems. Such a phenomenon seems to be of great importance for explaining the efficiency of electron transfer in electron transport chains without loss of energy. α-helix protein molecules could therefore be the ideal pathways of electron transport from a donor molecule to an acceptor molecule.

Fröhlich [183] first pointed out that the coupling of electrons with the displacement field of molecules in one-dimensional systems can result in a superconducting phenomenon without pairing. Fröhlich studied a one-dimensional model of the jellium-type metal, i.e., a homogeneous distribution of positive ion charges with a half-occupied electron conduction band, disregarding electron-electron interaction. By simply approximating the interaction of an electron with the displacement field by a sine wave, he showed the formation of a gap in the energy spectrum of the electrons. At low temperatures and in the absence of a current, the electrons occupy states below the gap only. However, the use of a rather simplified one-dimensional model of an electron system while neglecting the direct Coulomb interaction and the indirect one through the deformation of the one-dimensional lattice, makes Fröhlich's work less convincing. As was pointed out in this context by Bardeen [137], Fröhlich's work has been considered for along time as a mathematical study of a model having no real physical meaning.

The stability of electrosolitons, as well as that of soliton of Amide-I vibrations, is provided by the mutual compensation of nonlinear and dispersion effects. According to eqn. (26.21), the binding energy of a quasiparticle with deformation is proportional to the fourth power of the deformation parameter σ and inversely proportional to the square of the coefficient of elasticity κ. As ΔE increases, the electrosoliton becomes more stable. Large values of ΔE, which provide the stability of the soliton under the conduction band of a quasiparticle, dampens it strongly in the conduction band. This peculiarity is similar to that which occurs in the superconductivity of metals. Metals having a weak binding of electrons with the deformation of ionic lattices, such as silver and copper, possess rather a large conductivity, but do not demonstrate transition into the superconductivity state at low temperatures. Metals whose electrons are strongly coupled with a lattice deformation, e.g. plumbum, are bad conductors in the normal state but, however, transform into the superconducting state at low temperatures.

Therefore, the stability of electrosolitons is considerably large in soft-one-dimensional chains having a rather strong binding of electrons with the chain local deformation. The hydrogen-bonded chain of peptide groups of α-helix protein molecules refer to that class, and thus the α-helix segments of proteins are the ideal pathways of electron transport from donor molecules to the acceptor molecules.

One should note, however, that at a rather large value of ΔE there could be a complete localization of excitation at a single peptide group. Such an excitation could possess a very large effective mass and, in fact, be immobile in the chain.

26.1. Pairing of Two Excess Electrons in a One-Dimensional Molecular Chain

The possibility of autolocalization of an electron and its transport without loss of energy along one-dimensional molecular chains in the form of a soliton produce new properties of the electron motion in one-dimensional molecular structures [247,139].

In the foregoing we have considered the nonlinear equations describing the collective excitations arising from the interaction of one excess electron with displacements from the equilibrium positions of molecules in soft quasi-one-dimensional structures. Such an interaction results in electron autolocalization (electrosoliton formation) which stabilizes the motion of electron.

In order to study the electron conductivity of quasi-one-dimensional molecular crystals, it is necessary to consider the motion of many quasiparticles and, furthermore, to take into account their interaction and spin. The simplest way to formulate such a theory consists in constructing an approach which enables the description of the motion of two excess quasiparticles in a quasi-one-dimensional molecular chain which is able to accommodate these quasiparticles.

Such a problem is of particular interest. As is known, a pair of electrons with opposite spins is transferred from one molecule to another in the redox reactions occurring in living organisms. It is also known that the transport of electrons in the synthesis process of ATP molecules in conjugate membranes of mitochondria and chloroplasts is realized by pairs, but not individually [199]. We will demonstrate now that α-helix protein molecules could be the proper pathways for the transfer of electron pairs.

Two excess electrons in the singlet spin state form a combined potential well in a given chain of peptide groups of a protein molecule. This well is conditioned by the displacements of the equilibrium positions of peptide groups. The potential well created by the deformation of one electron, attracts the other electron which makes this well deeper, and so on. In such a way, a 'pairing' of electrons occurs. In this autolocalized state a given pair of electrons moves along the chain of peptide groups.

The simplest theory of pairing of excess electrons in a soft quasi-one-dimensional molecular chain was developed by Brizhik and the author [22] for the particular case of an infinite chain of peptide groups. We shall now sketch this approach.

The Hamiltonian of the system in question is written as

$$H = H_{el} + H_{vib} + H_{int}. \tag{26.22}$$

In eqn. (26.22), H_{el} is the Hamiltonian of two excess electrons in the field of the potential wells of the peptide groups in the underformed chain,

$$H_{el} = \sum_{j=1}^{2} \psi_j^* \left[\mathcal{E}_0 - \frac{\hbar^2}{2m^*} \frac{\partial^2}{\partial z^2} \right] \psi_j dz / a, \tag{26.23}$$

where $\mathcal{E}_0$ is the energy of the bottom of the conduction band of excess electrons, m^* is the effective mass of an electron given by eqn. (26.1); the function $\psi_j(z,t)$ with $j = 1,2$ is normalized by the condition (26.8a); and H_{vib} is the Hamiltonian of the coherent displacements $\beta(z,t)$ of the equilibrium positions of peptide groups,

$$H_{vib} = \frac{1}{2} M \int \left[\left(\frac{\partial \beta}{\partial t} \right)^2 + \rho^2 V_0^2 / a^2 \right] dz / a. \tag{26.24}$$

As in Section 26, the function $\rho(z,t) = -a\partial\beta/\partial z$ characterizes a decrease of distance between neighboring peptide groups, $V_0 = a\sqrt{\kappa/M}$ is the longitudinal sound velocity, and κ is the longitudinal elasticity of a given chain.

The deformation interaction of excess electrons with the peptide group chain is described by the Hamiltonian

$$H_{int} = -\sigma \int \rho(z,t) \sum_{j-=1}^{2} |\psi_j|^2 dz / a. \tag{26.25}$$

The following set of equations corresponds to the total Hamiltonian

$$\left[i\hbar \frac{\partial}{\partial t} - \mathcal{E}_0 + \frac{\hbar^2}{2m^*} \frac{\partial^2}{\partial z^2} + \sigma \rho(z,t) \right] \psi_j(z,t) = 0, \tag{26.26}$$

$$\left(\frac{\partial^2}{\partial t^2} - V_0^2 \frac{\partial^2}{\partial z^2} \right] \rho(z,t) + \frac{a^2 \sigma}{M} \frac{\partial^2}{\partial z^2} \sum_{j=1}^{2} |\psi_j(z,t)|^2 = 0. \tag{26.27}$$

We seek those solutions of this set which correspond to the movement of both electrons with a constant velocity V, which the average interelectronic separation,

al, being fixed (the adiabatic approximation). Let us define a new variable

$$\xi = (z - z_0 - Vt)/a \tag{26.28}$$

and the following function

$$\rho(z,t) = \rho(\xi), \tag{26.29a}$$

$$\psi_j(z,t) = \Phi_j(\xi)\exp\left\{\frac{i}{\hbar}\left[m^* zV - \left(\mathcal{E} + \frac{m^* V^2}{2}\right)t + \eta_0\right]\right\}. \tag{26.29b}$$

In eqn. (26.29b), $\mathcal{E}$ is the energy of an electron which interacts with the deformed chain, given relative to the coordinate frame moving with a velocity V, and η_0 is an arbitrary phase.

Equation (26.27) then transforms into the relation

$$\rho(\xi) = \sigma\left[\kappa(1-s^2)\right]^{-1}[\Phi^2(\xi) + \Phi^2(\xi + l)], \tag{26.30}$$

where $s = V/V_0$. Using the latter, one can rewrite eqn. (26.26) as follows

$$\left\{\frac{d^2}{\partial\xi^2} + \epsilon_l + 2g[\Phi^2(\xi) + \Phi^2(\xi + l)]\right\}\Phi(\xi) = 0, \tag{26.31}$$

where the abbreviated notations for dimensionless quantities,

$$g = m^* a^2 G/\hbar^2, \quad G = \sigma^2/\kappa(1-s^2), \tag{26.32}$$

$$\epsilon_l = 2m^* a^2(\mathcal{E} - \mathcal{E}_0)/\hbar^2, \tag{26.33}$$

are used. From eqn. (26.31) it follows that

$$\epsilon_l = \int\left\{\left[\frac{\partial\Phi}{\partial\xi}\right]^2 - 2g\Phi^2(\xi)[\Phi^2(\xi) + \Phi^2(\xi + l)]\right\}d\xi. \tag{26.34}$$

In agreement with eqn. (26.22) the total energy of two excess electrons, separated by a distance al and moving with a constant velocity V together with a chain deformation, is determined by the expression

$$E_l(V) = 2\left\{\frac{m^* V^2}{2} + \mathcal{E}_0 + \frac{\hbar^2}{2m^* a^2}\left[\epsilon_l + \frac{g(1+s^2)}{2(1-s^2)}\right] \times \right. \tag{26.35}$$

$$\left. \times \int[\Phi^2(\xi) + \Phi^2(\xi + l)]^2 d\xi\right]\Big\}.$$

Two electrons become independent at large interparticle distances al since, in the region of nonvanishing $\Phi(\xi)$, the function $\Phi(\xi+l) = 0$. Using eqn. (26.35) one obtains the expression for the total energy of two infinitely separated electrons linked with their local deformations,

$$E_\infty(V) = 2\left\{\frac{m^* V^2}{2} + \mathcal{E}_0 + \frac{\hbar^2}{2m^* a^2}\left[\epsilon_\infty + \frac{g(1+s^2)}{2(1-s^2)}\int \Phi_\infty^4(\xi)d\xi\right]\right\}, \tag{26.36}$$

where the values ϵ_∞ and $\Phi_\infty(\xi)$ are determined by the equation

$$\frac{d^2}{d\xi^2} + \epsilon_\infty + 2g\Phi_\infty^2(\xi)]\Phi_\infty(\xi) = 0. \tag{26.37}$$

One seeks the solution of eqn. (26.37) which obeys the normalization

$$\int \Phi^2(\xi)d\xi = 1.$$

Thus one finds

$$\Phi_\infty(\xi) = \frac{1}{2}\sqrt{g}\,\mathrm{sch}(g\xi/2) \tag{26.38}$$

with

$$\epsilon_\infty = -g^2/4. \tag{26.39}$$

Substituting these quantities into eqn. (26.36), one obtains the total energy of two independent electrosolitons

$$E_\infty(V) = 2\left\{\mathcal{E}_0 + \frac{m^* V^2}{2} - \frac{\hbar^2 g^2(1-5s^2)}{24m^* a^2(1-s^2)}\right\}. \tag{26.40}$$

Equation (26.40) is simplified at small velocities

$$E_\infty(V) \approx 2\left\{\epsilon_0 + \frac{1}{2}M_{\mathrm{sol}}V^2 - \hbar^2 g_0^2/24m^* a^2\right\}, \tag{26.41}$$

where

$$V \ll V_0, \; g_0 = m^* a^2\sigma^2/\kappa\hbar^2, \tag{26.42}$$

and

$$M_{\mathrm{sol}} = m^*(1 + M\sigma^4/6\kappa^3\hbar^2) \tag{26.43}$$

is the effective mass of a single electrosoliton.

At small distances, i.e., $l \approx 0$, two electrons in the singlet spin state move as a whole with an energy involving the energy of chain deformation,

$$E_0(V) = 2\left\{\mathcal{E}_0 + \frac{m^* V^2}{2} + \frac{\hbar^2}{2m^* a^2}\left[\epsilon_0 + \frac{2g(1+s^2)}{1-s^2}\int \Phi_0^4(\xi) d\xi\right]\right\}. \tag{26.44}$$

In the eqn. (26.44), ϵ_0 and Φ_0 obey the equation

$$\left[\frac{d^2}{d\xi^2} + \epsilon_0 + 4g\Phi_0^2(\xi)\right]\Phi_0(\xi) = 0. \tag{26.45}$$

An equation of this type for immobile pairs of electrons was also derived by Krivnov and Ovchinnikov [82] who did not, however, evaluate the energy of pairing and the effective mass of an electron pair by taking the chain deformation energy into account. The normalized solution of eqn. (26.45) has the form

$$\Phi_0(\xi) = \sqrt{g/2}\,\mathrm{sch}(g\xi), \tag{26.46}$$

with $\epsilon_0 = -g^2$. The total spin of the electrons vanishes at the state of pairing. This state of coupled electrons is called a *bisoliton*. In the bisoliton state, both the electrons move in the combined effective potential well,

$$U_{\uparrow\downarrow}(\xi) = -2g^2 J\,\mathrm{sch}^2(g\xi),$$

relative to the coordinate frame ξ. The radius of this well is twice as small as that of an isolated soliton.

The motion of an electron inside the well is described by the following function

$$\Phi_{\uparrow\downarrow}(\xi_1, \xi_2) = \frac{1}{2} g\,\mathrm{sch}(g\xi_1)\mathrm{sch}(g\xi_2)|\uparrow\rangle|\downarrow\rangle,$$

where $|\uparrow\rangle$ and $|\downarrow\rangle$ are spin-state functions. The average distance between the paired electrons in a given bisoliton is determined by the formula

$$L \approx a\Delta\xi \approx 2\pi a/g. \tag{26.46a}$$

If $g \approx 0.1-1$, L exceeds by more than ten times the distance between molecules. A bisoliton carries double the charge of an electron.

Substituting (26.46) into eqn. (26.44), one obtains the energy of paired electrons moving with velocity $V = sV_0$,

$$E_0(V) = 2\left[\mathcal{E}_0 + \frac{1}{2}mV^2 - \frac{\hbar^2 g^2(1-5s^2)}{6ma^2(1-s^2)}\right]. \tag{26.47}$$

The energy of pairing is given by

$$\Delta E(V) \equiv E_\infty(V) - E_0(V) = \frac{\hbar^2 g_0^2(1-5s^2)}{4ma^2(1-s^2)^2} \tag{26.48}$$

where the effective Coulomb repulsion is neglected. As follows from eqn. (26.48), a pairing becomes profitable energy wise at small electron velocities. For this regime of velocities, the energy can be represented as follows

$$\Delta E(V) \approx ma^2\sigma^4(1-2s^2)/4\kappa^2\hbar^2, \quad s^2 \ll 1. \tag{26.49}$$

The equations obtained tell us that a pairing must be broken by a strong electric field which results in a paired electron velocity which exceeds the threshold magnitude.

At small velocities, the energy (26.47) of paired electrons takes the form

$$E_0(V) = 2\mathcal{E}_0 - \frac{ma^2\sigma^4}{3\kappa^2\hbar^2} + \frac{1}{2}M_{ss}V^2, \tag{26.50}$$

where

$$M_{ss} = 2m(1 + 2M\sigma^4/3\kappa^3\hbar^2) \tag{26.51}$$

is the effective mass of paired electrons moving together with a local chain deformation. This exceeds twice the mass of two isolated electrosolitons. Therefore, as the velocity increases the kinetic energy of a pair increases more rapidly compared with that of two isolated electrosolitons. This fact explains the breakdown of pairing at large velocities.

One should point out that the energy of solution pairing with the further formation of bisolitons in the singlet spin state is independent of the mass M of heavier particles unlike that of the pairing predicted by the BCS theory [291]. In the context of the BCS theory, the average distance between paired electrons is about $3\times10^{-4}-4\times10^{-4}cm$ which is ten of thousands times greater than that between ions in a lattice. The BCS theory is developed only within the second-order perturbation treatment relative to a coupling parameter g. For this reason, it is justified for metals having $g \approx 10^{-3}-10^{-4}$.

The pairing of electrons in soft chains is provided by the nonlinear coupling of electrons with inertial chain deformation. The solutions of the nonlinear Schrödinger equation, describing the inertialess field with a self-interaction, refer to unbounded states. Using the inverse scattering method, Zakharov and Shabat [68] have shown that in this case the energy of two-soliton states is equal to the sum of single-soliton energies.

We neglect, as before, the Coulomb repulsion. Due to the screening of electrons, such repulsion in neutral systems is negligible. Hence this can be taken into account via the methods of perturbation theory. The minimum of the energy of paired quasiparticles corresponds to small nonvanishing values of l. Evaluating ϵ_0 and $\Phi_l(\xi)$ by means of eqns. (26.34) and (26.35), ignoring powers of l greater than l^2, one obtains

$$\epsilon_l = -g^2+4g^4l^2/15, \tag{26.52}$$

$$E_l(V) = E_0(V)+\frac{1}{2}l^2F(V). \tag{26.53}$$

In eqn. (26.53), $E_0(V)$ is given by eqn. (26.27), and the function $F(V)$ has the following form

$$E(V) = 4g^4\hbar^2(l-3s^2)/15m^*a^2(l-s^2). \tag{26.54}$$

The total energy, $W_l(V)$, of a pair of quasiparticles, taking into account the Coulomb repulsion, is equal to the sum of the energy $E_l(V)$, given by eqn. (26.53), and the energy, $U(l)$ of the Coulomb interaction of quasiparticles separated by al from each other,

$$W_l(V) = E_l(V)+U(l). \tag{26.55}$$

A quasiparticle carries an effective screened charge e^*_{eff} whose density along the chain axis is proportional to the probability $\Phi^2(\xi)$ and is uniformly distributed in the molecular plane of radius b across the chain. Hence, one obtains

$$U(l) = a^{-2}\int\Phi^2(x)\Phi^2(y+al)u(x-y)dxdy, \tag{26.56}$$

where

$$u(|x|) = e^2_{\text{eff}}\int\frac{dydz}{\sqrt{x^2+y^2+z^2}}.$$

Using cylindrical coordinates, one has

$$u(x) = \frac{2e_{\text{eff}}^2}{b^2}\int_0^b \frac{\rho d\rho}{\sqrt{x^2+\rho^2}} \approx \frac{2e_{\text{eff}}^2}{b^2}(\sqrt{a^2+b^2} - |x|). \tag{26.57}$$

Substituting eqn. (26.57) into eqn. (26.56) and choosing the function $\Phi(x)$ in the form of eqn. (26.38), and further integrating, one finds

$$U(l) \approx \frac{2e_{\text{eff}}^2}{b^2}(\sqrt{b^2+a^2l^2} - al).$$

Therefore, the energy (26.53) becomes

$$W_l(V) = E_0(V) + \frac{1}{2}F(V)l^2 + \frac{2\text{eeff}^2}{b^2}(\sqrt{b^2+a^2l^2} - al). \tag{26.58}$$

If the inequality

$$l_0 \equiv \left[\frac{e_{\text{eff}}^2}{aF}\right]^{1/3} > b/a \tag{26.59}$$

is valid, the energy (26.58) attains its minimum,

$$W_{l_0}(V) - E_0(V) + \frac{3}{2}(e_{\text{eff}}^4 F(V)/a^2)^{1/3} \tag{26.60}$$

at $l = l_0$. Hence, if Coulomb repulsion is taken into account, one can replace the condition (26.48) of the pairing of quasiparticles by

$$\Delta E(V) = E_\infty(V) - W_{l_0}(V) > 0, \tag{26.61}$$

from which it follows that at small velocities a pairing is profitable energy wise, if the dimensionless coupling constant g is greater than some critical value,

$$g_{\text{cr}} \equiv (4e_{\text{eff}}/bJ)^{1/2}. \tag{26.62}$$

If this condition is satisfied, the bisolitons formed transmit pairs of quasiparticles without energy loss along the chains of molecules. Such a process of electron pair transport in the singlet spin state appears to take place even at a temperature of $300\,K$ in α-helix protein molecules of living organisms under the synthesis of ATP molecules in conjugate membranes of mitochondria and chloroplasts [199], and in some redox reactions.

27. The Interaction of an Electron with Nondipole Optical Phonons

We begin by suggesting the possibility of nondipole intermolecular vibrations with frequency Ω_0 in molecules constituting a given molecular chain. The interaction of neighboring molecules in the chain results in optical phonons with frequencies

$$\Omega(k) = (\Omega_0^2 + Dk^2)^{1/2}, \tag{27.1}$$

depending on the wavenumber k. Such a dependence characterizes the spatial dispersion of the optical phonons. A parameter D having the dimension of velocity determines the type of dispersion; namely, a positive D ($D>0$) corresponds to positive dispersion, and a negative D ($D<0$) to negative dispersion.

The group velocity of the displacement of wave packets of optical phonons

$$v_{gr} = kD/(\Omega_0^2 + Dk^2)^{1/2} \tag{27.1a}$$

is just $(|D|)^{1/2}$ in the limit of rather short wavelengths ($k \to \infty$). In a case of negative dispersion, the group velocity is directed contrary to the direction of propagation of a plane wave with wavenumber k.

Within the continuum approach, the following Hamiltonian corresponds to optical phonons having the dispersion (27.1)

$$H_{\text{ph}} = (2a)^{-1} \int \left\{ \left[\frac{\partial u}{\partial t} \right]^2 + \Omega_0^2 u^2 + D \left[\frac{\partial u}{\partial z} \right]^2 \right\} dz, \tag{27.2}$$

where $u(z,t)$ is the relative intramolecular displacement of atoms in molecules, M is an appropriate reduced mass, and the integer z/a is the number of molecules in the chain.

If the displacements $u(z,t)$ are not coupled with a change of internal electric dipole moments of molecules, their interaction with an electron is determined by the local deformation potential. Within the continuum approach, the latter takes the form

$$H_{\text{int}} = \chi a^{-1} \int u(z,t) |\psi(z,t)|^2 dz, \tag{27.3}$$

where χ is a parameter of the deformation interaction. This interaction appears to be negligible in real systems. However, in some cases, the interaction with optical phonons is of greater importance compared with the case of acoustic phonons. We now consider the work by Enolskii and the present author [54] in order to study this case. The short-range interaction of an electron with nondipole optical vibrations of the chains of diatomic molecules was first studied by Holstein [201].

The Hamiltonian of the chain which takes on an electron is given by a sum of the Hamiltonians (26.2), (27.2), and (27.3). Hence, the system is described by the following equations of motion

$$\frac{\partial^2 u}{\partial t^2}+\Omega_0^2 u-D\frac{\partial^2 u}{\partial z^2}+\chi M^{-1}|\psi(z,t)|^2 = 0, \tag{27.4}$$

$$\left[i\hbar\frac{\partial}{\partial t}-\mathcal{E}_0+\frac{\hbar^2}{2m}\frac{\partial^2}{\partial z^2}-\chi u(z,t)\right]\psi(z,t) = 0, \tag{27.5}$$

where m is the positive effective mass of a free quasiparticle (electron) in its energy band the energy of which at the bottom is just $\mathcal{E}_0$.

In a case of the positive dispersion of optical phonons, i.e., $D>0$, a given quasiparticle can be bounded with the field of virtual optical phonons. Due to this binding, if the latter is, of course, profitable energy wise, the combined collective excitation, or the dressed quasiparticle, can move along the chain. If a quasiparticle is an excess electron in the conduction band, the localized state generated carries the electric charge of an electron along the chain. For this reason, it is convenient to consider this state as an electrosoliton.

To describe the stationary state of motion of the dressed quasiparticle, or electrosoliton, with velocity V, one can define the following real functions, $u(\xi)$ and $\varphi(\xi)$, depending on the dimensionless variables

$$\xi = (z-z_0-Vt)/a, \quad V = k\hbar/m, \tag{27.6}$$

by means of the equations

$$u(z,t) = u(\xi), \quad \psi(z,t) = \varphi(\xi)\exp[i(kz-\omega t)]. \tag{27.7}$$

Stationary motion takes place iff the velocity of the dressed quasiparticle is less than the velocity $D^{1/2}$ which characterizes the positive dispersion of optical phonons and the maximum velocity (4.7) of a free quasiparticle in its conduction band. We suggest, henceforth, that these necessary conditions are both satisfied.

Substituting the functions, given by eqn. (27.7), into the set of eqns. (27.4) and (27.5), one obtains, relative to the moving coordinate frame,

$$\left[\frac{\hbar^2}{2m}\frac{d^2}{d\xi^2}+\Lambda-\chi u(\xi)\right]\varphi(\xi) = 0, \tag{27.8}$$

$$\left[\nu\frac{d^2}{d\xi^2}+1\right]u(\xi)-\chi(M\Omega_0^2)^{-1}\varphi^2(\xi) = 0 \tag{27.9}$$

where the following abbreviated notations,

$$\Lambda \equiv \hbar\omega-\mathcal{E}_0-\frac{1}{2}mV^2, \tag{27.10}$$

$$\nu \equiv (D-V^2)/a^2\Omega_0^2 \geqslant 0, \tag{27.11}$$

are used. In eqn. (27.10), $\hbar\omega$ determines the energy of a quasiparticle interacting with the displacement field $u(\xi)$ moving along the chain with velocity V.

We are now interested in the autolocalized states of a quasiparticle within the coordinate frame ξ. For this reason we search for the solutions which satisfy the boundary conditions

$$u(\pm\infty) = \varphi(\pm\infty) = 0, \tag{27.12}$$

and the normalization condition

$$\int\varphi^2(\xi)d\xi = 1. \tag{27.12a}$$

The constrained solutions of eqn. (27.9) are given by the integral

$$u(\xi) = -\chi(M\Omega_0^2)^{-1}\int G(\xi_1-\xi)\varphi^2(\xi_1)d\xi_1, \tag{27.13}$$

where the Green's function is defined as follows

$$G(\xi) = \lim_{\delta\to 0}(2\pi)^{-1}\int\frac{\exp(iq\xi)dq}{1+\gamma(q+i\delta)^2}. \tag{27.14}$$

Green's function, introduced in eqn. (27.13), describes the nonlocal interaction of a quasiparticle with nondipole optical phonons. The origin of nonlocality is due to the effect of the spatial dispersion ($D\neq 0$) of optical phonons and the time retardation effect of their motion ($V\neq 0$). These effects partially cancel each other. But if $V^2 = D$ and $\gamma = 0$, i.e., both effects are completely compensated, Green's function $G(\xi)$ reduces to a delta-function. In general, the velocity V of motion does not exceed the maximum group velocity $D^{1/2}$ of the phonons. Hence, $\gamma \geqslant 0$ and Green's function becomes as follows

$$G(\xi) = \begin{cases}(2\gamma)^{-1}\exp(-|\xi|/\gamma) & \text{if} \quad \gamma>0,\\ \delta(\xi) & \text{if} \quad \gamma = 0.\end{cases} \tag{27.14a}$$

In this case, the solutions of the type (27.13) obey the necessary boundary conditions (27.12).

In the opposite case, when the velocity of a quasiparticle is greater than the limiting velocity of optical phonons, i.e., when the inequality $V > D^{1/2}$ holds, Green's function takes the form

$$G(\xi) = \begin{cases} (|\gamma|)^{1/2}\sin(-|\xi|/\sqrt{|\gamma|}) & \text{if} \quad \xi > 0, \ \gamma < 0, \\ 0 & \text{if} \quad \xi \geqslant 0, \ \gamma < 0. \end{cases}$$

Hence, the solution of eqn. (27.13) is

$$u(\xi) = \frac{\chi}{M\Omega_0^2}\left[\sin\xi \int_{-\infty}^{\xi} \varphi^2(\xi_1)\cos\xi_1 d\xi_1 - \cos\xi \int_{-\infty}^{\xi} \varphi^2(\xi_1)\sin\xi_1 d\xi_1\right].$$

This function possesses an oscillating nonvanishing asymptotic at

$$u(\xi)_{\xi\to\infty} \to -\frac{\chi}{M\Omega_0^2}\sin\xi \int_{-\infty}^{\infty} \varphi^2(\xi_1)\cos\xi_1 d\xi_1.$$

Therefore, the autolocalized quasiparticle states occur if $V \leqslant D^{1/2}$ only. This is a unique case when the local deformation follows the displacement of the localized quasiparticle.

Substituting the function (27.13) into eqn. (27.8), one obtains the integro-differential equation for the function $\varphi(\xi)$ which determines the distribution of a quasiparticle with mass m along the chain,

$$\left\{\frac{\hbar^2}{2ma^2} + \frac{d^2}{d\xi^2} + \Lambda + \frac{\chi}{M\Omega_0^2}\int G(\xi-\xi_1)\varphi^2(\xi_1)d\xi_1\right\}\varphi(\xi) = 0. \qquad (27.15)$$

We recall that the motion of a quasiparticle, interacting locally with acoustic phonons, is described by the nonlinear differential equation (26.7).

Solving eqn. (27.15), with the help of eqn. (27.13) one can find the function $u(\xi)$ giving the displacement field of atoms in molecules constituting the chain. According to eqns. (27.10) and (27.15), the energy $\hbar\omega$ of a quasiparticle interacting with the deformation field is as follows

$$\hbar\omega = \mathcal{E}_0 + \frac{1}{2}mV^2 - \frac{\hbar^2}{2ma^2}\int \varphi(\xi)\frac{d^2}{d\xi^2}\varphi(\xi)d\xi + \qquad (27.16)$$

$$+\frac{\chi^2}{2M\Omega_0^2}\int \varphi^2(\xi)G(\xi-\xi_1)\varphi^2(\xi_1)d\xi d\xi_1.$$

Hence, for the known function $u(\xi)$, the energy of atomic displacements in the chain is

$$\mathcal{E}_{\rm ph} = \frac{M}{2a}\int\left[(V^2+D)\left[\frac{\partial u}{\partial \xi}\right]^2+\Omega_0^2 u^2(\xi)\right]d\xi. \tag{27.17}$$

The total energy, $E(V)$, of the dressed quasiparticle is equal to the following sum

$$E(V) = \hbar\omega+\mathcal{E}_{\rm ph}. \tag{27.18}$$

The explicit form of the equations derived is considerably simplified in the case when the particle velocity V is just the limiting one, $D^{1/2}$, i.e., $\gamma = 0$. Then the interaction becomes local, and, in accordance with eqn. (27.13), the displacement function $u(\xi)$ is directly related to the probability $\varphi(\xi)$ of the spatial distribution of the quasiparticle,

$$u(\xi) = -\frac{\chi}{M\Omega_0^2}\varphi^2(\xi). \tag{27.19}$$

The function $\varphi(\xi)$ satisfies the following rather simple nonlinear differential equation

$$\left[\frac{\hbar^2}{2ma^2}\frac{d^2}{d\xi^2}+\Lambda+\frac{\chi^2}{M\Omega_0^2}\varphi^2(\xi)\right]\varphi(\xi) = 0 \tag{27.20}$$

whose normalized solution has the bell-like form

$$\varphi(\xi) = (\tfrac{1}{2}Q)^{1/2}\mathrm{sech}(Q\xi) \tag{27.21}$$

where

$$Q = a^2\chi^2/2\hbar^2\Omega_0^2 \text{ and } \Lambda = -\hbar^2 Q^2/2ma^2. \tag{27.22}$$

The field of displacements accompanying a quasiparticle, is given by the function

$$u(\xi) = -\frac{a^2\chi^2}{2M\hbar^3\Omega_0^4}\mathrm{sech}^2(Q\xi). \tag{27.23}$$

As is known, the spatial dispersion of realistic chains is negligible, and the strong inequality

$$D \ll a^2\Omega_0^2$$

takes place. The parameter γ, given by eqn. (27.11), is always small, $\gamma \ll 1$, since the dressed quasiparticle moves with velocity $V \leqslant D^{1/2}$. In this case, Green's function (27.14) of the integro-differential equation (27.15) can be represented by the following series expansion

$$G(\xi) = \left[1+\gamma\frac{d^2}{d\xi^2}+\gamma^2\frac{d^2}{d\xi^4}+...\right]\delta(\xi). \tag{27.24}$$

In only the first terms on the right-hand side of (27.24) are taken into account, eqn. (27.15) is transformed into the modified Schrödinger equation,

$$\frac{d^2\varphi}{d\xi^2}+4\alpha\varphi^3+4\alpha\gamma\varphi\frac{d^2\varphi}{d\xi^2}-\lambda^2\varphi = 0, \tag{27.25}$$

with the dimensionless parameters

$$\alpha = ma^2\chi^2/2\hbar^2M\Omega_0^2, \;\; \lambda^2 = -2ma^2\Lambda/\hbar^2. \tag{27.26}$$

Equation (27.25) possesses implicitly the exact soliton-type solution

$$\xi = (2\lambda^{-1})\ln\left[\frac{|A-B|}{A+B}\right]+C\gamma^{1/2}, \;\; \gamma>0, \tag{27.27}$$

where the abbreviated notations,

$$A^2 \equiv 1+8\alpha\gamma\varphi^2, \;\; B^2 \equiv 1-2\alpha\varphi^2\lambda^{-2}, \tag{27.28}$$

$$C \equiv 2\mathrm{arctg}(A/2\lambda B\gamma^{1/2}),$$

are introduced. As expected, this equation reduces to the nonlinear Schrödinger equation for vanishing γ. Inverting eqn. (27.20) assuming $\gamma = 0$, and using the normalization condition, one finds

$$\varphi(\xi) = \left[\frac{1}{2}\alpha\right]^{1/2}\mathrm{sech}(\alpha\xi).$$

It follows from eqn. (27.27) that

$$\xi \approx (2\lambda)^{-1}\ln\left\{\frac{(1-B)[1-4\gamma\lambda^2]}{1+B}\right\} \tag{27.29}$$

for nonvanishing positive γ satisfying the inequality $8\alpha\gamma\varphi^2 \ll 1$. Resolving this equation with respect to $\varphi^2(\xi)$, one finds within the term $(\gamma\lambda)^2$

$$\varphi^2(\xi) = \frac{\lambda^2}{2\alpha}\text{sech}^2(\lambda\xi)[1+4\lambda^2\gamma\text{th}^2(\lambda\xi)]. \tag{27.30}$$

Using the normalization condition, one has

$$\lambda = \alpha(1-4\gamma\alpha^3/3). \tag{27.31}$$

Therefore, the dimension of the soliton is given by the following expression in terms of z,

$$\Delta z \approx 2\pi a/|\lambda| \approx 2\pi a\alpha^{-1}(1+4\gamma\alpha^2/3). \tag{27.32}$$

The continuum approach is valid if the inequality

$$\Delta z \approx 2\pi\alpha^{-1} = \frac{\pi\chi^2 ma^3}{\hbar^2 M\Omega_0^3} \gg a \tag{27.33}$$

is satisfied.

The solution (27.30) enables the function $u(\xi)$ to be obtained which characterizes the intramolecular displacements of atoms,

$$u(\xi) = -\frac{\chi\lambda^2}{2\alpha M\Omega_0^2}\text{sech}^2(\lambda\xi)\{1+2\gamma\lambda^2[4-5\text{sech}^2(\lambda\xi)]\}, \tag{27.34}$$

and the evaluation of the total energy and momentum of slowly moving ($V^2<D$) soliton, respectively,

$$E(V) = E(0)+\frac{1}{2}M_{\text{sol}}V^2, \quad P(V) = M_{\text{sol}}V. \tag{27.35}$$

In eqn. (27.35), the effective mass M of the soliton is determined by the relation

$$M_{\text{sol}} = m[1+8\hbar^2\alpha^4/15m^2\Omega_0^2a^4] \tag{27.36}$$

and the energy of the rest soliton takes the value

$$E_0 = \mathcal{E}_0-\frac{\hbar^2\alpha^2}{6ma^2}(1-8\alpha^2D/5a^2\Omega_0^2). \tag{27.37}$$

The value of α is given by eqn. (27.26).

Hence, the origin of the soliton energy band is separated from the energy of the bottom of the conduction and by

$$\Delta E = -\frac{\hbar^2\alpha^2}{6ma^2}(1-8\alpha^2D/5a^2\Omega_0^2),$$

which characterizes the condition of stability of the dressed quasiparticle

$$\Delta E < 0. \tag{27.38}$$

For nonvanishing D, an increase of V^2 from to D results in a rather small contraction of its linear dimension, in accordance with (27.32).

Taking the two first terms in eqn. (27.24) into account, one can obtain the solutions of eqn. (27.16) by minimizing the following functional

$$\mathcal{J}(\varphi) = -\frac{\hbar^2}{2ma^2}\int \varphi(\xi)\left[\frac{d^2\varphi}{d\xi^2} + 2\alpha\varphi\left(\varphi^2 + \gamma\frac{d^2\varphi^2}{d\xi^2}\right)\right]d\xi \tag{27.39}$$

and imposing the normalization condition for $\varphi(\xi)$. One can use a normalized trial function of the form

$$\varphi(\xi) = (\frac{1}{2}Q)^{1/2}\cosh^{-1}(Q\xi), \tag{27.40}$$

depending on the single variational parameter Q. Substituting eqn. (27.40) into (27.39), one derives a cubic polynomial

$$\mathcal{J}(Q) = \frac{\hbar^2}{3ma^3}\left[\frac{4}{5}\gamma\alpha Q^3 + \frac{1}{2}Q^2 - \alpha Q\right] \tag{27.41}$$

whose minimum is

$$Q = Q_0 \equiv \alpha\left(1 - \frac{12}{5}\gamma\alpha^2\right). \tag{27.42}$$

Note that Q_0 differs slightly from its exact value given by (27.22).

Therefore, autolocalization of a quasiparticle (electron) in a molecular chain with its interaction with optical phonons possessing the dispersion (27.1) is realized if its velocity V is less than $D^{1/2}$, i.e., it has a velocity than less than the limiting velocity of wave packets composed of optical phonons. In the opposite case, $V > D^{1/2}$, the displacements of atoms in molecules are unable to follow the quasiparticle motion. In the latter case, its motion is described by stationary plane waves, or nonstationary packets of stationary states of a free quasiparticle in its energy band.

At velocities $V < D^{1/2}$, the set of equations (27.8) and (27.9) possess plane-wave type solutions, $\varphi(\xi) = N^{-1/2}$, $u(\xi) = -\chi/NM\Omega_0^2$ (the chain length is equal to aN)

together with the autolocalized solutions. The energy of the former solutions moving with velocity $V < D^{1/2}$, is

$$\hbar\omega = \mathcal{E}_0 + \frac{1}{2}mV^2 + \alpha^2 / M\Omega_0^2 N^2.$$

For rather large $N \gg 1$, this coincides exactly with the energy of a free quasiparticle of effective mass m in its energy band. Such states are metastable since, if $\chi \neq 0$, the localized quasiparticle can migrate with a considerably small energy.

Chapter 8

SPACE-PERIODIC EXCITATIONS IN NONLINEAR SYSTEMS

28. Mobile Periodic Excitations in One-Dimensional Systems

In the preceding Chapters we were interested basically in the study of mobile and isolated autolocalized excitations in soft one-dimensional molecular systems. Such excitations are described by nonlinear differential equations. Together with two-component solitons these equations possess single-component solutions corresponding to collective states without a local deformation, i.e., to the excitations of the band electron or exciton types (if they do not carry an electric charge). Unlike the case of solutions, the excitation of band states is accompanied by a rather small homogeneous displacement of molecules from their equilibrium positions. The band excitations are described by plane waves, and in one-dimensional systems they become metastable at small velocities since their energies usually exceed those of solitons.

However, the same nonlinear equations possess a third type of solutions which correspond to the spatially periodic distributions of electrons or excitons moving along a chain with a constant velocity. We describe these briefly as quasiparticles. The inhomogeneous distribution of quasiparticles is accompanied by a consistent variation of intermolecular distances, i.e., with a periodic distribution of the local deformation. For this reason, these excitations are two-component too, since they are affected by two fields: the field of quasiparticles and that of deformations.

We shall now consider such solutions following the paper by Pestryakov and the present author [52]. As a model we consider an infinite chain of molecules with mass M separated by a distance a from each other. We suggest that this chain may attach excess quasiparticles (see Section 26). Assuming the molecules to be 'frozen' at the equilibrium positions na, quasiparticles are characterized by an effective mass m in the long-wave limit. We also suggest that these particles do not interact with each other, and that their local deformation interaction with the chain is rather strong, so the motion of the particles may be accompanied by displacements of the equilibrium positions of molecules.

All energies refer to the energy of the bottom of the energy band of free quasiparticles. In the long-wave, $na \to x$, the Hamiltonian density of the system in question has the following form

$$\mathcal{H}(x,t) = \frac{\hbar^2}{2m}\left|\frac{\partial\psi}{\partial x}\right|^2 - 2\sigma\varphi|\psi|^2 + \frac{M}{2a}\left[V_0^2\rho^2 + \left(\frac{\partial u}{\partial t}\right)^2\right]. \qquad (28.1)$$

In eqn. (28.1), $u(x,t)$ is the displacement of molecules from their equilibrium positions expressed in terms of the coherent states of virtual phonons,

$$\rho(x,t) = -\frac{\partial u(x,t)}{\partial x} \qquad (28.2)$$

represents a decrease of distance, in units of a, between neighboring molecules, V_0 is the longitudinal sound velocity in the chain without quasiparticles, $|\psi(x,t)|^2$ is the linear density of quasiparticles at the point x and at time t, and σ is the energy of the local deformation interaction of quasiparticles with molecular displacements.

The system under study, described by the Hamiltonian density (28.1), possesses the following integrals of motion;

$$(I) \int|\psi(x,t)|^2 dx = \text{const} \qquad (28.3)$$

is the number of independent quasiparticles in the system;

$$(II)\ \ni = \int\mathcal{H}(x,t)dx \qquad (28.4)$$

is the excited state energy;

$$\mathcal{P} = \int\left\{\frac{\hbar}{2i}\left[\psi^*\frac{\partial\psi}{\partial x} - \psi\frac{\partial\psi^*}{dx}\right] - \frac{M}{a}\frac{\partial u}{\partial t}\frac{\partial u}{\partial x}\right\}dx \qquad (28.5)$$

is the excited state momentum.

From the Hamiltonian (28.1) we obtain the set of differential equations:

$$i\hbar\frac{\partial\psi}{\partial t} + \frac{\hbar^2}{2m}\frac{\partial^2\psi}{dx^2} + \sigma\rho\psi = 0, \qquad (28.6a)$$

$$\left[\frac{\partial^2}{\partial t^2}-V_0^2\frac{\partial^2}{\partial x^2}\right]\beta-\frac{a\sigma}{M}\frac{\partial}{\partial x}(|\psi|^2) = 0. \tag{28.6b}$$

Using the invariance of the Hamiltonian (28.1) relative to the choice of the origin of the spatial reference frame, one can seek the solution of the set of equations (28.6) in the form

$$\psi(x,t) = \frac{1}{\sqrt{a}}\Phi(\xi)\exp\left\{\frac{i}{\hbar}\left[mxV-\left[\mathcal{E}+\frac{mV}{2}\right]f+\eta_0\right]\right\}, \tag{28.7a}$$

$$\beta(x,t) = \beta(\xi), \quad \rho(x,t) = \rho(\xi), \tag{28.7b}$$

where

$$\xi = (x-x_0-Vt)/a \tag{28.8}$$

and η_0 is an arbitrary phase. In eqn. (28.7a), the quantity $\mathcal{E}$ determines the energy of quasiparticles in the field of the deformation wells, $\sigma\rho(\xi)$, moving with velocity V. Substituting the function (28.7a) into eqn. (28.6a), one obtains the equation for the envelope $\Phi(\xi)$ of the function (28.7a),

$$\left[\frac{\hbar^2}{2ma^2}\frac{d^2}{d\xi^2}+\mathcal{E}+\sigma\rho(\xi)\right]\Phi(\xi) = 0. \tag{28.9}$$

Integrating eqn. (28.6b) under the condition $\partial\beta/\partial x = 0$ over all regions where $|\psi(x,t)|^2 = 0$, one finds

$$\rho(\xi) = \sigma\Phi^2(\xi)/[MV_0^2(1-s^2)], \quad s = V/V_0. \tag{28.10}$$

Due to the presence of the factor $(1-s^2)$ in the denominator of eqn. (28.10), the approach developed is justified for those velocities V of excitation motion which are considerably less than the longitudinal sound velocity V_0. As $s^2\rightarrow 1$, the continuum and harmonic approaches used in deriving the Hamiltonian (28.1) become invalid. However, as shown in Chapter V, this difficulty is overcome by including anharmonicity.

Substituting the formula (28.10) into eqn. (28.9), one transforms the latter into the nonlinear Schrödinger equation for the envelope function $\Phi(\xi)$:

$$\left[\frac{d^2}{d\xi^2}+\epsilon+2g\Phi^2(\xi)\right]\Phi(\xi) = 0, \tag{28.11}$$

where

$$\epsilon = 2ma^2\mathcal{E}/\hbar^2, \quad g = ma^2G/\hbar^2, \tag{28.12}$$

and

$$G = \sigma^2[MV_0^2(1-s^2)]^{-1}. \tag{28.12a}$$

Substituting further (28.7a) and (28.10) into eqn. (28.4) and using (28.2), (28.8), and the relation $\partial u/\partial t = V\rho$, one obtains the expression for the total excitation energy per particle which involves the deformation chain energy (within the laboratory frame)

$$\ni(V) = \frac{1}{2}mV^2 + \frac{\hbar^2}{2ma^2}\left[\epsilon + (1+s^2)g\int\Phi^4(\xi)d\xi\right], \tag{28.13}$$

where ϵ is given by the following formula

$$\epsilon = \int\left\{\left|\frac{d\varphi}{d\xi}\right|^2 + 2g\Phi^4\right\}d\xi \tag{28.13a}$$

in agreement with eqn. (28.11). The expression (28.5) for momentum is transformed similarly,

$$\mathcal{P} = V\left\{m + \frac{g\hbar^2}{ma^2V_0^2}\int\Phi^4 d\xi\right\}. \tag{28.14}$$

If a single quasiparticle exists in the system, eqn. (28.11) has the solution

$$\Phi_{\rm sol}(\xi) = \frac{1}{2}\sqrt{g}\,{\rm sch}(g\xi/2), \quad \epsilon_{\rm sol} = -g^2/4 \tag{28.15}$$

which decreases exponentially as $|\xi| \to \infty$ and obeys the normalization condition

$$\int\Phi^2(\xi)d\xi = 1. \tag{28.16}$$

This function describes the movement of a single-soliton excitation with velocity V.

Using the explicit form of the function (28.16), one finds

$$\int_{-\infty}^{\infty}\Phi_{\rm sol}^4(\xi)d\xi = g/6.$$

Substituting this value and $\epsilon_{\rm sol}$ into eqns. (28.13) and (28.14), one obtains the energy and momentum of the soliton,

$$\ni_{sol}(V) = \frac{1}{2}mV^2 - \hbar^2 g^2/(1-2s^2)24ma^2, \tag{28.17}$$

$$\mathcal{P}_{sol}(V) = V[m + \hbar^2 g^2/6ma^2 V_0^2(1-s^2)]. \tag{28.18}$$

Due to the fact that quasiparticle motion is accompanied by a local chain deformation

$$\rho_{sol} = 2\sigma g \operatorname{sch}^2(g\xi/2)/8MV_0^2(1-s^2), \tag{28.19}$$

the effective mass of the whole excitation (soliton) exceeds significantly the mass m of a quasiparticle in the conduction band and becomes equal to

$$m_{sol} = m\left\{1 + \frac{\hbar^2}{6Ma^2 V_0^2(1-s^2)}\right\}. \tag{28.20}$$

Let us now study the periodic solutions of the set of equations (28.6). We suggest that, in terms of x, the period is equal to aL, i.e., the conditions

$$|\psi(x,t)|^2 = |\psi(x+aL,t)|^2, \quad \rho(x,t) = \rho(x+aL), \tag{28.21}$$

apply. From eqn. (28.7) there follows the condition of periodicity

$$\Phi(\xi+L) = \Phi(\xi). \tag{28.22}$$

We seek the solutions $\Phi(\xi)$ of eqn. (28.11) which satisfy the conditions (28.22) of periodicity and normalization of a single particle per period,

$$\int_0^L \Phi^2(\xi)d\xi = 1. \tag{28.23}$$

In Section 29, we demonstrate that these solutions are expressed in terms of two Jacobian elliptic functions $dn(u,k)$ and $cn(u,k)$ depending on a variable u as the function of ξ. Each function is characterized by the modulus k having a range in the interval (0,1). Notice that the elementary data on the Jacobian elliptic functions are presented in the next Section.

To find in a simple way the explicit form of these solutions of eqn. (28.11), one can use the differential relations to which the Jacobian elliptic functions, $dn(u,k)$ and $cn(u,k)$, are satisfied (see Section 29):

$$\frac{d^2}{du^2}\mathrm{dn}(u,k) = (2-k)^2\mathrm{dn}(u,k) - 2\mathrm{dn}^3(u,k), \tag{28.24}$$

$$\frac{d^2}{du^2}\mathrm{cn}(u,k) = (2k^2-1)\mathrm{cn}(u,k) - 2k^2\mathrm{cn}^3(u,k). \tag{28.25}$$

Applying these equations, one can check that the function

$$\Phi_d(\xi,k) = A\,\mathrm{dn}(\alpha\xi,k) \tag{28.26a}$$

$$\Phi_a(\xi,k) = B\,\mathrm{cn}(\beta\xi,k) \tag{28.26b}$$

satisfy eqn. (28.11) under the following constraints:

$$\epsilon_d = -\alpha^2(2-k^2), \; g = (\alpha/A)^2, \tag{28.27a}$$

$$\epsilon_c = -\beta^2(2k^2-1), \; g = (k\beta/B)^2. \tag{28.27b}$$

The Jacobian functions $dn(u,k)$ and $cn(u,k)$ are periodic functions of their argument u. The solutions of type (28.26) are called cnoidal waves. The constants A, B, α, β and the modulus k of the elliptic cnoidal waves. The constants A, B, α, β and the modulus k of the elliptic functions, appearing in eqn. (28.26), are determined by the period L relative to the dimensionless variable ξ and by the normalization condition. Now we study the solutions of the types (28.26a) and (28.26b) separately.

28.1. First Type of Periodic Excitations

The solutions corresponding to the Jacobian elliptic function $\mathrm{dn}(u,k)$ are defined via inverting the integral (see Section 29)

$$u = \int_l^{\mathrm{dn}(u,k)} \frac{dt}{\sqrt{(1-t^2)(t^2-k_l^2)}}, \quad k_1 = \sqrt{1-k^2}, \tag{28.28}$$

where k_1 is the complementary modulus. In the case under consideration the independent variable is $u = \alpha\xi$. The normalization of the function (28.26a) per period is reduced to the following relation

$$l = A^2\int_0^L \mathrm{dn}^2(\alpha\xi) = \frac{A^2}{\alpha}\int_0^{\alpha L} \mathrm{dn}^2(u,k)du. \tag{28.29}$$

The period of the function $\mathrm{dn}(u,k)$ relative to u is equal to $2K(k)$, where $K(k)$ is the complete elliptic integral of the first kind. The value L is the period of the function (28.26a) if the relation

$$\alpha L = 2K(k) \tag{28.30}$$

holds. In that case, the normalization condition (28.29) takes the form

$$2A^2E(k) = \alpha, \tag{28.31}$$

where $E(k)$ is the complete elliptic integral of the second kind,

$$E(k) \equiv \int_0^{K(k)} \mathrm{dn}^2(u,k)du. \tag{28.32}$$

From eqns. (28.27a) and (28.31) it follows that

$$\alpha = \frac{g}{2E(k)}, \quad A = \frac{\sqrt{g}}{2E(k)}, \quad \epsilon_d = -\frac{g^2(2-k^2)}{4E^2(k)}. \tag{28.33}$$

Substituting this value of α into eqn. (28.31), one obtains the equation

$$gL = 4E(k)K(k), \tag{28.34}$$

which enables the modulus k to be determined for a given spatial period L and nonlinearity parameter g.

Small values of k_1 correspond to large periods. If $k_1^2 \ll 1$, the functions $K(k)$ and $E(k)$ have the following asymptotics (see Section 29):

$$K(k) = \Lambda + \frac{1}{4}(\Lambda - 1)k_1^2 + ... \tag{28.35}$$

$$E(k) = 1 + \frac{1}{2}\left(\Lambda - \frac{1}{2}\right)k_1^2 + ... \tag{28.36}$$

where

$$\Lambda \equiv \ln\left(\frac{4}{k_1}\right). \tag{28.37}$$

Determining the modulus k for given gL by means of eqn. (28.34), one obtains the function

$$\Phi_d(\xi,k) = \frac{1}{2}\sqrt{g}E^{-1}(k)\mathrm{dn}\left(\frac{g\xi}{2E(k)},k\right) \tag{28.38}$$

in terms of the dimensionless variable ξ. Using eqn. (28.10), one finds the second component of the soliton, the deformation field

$$\rho_d(\xi) = \frac{\sigma g}{4MV_0^2(1-s^2)E^2(k)}\mathrm{dn}^2\left[\frac{g\xi}{2E(k)},k\right]. \tag{28.38a}$$

Limiting values $k \to 1$ and $E(k) \to 1$ correspond to the infinitely large period $L \to \infty$. In this case, the function (28.38) transforms into the form

$$\Phi_d(\xi,k) = \Phi_s(\xi) = \frac{1}{2}\sqrt{g}\,\mathrm{sch}(g\xi/2), \quad k \approx 1. \tag{28.39}$$

This function describes the motion of a single excitation of the solitonic type with velocity V (28.16). For $k_1^2 \ll 1$, the function (28.38) is approximated by the hyperbolic function

$$\Phi_d(\xi,k) = \frac{1}{2}\sqrt{g}\left\{1+\frac{k_1^2}{4}(u+\mathrm{sh}u\mathrm{ch}u)\mathrm{th}u\right\}\mathrm{sch}u, \tag{28.40}$$

with $u \equiv g\xi/2$. In the opposite limiting case of small values of $k, E(k) \approx \pi/2$ and the function becomes

$$\Phi_d(\xi,k) \simeq \frac{1}{2}\sqrt{g}\left[1-\frac{1}{2}k^2\sin^2 u\right], \quad k^2 \ll 1, \quad u = g\xi/\pi. \tag{28.40a}$$

Now we evaluate the energy and momentum per period L of the excitation moving with velocity $V < V_0$. Using eqn. (28.26a) and (28.33), and substituting into eqns. (2813a) and (28.14) the integral

$$\int_0^L \Phi_d^4(\xi,k)d\xi = gQ_d(k)/4E(k), \tag{28.41}$$

where

$$Q_d(k) \equiv \int_0^{K(k)} \mathrm{dn}^4(u,k)\mathrm{du} = \frac{1}{3}[2(2-k^2)E(k)-(1-k^2)K(k), \tag{28.42}$$

one obtains the total energy and momentum of the two-component excitation which is characterized by the modulus k,

$$\ni_{d,k}(V) = \frac{1}{2}mV^2 - \frac{\hbar^2 g^2}{8ma^2E(k)}\left[\frac{2-k^2}{E(k)}-(1+s^2)Q_d(k)\right], \tag{28.43}$$

$$\mathcal{P}_{d,k}(V) = V\left[m+\frac{\hbar^2 g^2 Q_d}{4ma^2V_0^2(1-s^2)E(k)}\right]. \tag{28.44}$$

For sufficiently large L, the complementary modulus k_1 becomes small, and one can use the asymptotics given by (28.35) and (28.36). Then eqn. (28.34) is replaced by the approximate one

$$gL \approx 4\ln\left[\frac{4}{k_1}\right], \quad \text{or} \quad k_1^2 \approx 16\exp(-gL/2). \tag{28.45}$$

Within this approximation, the function $Q_d(k)$ has the value

$$Q_d(k) \approx \frac{2}{3}[1-4gL\exp(-gL/2)], \quad k_2^2 \ll 1. \tag{28.45a}$$

In the limit $L \to \infty$, $\lim Q_d = 2/3$. Hence, the limiting values of (28.43) and (28.44) are just the corresponding ones, (28.17) and (28.17a), for a single soliton.

28.2. Second Type of Periodic Excitations

Let us now study the periodic solutions corresponding to the function (28.26b), whose normalization condition is reduced to the following

$$1 = B^2\int_0^L \mathrm{cn}^2(\beta\xi)d\xi = \frac{B^2}{\beta}\int_0^{\beta L} \mathrm{cn}^2(u,k)du. \tag{28.46}$$

The Jacobian function $\mathrm{cn}(u,k)$ is periodic relative to the variable u with period $4K(k)$. Thus, the condition of periodicity, eqn. (28.20), is satisfied if the following relation

$$\beta L = 4K(k) \tag{28.47}$$

holds. Using the integral

$$\int_0^{4K(k)} \mathrm{cn}^2(u,k)du = 4\delta(k)/k^2, \tag{28.48}$$

where

$$\delta(k) \equiv k^2\int_0^{K(k)} \mathrm{cn}^2(u,k)du = E(k)-k_1^2K(k), \tag{28.48a}$$

one can obtain

$$\beta k^2 = 4B^2\delta(k). \tag{28.49}$$

At small values of the complementary modulus, one has

$$\delta(k) \approx E(k) \approx 1, \quad k_1^2 \ll 1.$$

From eqns. (28.27b) and (28.49) it follows

$$\beta = g/4\delta(k), \quad B = k\sqrt{g}/4\delta(k), \tag{28.50}$$

$$\epsilon_c = -g^2(2k^2-1)/16\delta^2(k). \tag{28.51}$$

Substituting this value of β into eqn. (28.49), one obtains the relation

$$gL = 16\delta(k)K(k), \tag{28.52}$$

which determines the modulus of the Jacobian function (28.26b) taking now the final form

$$\Phi_c(\xi,k) = \frac{k\sqrt{g}}{4\delta(k)}\text{cn}\left[\frac{g\xi}{4\delta(k)},k\right] \tag{28.53}$$

The square of the function $\Phi_c(\xi,k)$ characterizes the probability of distribution of a quasiparticle. The quasiparticle motion is accompanied by the deformation field which is given by the function

$$\rho_c(\xi) = \frac{\sigma g k^2}{16MV_0^2(1-s^2)\delta^2(k)}\text{cn}^2\left[\frac{g\xi}{4\delta(k)},k\right] \tag{28.53a}$$

in agreement with eqn. (28.10).

To evaluate the energy of the state described by the function (28.53), one should substitute eqn. (28.51) and the following integral

$$\int_0^L \Phi_c^4(\xi,k)d\xi = gQ_c(k)/16\delta^3(k), \tag{28.54}$$

where

$$Q_0(k) \equiv k^4 \int_0^{K(k)} \text{cn}^4(u,k)du = \frac{1}{3}[2(2k^2-1)E(k)-k_1^2(2-3k^2)K(k)] \tag{28.55}$$

into the formula (28.13). Then one obtains the total excitation energy

$$\ni_c(V) = \frac{1}{2}mV^2 - \frac{\hbar^2 g^2}{32ma^2\delta(k)}\left\{(2k^2-1) - \frac{(1+s^2)Q_c(k)}{\delta(k)}\right\}. \tag{28.56}$$

Since $Q_c(k) > 0$, it follows that the negative energies correspond to those states having a modulus k only which satisfies the inequality

$$(2k^2-1)\delta(k) > Q_c(k), \tag{28.57}$$

or

$$k^2 > \frac{1}{2} \tag{28.57a}$$

for vanishing V.

For small values of the complementary modulus, $k_1^2 \ll 1$, the functions $\delta(k) \approx E(k) \approx 1$, and hence, $Q_c(k) \approx \frac{2}{3}(1-2k_1^2)$. The total energy takes now the magnitude

$$\ni_c(V) = \frac{1}{2}mV^2 - \frac{\hbar^2 g^2(1-2k_1^2)(1-2s^2)}{96ma^2}. \tag{28.58}$$

Finally, the wave function is expressed in terms of the hyperbolic functions,

$$\Phi_c(\xi,k) = \frac{1}{4}\sqrt{g}\left[1+\frac{k_1^2}{4}(u-\text{sh}u\text{ch}u)\text{th}u\right]\text{sch}u,\quad u = g\xi/4,\quad k_1^2 \ll 1. \tag{28.59}$$

In this limiting case, which corresponds to an infinite limit, $k_1^2 \approx 0$ and the function (28.59) describes the autolocalized state of the soliton type.

Therefore, the periodic excitations described by the functions which satisfy the conditions (28.20), may arise in a molecular chain. Such excitations are called *cnoidal waves*, since they are expressed in terms of Jacobian elliptic functions. These functions are characterized by a modulus k in the range (0,1). If $k^2 > 1/2$, two excitation branches correspond each to a value of k. These branches are described by the functions $\Phi_d(\xi,k)$ and $\Phi_c(\xi,k)$ with differing energies given by eqns. (28.43) and (28.56), respectively.

One should recall that the states of the system described by the functions $\Phi_d(\xi,\tau)$ and $\Phi_c(\xi,\tau)$ are two-component. The squares of these functions give the probability of the distribution of quasi-particles whose motion is accompanied by the movement of the deformation fields $\rho_d(\xi,\tau)$ and $\rho_c(\xi,\tau)$ determined by equations of the type (28.10).

At rather large periods L, when $gL \gg 1$ and $k_1^2 \ll 1$, the states corresponding to the wave functions $\Phi_c(\xi,k)$, are characterized by larger values of energy compared with those described by $\Phi_d(\xi,k)$. This is explained by the smaller degree of the local deformation in the neighborhood of the particle density maximum for the states $\Phi_c(\xi,k)$ in comparison with the states $\Phi_d(\xi,k)$.

The energy of excited states is lower than the bottom of the one-particle band for $V = 0$ if the spatial period $gL \rightarrow aL$ exceeds a somewhat critical value. In other words, the energetically preferable and spatially periodic configurations of the autolocalized states can be realized only for comparatively small densities of particles, $|\psi(\xi,t)|^2 \approx (aL)^2 \ll 1$. A study of the possibility of their clusterization requires an account of interparticle interaction.

29. Elementary Data on Jacobian Elliptic Functions

In order to facilitate the reading of the present book, we present here elementary data on three Jacobian elliptic functions which are used throughout the book. A complete description of their properties are given elsewhere [2,14,110,127,128].

Three Jacobian elliptic functions, $\mathrm{sn}(u,k)$, $\mathrm{cn}(u,k)$, and $\mathrm{dn}(u,k)$, depend on a variable u and modulus k taking all values in the interval

$$0 \leqslant k \leqslant 1. \tag{29.1}$$

Each elliptic function is defined by inverting the corresponding elliptic integral of the first kind

$$u = \int_0^{\mathrm{sn}(u)} \frac{dt}{[(1-t^2)(1-k^2)]^{1/2}} = \int_0^{\mathrm{cn}(u)} \frac{dt}{[(1-t^2)(k_1^2+k^2t^2)]^{1/2}} = \tag{29.2}$$

$$= \int_0^{\mathrm{dn}(u)} \frac{dt}{[(1-t^2)(t^2-k^2)]^{1/2}},$$

where $k_1 = (1-k^2)^{1/2}$ is the complementary modulus of the elliptic function.

Jacobian elliptic functions are single-valued analytic functions of a complex variable u except for their poles. They are single-valued functions of a modulus k, too. The functions $\mathrm{cn}(u,k)$ and $\mathrm{dn}(u,k)$ are even, and the function $\mathrm{sn}(u,k)$ is odd relative to the variable u, that is,

$$\mathrm{cn}(-u,k) = \mathrm{cn}(u,k), \quad \mathrm{dn}(-u,k) = \mathrm{dn}(u,k), \tag{29.3}$$

$$\mathrm{sn}(-u,k) = -\mathrm{sn}(u,k).$$

These functions are interrelated as follows

$$\mathrm{sn}^2(u,k)+\mathrm{cn}^2(u,k) = 1, \quad k^2\mathrm{sn}^2(u,k)+\mathrm{dn}^2(u,k) = 1, \tag{29.4}$$

$$\mathrm{dn}^2(u,k)-k^2\mathrm{cn}(u,k) = k_1^2,$$

$$k_1^2-\mathrm{dn}^2(u,k) = -k^2\mathrm{cn}^2(u,k) = k^2[\mathrm{sn}^2(u,k)-1]$$

and take the following particular values

$$\mathrm{sn}\left[\frac{1}{2}K,k\right] = (1+k_1)^{-1/2}, \quad \mathrm{cn}\left[\frac{1}{2}K,k\right] = [k_1/(1+k_1)]^{1/2}, \tag{29.5}$$

$$\mathrm{dn}\left[\frac{1}{2}K,k\right] = k_1^{1/2}, \quad \mathrm{sn}(K,k) = 1, \quad \mathrm{cn}(K,k) = 0, \quad \mathrm{dn}(K,k) = k_1$$

where $K = K(k)$ is the complete elliptic integral of the first kind (for the definition see below).

Introducing the abbreviated notations

$$s \equiv \mathrm{sn}(u,k), \quad c \equiv \mathrm{cn}(u,k), \quad d \equiv \mathrm{dn}(u,k)$$

and designating differentiation with respect to u by a prime, one can represent derivatives of elliptic functions in the form

$$s' = cd, \quad c' = -sd, \quad d' \equiv -k^2sc, \quad (s')^2 = (1-s^2)(1-k^2s^2), \tag{29.6}$$

$$(c')^2 = (1-c^2)(k^2c^2+k_1^2), \quad (d')^2 = (1-d^2)(d^2-k_1^2), \quad s'' = -s(d^2+k^2c^2),$$

$$c'' = -c(d^2-k^2s^2), \quad d'' = -k^2d(c^2-s^2).$$

Jacobian elliptic functions satisfy the following differential equations

$$s''+(1+k^2)s-2k^2s^3 = 0, \quad c''+(1-2k^2)c+2k^2c^3 = 0, \tag{29.7}$$

$$d''+(k^2-2)d+2d^3 = 0.$$

For $k_1^2 \ll 1$, the asymptotics of elliptic functions are expressed in terms of hyperbolic functions

$$\mathrm{sn}(u,k) = \mathrm{th}u+\frac{1}{4}k_1^2(\mathrm{sh}u\,\mathrm{ch}u-u)\mathrm{sch}^2u, \tag{29.8}$$

$$\mathrm{cn}(u,k) = \mathrm{sch}u-\frac{1}{4}k_1^2(\mathrm{sh}u\,\mathrm{ch}u-u)\mathrm{sch}u\,\mathrm{th}u,$$

$$\mathrm{dn}(u,k) = \mathrm{sch}u+\frac{1}{4}k_1^2(\mathrm{sh}u\,\mathrm{ch}u+u)\mathrm{sch}u\,\mathrm{th}u.$$

For $k^2 \ll 1$, the Jacobian elliptic functions are represented through the trigonometric functions

$$\mathrm{sn}(u,k) = \sin u - \frac{1}{4}k^2(u - \sin u \cos u)\cos u, \tag{29.9}$$

$$\mathrm{cn}(u,k) = \cos u + \frac{1}{4}k^2(u - \sin u \cos u)\sin u,$$

$$\mathrm{dn}(u,k) = 1 - \frac{1}{2}k^2 \sin^2 u.$$

For small values of the variable u, $u \ll 1$, one has the following power series expansions

$$\mathrm{sn}(u,k) = u + (1+k^2)u^2/3! + (1+14k^2+k^4)u^5/5! + \ldots \tag{29.10}$$

$$\ldots \mathrm{cn}(u,k) = 1 - \frac{u^2}{2!} + (1+4k^2)u^4/4! - \ldots \mathrm{dn}(u,k) =$$

$$= 1 - k^2u^2/2! + k^2(4+k^2)/u^4/4! + \ldots$$

Jacobian elliptic functions are doubly-periodic functions of the variable u. Their periods on the real axis are expressed in terms of the complete elliptic integral of the first kind,

$$K(k) = \int_0^{\pi/2} \frac{d\varphi}{[1-k^2\sin^2\varphi]^{1/2}} = \int_0^1 \frac{dt}{[(1-t^2)(1-k^2t^2)]^{1/2}} \tag{29.11}$$

through the following relations

$$\sin(u+4K) = -\mathrm{sn}(u+2K) = \mathrm{sn}(u), \tag{29.12}$$

$$\mathrm{cn}(u+4K) = -\mathrm{cn}(u+2K) = \mathrm{cn}(u), \quad \mathrm{dn}(u+2K) = \mathrm{dn}(u)$$

where the modulus k is simplified. Therefore, the functions $\mathrm{sn}(u)$ and $\mathrm{cn}(u)$ have a period of $4K$, while the period of $\mathrm{dn}(u)$ is less and equal to $2K$. The following relations are also valid

$$\mathrm{sn}(u+K) = \mathrm{cn}(u)/\mathrm{dn}(u), \tag{29.13}$$

$$\mathrm{cn}(u+K) = -\frac{k_1\mathrm{sn}(u)}{\mathrm{dn}(u)}, \quad \mathrm{dn}(u+K) = \frac{k_1}{\mathrm{dn}(u)}.$$

The complete elliptic integral of the first kind as a function of the complementary modulus k_1 possesses a specific notation

$$K_1(k) = K(k_1). \tag{29.14}$$

Thus, one has

$$K_1(k) = \int_0^{\pi/2} [1-k_1^2 \sin^2\varphi]^{-1/2} d\varphi. \tag{29.15}$$

The imaginary periods of the Jacobian elliptic functions are determined through the integral (29.15)

$$\mathrm{sn}(u+2iK_1) = \mathrm{sn}(u),\ \ \mathrm{cn}(u+2iK_1) = -\mathrm{cn}(u),\ \ \mathrm{cn}(u+4iK_1) = \mathrm{cn}(u), \tag{29.16}$$

$$\mathrm{dn}(u+2iK_1) = -\mathrm{dn}(u),\ \ \mathrm{dn}(u+4iK_1) = \mathrm{dn}(u).$$

Therefore, the functions $\mathrm{cn}(u)$ and $\mathrm{dn}(u)$ have the period $4iK_1$, while the period of $\mathrm{sn}(u)$ is smaller and is $2iK_1$.

Using the formula

$$E(u,k) = \int_0^u \mathrm{dn}^2(u',k)du' \tag{29.17}$$

one can define the elliptic integral of the second kind. This integral is a single-valued and odd function of the variable u,

$$\frac{dE(u,k)}{du} = \mathrm{dn}^2(u,k),\ \ E(0,k) = 0. \tag{29.18}$$

For $u = K$, the elliptic integral of the second kind is called the complete elliptic integral of the second kind

$$E(K,k) = \int_0^K \mathrm{dn}^2(u,k)du \equiv E(k). \tag{29.19}$$

The complete elliptic integrals satisfy the following relations

$$E(u+2nK) = E(u)+2nE \ \text{ if } n \text{ is an integer} \tag{29.20}$$

$$E(u+K)-E(u) = E-k^2\mathrm{sn}u\mathrm{cn}u/\mathrm{dn}u,\ \ E-K = kdE/dk. \tag{29.21}$$

Other elliptic integrals of the second kind

$$\mathrm{Sn}(u,k) = \int_0^u \mathrm{sn}^2(u',k)du',\ \ \mathrm{Cn}(u,k) \equiv \int_0^u \mathrm{cn}^2(u',k)du'$$

are expressed via the basic integral of the second kind $E(u,k)$ through the relations

$$k^2\mathrm{Sn}(u,k) = u - E(u,k), \quad k^2\mathrm{Cn}(u,k) = E(u,k) - k_1^2 u. \tag{29.22}$$

The complete elliptic integrals of the first and second kinds have the following asymptotics:
(i) $k^2 \ll 1$

$$K(k) = \frac{\pi}{2}\left\{1 + \left[\frac{1}{2}\right]^2 k^2 + \left[\frac{1\cdot 3}{2\cdot 4}\right]^2 k^4 + \ldots\right\}, \tag{29.23}$$

$$E(k) = \frac{\pi}{2}\left\{1 - \left[\frac{1}{2}\right]^2 k^2 - \left[\frac{1\cdot 3}{2\cdot 4}\right]^2 k^4/3 + \ldots\right\}$$

(ii) $k_1^2 \ll 1$

$$K(k) = \Lambda + (\Lambda - 1)k_1^2/4 + \ldots E(k) = 1 + \frac{1}{2}(\Lambda - 1)k_1^2 + \ldots, \tag{29.24}$$

where

$$\Lambda \equiv \ln(4/k_1^2). \tag{29.25}$$

Hence, the integral $K(k)$ increases monotonically in the interval $0 \leqslant k^2 \leqslant 1$. At $k = 0$, it becomes equal to $\pi/2$. Further, it approaches logarithmically infinity. The complete elliptic integral of the second kind $E(k)$ decreases monotonically in the same interval, from $\pi/2$ to unity.

To conclude this Section we give the integrals of the elliptic functions $\mathrm{cn}(u,k)$ and $\mathrm{dn}(u,k)$:

$$\int_0^{NK} \mathrm{cn}^4(u,k)du = N[k_1^2(3k_1^2 - 1)K(k) + 2(1 - 2k_1^2)E(k)](3k^4)^{-1}, \tag{29.26}$$

$$\int_0^{NK} \mathrm{dn}^4(u,k)du = N[2(2 - k^2)E(k) - k_1^2 K(k)](3)^{-1}. \tag{29.27}$$

30. Periodic Solutions of the Nonlinear Schrödinger Equation. Cnoidal Waves

Periodic solutions of nonlinear one-dimensional differential equations are expressed in terms of Jacobian elliptic functions, $\mathrm{sn}(u,k)$, $\mathrm{cn}(u,k)$, and $\mathrm{dn}(u,k)$. The first periodic solutions in terms of $\mathrm{cn}(u,k)$ were obtained by Korteweg and de Vries [213] in their study of gravitational waves on shallow water. Korteweg and de Vries called these waves cnoidal to emphasize that they are expressed through the elliptic cosine.

At present, the term 'cnoidal wave' is reserved also for periodic solutions of other nonlinear equations if they are expressed in terms of Jacobian elliptic functions.

We now deal with the cnoidal waves which are periodic solutions of the nonlinear Schrödinger equation. So, let $\Phi(\xi)$ be the periodic, i.e.,

$$\Phi(\xi+L) = \Phi(\xi) \tag{30.1}$$

solution of the following nonlinear Schrödinger equation,

$$\frac{d^2\Phi}{d\xi^2}+\epsilon\Phi+2g\Phi^3 = 0, \tag{30.2}$$

which corresponds to a positive g and negative energies, $\epsilon<0$. Introducing the quantity

$$\nu = -\epsilon/2g>0 \tag{30.3}$$

and multiplying eqn. (30.2) on the left by $d\Phi/d\xi$, and further integrating, one obtains the equation

$$\left[\frac{d\Phi}{d\xi}\right]^2 = g[2\xi\Phi^2-\Phi^4-C], \tag{30.4}$$

where C is an integration constant. Different solutions correspond to positive and negative values of C.

30.1. Cnoidal Waves of the First Kind

Let us consider the solutions of the first kind corresponding to positive C. At the start, we denote the maximum and minimum values of the function $\Phi(\xi)$ by Φ_0 and Φ_1, respectively. Thus, one obtains

$$\Phi_0^2 = \nu + \sqrt{\nu^2 - C}, \quad \Phi_1^2 = \nu - \sqrt{\nu^2 - C}, \tag{30.5}$$

or

$$\Phi_0^2 + \Phi_1^2 = 2\nu, \quad \Phi_0^2\Phi_1^2 = C > 0. \tag{30.5a}$$

Then eqn. (30.4) can be rewritten as follows

$$\frac{d\Phi}{d\xi} = \eta\sqrt{g}\sqrt{(\Phi_0^2 - \Phi^2)(\Phi^2 - \Phi_1^2)}, \quad \eta = \pm 1.$$

Integrating, one has

$$\eta\sqrt{g}\,\xi\Phi_0 = \int_0^{\Phi/\Phi_0} \frac{dt}{\sqrt{(1-t^2)(t^2-k_1^2)}}, \tag{30.6}$$

where

$$\Phi(0) = \Phi_0, \quad k_1^2 = (\Phi_1/\Phi_0)^2 \leqslant 1, \quad t = \Phi/\Phi_0. \tag{30.7}$$

An inversion of the integral (30.6) defines the Jacobian elliptic function dn (see Section 29)

$$\Phi(\xi) = \Phi_0 \mathrm{dn}(\sqrt{g}\,\Phi_0\xi, k), \tag{30.8}$$

where the modulus k is expressed via a complementary modulus $k_1 = \Phi_1/\Phi_0$ as

$$k = \sqrt{1-k_1^2}. \tag{30.8a}$$

The complementary modulus k_1 is determined by the ratio of the minimum value of $\Phi(\xi)$ to its maximum. The latter is obtained from the normalization condition of $\Phi(\xi)$ per period,

$$1 = \int_0^L \Phi^2(\xi)d\xi = \frac{\Phi_0}{\sqrt{g}} \int_0^{\Phi_0 L\sqrt{g}} \mathrm{dn}^2(u,k)du. \tag{30.9}$$

The upper limit of the integral must be equal to the period $2K(k)$ of the function $\mathrm{dn}(u,k)$. Using the relation (see Section 29)

$$\int_0^{2K(k)} \mathrm{dn}^2(u,k)du = 2E(k), \tag{30.10}$$

one finds

$$\Phi_0 = \sqrt{g}/2E(k).$$

Therefore, the function (30.8) takes finally the form of a cnoidal wave,

$$\Phi(\xi) = \frac{\sqrt{g}}{2E(k)}\mathrm{dn}\left[\frac{g\xi}{2E(k)},k\right]. \tag{30.11}$$

Applying eqns. (30.3), (30.5a), and (30.7), one determines the energy

$$\epsilon = -g^2(1+k_1^2)/4E^2(k). \tag{30.12}$$

30.2. Cnoidal Waves of the Second Kind

The second solution of eqn. (30.1) is obtained by assuming the integration constant in eqn. (30.4) to be negative. In this case eqn. (30.4) takes the form

$$\left[\frac{d\Phi}{d\xi}\right]^2 = g[2\gamma\Phi^2-\Phi^4+C_1], \quad C_1>0. \tag{30.13}$$

Introducing the quantities,

$$t = \Phi/\Phi_0, \quad \Phi_1^2/\Phi_0^2 = -k_1^2/k^2, \tag{30.14}$$

where Φ_0 and Φ_1 are defined by means of the relations,

$$\Phi_0^2\Phi_1^2 = -C_1<0, \quad \Phi_0^2-\Phi_1^2 = 2\gamma, \tag{30.15}$$

one can transform eqn. (30.13) as follows

$$\frac{d\Phi}{d\xi} = \eta\sqrt{g}\frac{\Phi_0}{k}\sqrt{(1-t^2)(t^2k^2+k_1^2)}, \quad \eta = \pm 1. \tag{30.16}$$

Integrating this, one obtains

$$\int_1^{\Phi/\Phi_0}\frac{dt}{\sqrt{(1-t^2)(t^2k^2+k_1^2)}} = \eta\frac{\xi\Phi_0}{k}\sqrt{g}. \tag{30.17}$$

An inversion of this integral defines the second Jacobian elliptic function

$$\Phi_0(\xi,k) = \Phi_0\mathrm{cn}\left[\frac{\xi\Phi_0}{k}\sqrt{g},k\right] \tag{30.18}$$

of modulus k. The magnitude of Φ_0 is obtained from its normalization per period L.

The upper limit of the integral must be equal to the period $4K(k)$ of the Jacobian function $\mathrm{cn}(u,k)$, i.e., the relation

$$L\Phi_0\sqrt{g}/k = 4K(k) \tag{30.19}$$

must hold. Using the relation

$$\int_0^{4K(k)} \mathrm{cn}^2(u,k)du = 4\delta(k)/k^2,$$

where the function $\delta(k)$ is given by eqn. (28.48), one obtains

$$\Phi_0 = k\sqrt{g}/4\delta(k). \tag{30.20}$$

Therefore, the cnoidal wave of the second kind (30.18) has finally the form

$$\Phi_0(\xi,k) = \frac{k\sqrt{g}}{4\delta(k)}\mathrm{cn}\left[\frac{g\xi}{4\delta(k)},k\right]. \tag{30.21}$$

Substituting eqn. (30.20) into eqn. (30.19), one obtains the following

$$Lg = 16K(k)\delta(k), \tag{30.21a}$$

which determines the modulus of the function (30.21).

By virtue of eqns. (30.3), (30.14), and (30.15), one obtains the quasiparticle energy

$$\epsilon_c = -g^2(2k^2-1)/16\delta^2(k). \tag{30.22}$$

From eqn. (30.21), it follows that the cnoidal waves (30.21) describe the states with negative energies if the modulus obeys the inequality $k^2 > 1/2$.

30.3. Solutions of the Nonlinear Schrödinger Equation with a Negative Nonlinearity Parameter

If the nonlinearity parameter g is negative, eqn. (30.2) has periodic solutions for the positive energies $\epsilon > 0$ only. To investigate this case, we introduce the following subsidiary quantities,

$$g_1 = -g > 0, \;\; \gamma = \epsilon/2g_1 > 0. \tag{30.23}$$

Multiplying eqn. (30.2) by $\Phi(\xi)$ and integrating further, we obtain

$$\left[\frac{d\Phi}{d\xi}\right]^2 = g_1[\Phi^4 - 2\gamma\Phi^2 + d], \;\; d > 0. \tag{30.24}$$

Introducing the quantities,

$$\Phi_0^2 = \gamma + \sqrt{\gamma^2 - d}, \quad \Phi_1^2 = \gamma - \sqrt{\gamma^2 - d} \tag{30.25}$$

one can transform eqn. (30.24) into the following

$$\frac{d\Phi}{d\xi} = \eta\sqrt{g_1}\sqrt{(\Phi_0^2 - \Phi^2)(\Phi_1^2 - \Phi^2)}, \quad \eta = \pm 1. \tag{30.26}$$

Assuming further that $\Phi(\xi)/\Phi_1 = t$ and $\Phi_1/\Phi_0 = k$, and integrating eqn. (30.26) over the interval $(0,\xi)$, and imposing finally the condition $\Phi(0) = 0$, one obtains

$$\int_0^{\Phi(\xi)/\Phi_1} \frac{dt}{\sqrt{(1-t^2)(1-k^2t^2)}} = \eta\xi\sqrt{g_1}\Phi_0. \tag{30.27}$$

Taking into account that the third Jacobian function $\mathrm{sn}(u,k)$ is defined by inverting the integral

$$\int_0^{\mathrm{sn}(u,k)} \frac{dt}{\sqrt{(1-t^2)(1-k^2t^2)}} = u, \tag{30.28}$$

one can check up that from eqn. (30.27) it follows

$$\Phi_s(\xi) = k\Phi_0 \mathrm{sn}[\sqrt{g_1}\Phi_0\xi, k]. \tag{30.29}$$

To obtain the final form of the cnoidal function $\Phi(\xi)$, one should determine Φ_0 by using its normalization condition per period L,

$$1 = \int_0^L \Phi_s^2(\xi)d\xi = \frac{k^2\Phi_0}{\sqrt{g_1}} \int_0^{\sqrt{g_1}\Phi_0 L} \mathrm{sn}^2(u,k)du. \tag{30.30}$$

Let us take into account that the period of the Jacobian function $\mathrm{sn}(u,k)$ is equal to $4K(k)$. One can easily prove the following relation:

$$\int_0^{4K(k)} \mathrm{sn}^2(u,k)du = \frac{4}{k^2}\delta_s(k), \tag{30.31}$$

where

$$\delta_s(k) \equiv K(k) - E(k). \tag{30.32}$$

Then assuming that

$$\sqrt{g_1}\Phi_0 L = 4K(k), \tag{30.33}$$

from eqn. (30.30) one finds the value

$$\Phi_0 = \sqrt{g_1}/4\delta_s(k). \tag{30.34}$$

Substituting eqn. (30.34) into (30.33), one obtains the equation

$$g_1 L = 16K(k)\delta_s(k), \tag{30.35}$$

which determines the modulus of the elliptic function (30.29) in terms of $g_1 L$. Using eqn. (30.34), the final form for the wave function (30.29) is obtained:

$$\Phi_s(\xi) = \frac{k\sqrt{g_1}}{4\delta_s(k)} \mathrm{sn}\left[\frac{g_1\xi}{4\delta_s}, k\right]. \tag{30.36}$$

Taking into account the relation (see Section 29)

$$\frac{d^2 s}{du^2} = -(1+k^2)s + 2s^2k^2, \quad s \equiv \mathrm{sn}(u,k), \tag{30.37}$$

and using eqn. (30.1), one finds the quasiparticle energy

$$\epsilon_s = \frac{g_1^2(1+k^2)}{16\delta_s^2} > 0. \tag{30.38}$$

Therefore, for a negative nonlinearity parameter, the energy which is above the bottom of the energy band of quasiparticles described by plane waves without a local chain deformation, corresponds to the cnoidal wave (30.36) with an arbitrary value of the modulus in the interval (0,1). In this relation, all autolocalized states of this type are unstable as far as their energy is concerned.

31. Periodic Solutions for Some Quasiparticles in a Quasi-One-Dimensional Molecular Chain

In Subsection 26.1, we considered soft molecular chains which are able to accommodate excess quasiparticles, electrons or holes, thus initiating local chain deformations. We also demonstrated that these quasiparticles are paired in a singlet spin state. Together with a local deformation, the paired quasiparticles having a total spin of zero move as a single entity along a molecular chain without energy loss.

In order to describe theoretically the motion of many excess quasiparticles coming to the chain from donor molecules, one should take into account the non-linear nature of bonds between quasiparticles and of their interaction with coherent displacements of equilibrium positions in the chain. As a model, Brizhik and the author [21] deal with an infinite chain of molecules with mass M separated by equal distances from each other, this chain being able to accommodate a few paired quasiparticles.

Let us assume that the molecules are rigidly fastened at the equilibrium positions na. Then, within the long-wave limit, each quasiparticle is characterized by an effective mass m. It is also suggested that their deformation interaction with molecules of the chain is strong enough to initiate local displacements of the molecular equilibrium positions.

Referring all energies to the bottom of the energy band of quasiparticles in the unperturbed chain, one can write the linear density of the Hamiltonian of the chain having an excess of quasiparticles within the continuum approach as follows:

$$\mathcal{H}(x,t) = \sum_{j=1,2} \frac{1}{2}\left[\frac{\hbar^2}{2m}\left|\frac{\partial\psi_j}{\partial x}\right|^2 - \sigma\gamma|\psi_j|^2\right] + \frac{M}{2a}\left[V_0^2\gamma^2 + \left(\frac{\partial\beta}{\partial t}\right)^2\right]. \tag{31.1}$$

In eqn. (31.1), the index j denotes the state of quasiparticles describing two possible orientations of spin. The coordinate wave $\psi_j(x,t)$ satisfies the normalization condition

$$\int|\psi_j(x,t)|^2 dx = N/2, \quad j = 1,2, \tag{31.2}$$

where N is the number of excess quasiparticles in a rather lengthy chain, $N_0 a \gg a$ and $N < N_0$. The integration is taken over the whole length aN_0. Here, $\beta = \beta(x,t)$ is the displacement of the equilibrium positions of molecules expressed in terms of the coherent states of virtual phonons; the function

$$\gamma(x,t) = -\partial\beta/\partial x \tag{31.3}$$

describes a decrease in the distance between neighboring molecules, in units of a; $V_0 = a\sqrt{\kappa/M}$ is the longitudinal sound velocity in the chain without excess quasiparticles, κ is the longitudinal coefficient of elasticity of the chain, and σ is the energy of the local interaction of quasiparticles with molecular displacements.

The following set of equations follows from the Hamiltonian (31.1)

$$\left[i\hbar\frac{\partial}{\partial t}+\frac{\hbar^2}{2m}\frac{\partial^2}{\partial x^2}+\sigma\gamma\right]\psi_j = 0, \; j = 1,2, \tag{31.4}$$

$$\left[\frac{\partial^2}{\partial t^2}-V_0^2\frac{\partial^2}{\partial x^2}\right]\beta-\frac{a\sigma}{M}\frac{\partial}{\partial x}(|\psi_1|^2+|\psi_2|^2) = 0. \tag{31.5}$$

We have already shown (Subsection 26.1) that the coordinate functions $\psi_1(x,t)$ and $\psi_2(x,t)$ are just the same in the state of pairing. Removing the subscript j of the function $\psi_j(x,t)$ and applying the invariance of the Hamiltonian (31.1) relative to the choice of the origin of the spatial coordinate frame, one seeks the solutions of eqns. (31.4) and (31.5) in the form

$$\psi(x,t) = \frac{1}{\sqrt{a}}\Phi(\xi)\exp\left\{\frac{i}{\hbar}\left[mxV-\left[\mathcal{E}+\frac{mV^2}{2}\right]t\right]\right\}, \tag{31.6}$$

$$\beta(x,t) = \beta(\xi), \tag{31.7}$$

where

$$\xi = (x-x_0-Vt)/a. \tag{31.8}$$

The quantity $\mathcal{E}$ appearing in eqn. (31.6) determines the energy of a quasiparticle in the field of the deformation wells $\sigma\gamma(\xi)$ moving with a velocity V.

Taking eqn. (31.6) into account and integrating eqn. (31.5) under the condition $\gamma(\xi)$ 0 everywhere in the chain where $\Phi(\xi)$ vanishes, one obtains

$$\gamma(\xi) = 2\sigma\Phi^2(\xi)/MV_0^2(1-s^2), \;\; s = V/V_0. \tag{31.9}$$

Due to the existence of the factor $(1-s^2)$ in the denominator of this expression, the theory developed is justified only if the velocity V of a quasiparticle is considerably less than that of the longitudinal sound velocity V_0. As was demonstrated in Subsection 26.1, the same condition is necessary to explain the pairing of quasiparticles. So, we suggest henceforth that the inequality $s^2 \ll 1$ holds.

Using eqns. (31.6) and (31.4) one finds the equation for the envelope $\Phi(\xi)$ of the function (31.6),

$$\left[\frac{\hbar^2}{2ma^2}\frac{d^2}{d\xi^2}+\mathcal{E}+\sigma\gamma(\xi)\right]\Phi(\xi) = 0.$$

Substituting further eqn. (31.8) in this equation, one can transform it into the nonlinear Schrödinger equation

$$\left[\frac{d^2}{d\xi^2}+\epsilon+4g\Phi^2(\xi)\right]\Phi(\xi) = 0 \tag{31.10}$$

where

$$\epsilon = 2ma^2\mathcal{E}/\hbar^2,\ \ g = ma^2\sigma^2/M\hbar^2V_0^2(1-s^2). \tag{31.11}$$

We demonstrated in Subsections 30.1 and 30.2 that eqns. (31.10) admits two periodic solutions. In the case under study, in each period aL there are two paired quasiparticles in the singlet spin state. The coordinate wave functions of quasiparticles and the function characterizing a chain deformation must satisfy the conditions of periodicity,

$$|\psi(x,t)|^2 = |\psi(x+aL,t)|^2,\ \ \gamma(x,t) = \gamma(x+aL,t). \tag{31.12}$$

According to (31.6), from these relations there follows the condition of periodicity for the envelope

$$\Phi(\xi+L) = \Phi(\xi). \tag{31.13}$$

The first type of the periodic collective excitations of eqn. (31.10), which satisfy the conditions of periodicity (31.13) and normalization,

$$\int_0^L \Phi^2(\xi)d\xi = 1, \tag{31.14}$$

is expressed in terms of the Jacobian elliptic function $du(u,k)$,

$$\Phi_d(\xi,k) = A\,\mathrm{dn}(u,k),\ \ u \equiv \alpha\xi. \tag{31.15}$$

For a given modulus k in the interval (0,1), the function (31.15) satisfies eqn. (31.10) if the relations

$$\epsilon(k) = -\alpha^2(2-k^2),\ \ \alpha = a\sqrt{2g} \tag{31.16}$$

hold. Equation (31.14), which determines the normalization of function, takes the form

$$1 = A^2\int_0^L dn^2(\alpha\xi,k)d\xi = \frac{A^2}{\alpha}\int_0^{aL} dn^2(u,k)du. \tag{31.17}$$

Since the period of the function $dn(u,k)$ relative to u is equal to $2K(k)$, where $K(k)$ is the complete elliptic integral of the first kind, the relation

$$\alpha L = 2K(k) \tag{31.18}$$

must hold. If this is the case, from eqn. (31.17) it follows that

$$\alpha = 2A^2 E(k), \tag{31.19}$$

where

$$E(k) \equiv \int_0^{K(k)} dn^2(u,k)du$$

is the complete elliptic integral of the second kind. Using eqns. (31.16) and (31.19), one finds

$$A = E^{-1}(k)\sqrt{g/2}, \quad \alpha = g/E(k), \tag{31.20}$$

$$\epsilon(k) = -g^2(2-k^2)/E^2(k). \tag{31.21}$$

Substituting α given by eqn. (31.20) into eqn. (31.18), one obtains the following equation for the modulus k in terms of gL,

$$gL = 2E(k)K(k). \tag{31.22}$$

The wave function

$$\Phi_d(\xi,k) = \sqrt{g/2}E^{-1}(k)\mathrm{dn}\left[\frac{g\xi}{E(k)},k\right] \tag{31.23}$$

corresponds to a fixed value of k.

For given g, small values of the complementary modulus $k_1 = \sqrt{1-k^2}$ correspond to large periods, or to small densities of excess electrons. If $k_1^2 \ll 1$, the asymptotics of the complete elliptic integrals is the following,

$$E(k) \approx 1, \quad K(k) \approx \ln\left[\frac{4}{k_1}\right], \quad k_1^2 \ll 1. \tag{31.24}$$

Hence, eqn. (31.22) is transformed as follows

$$gL \approx 2\ln(4/k_1). \tag{31.25}$$

For small values of k_1, the Jacobian elliptic function is approximated by the hyperbolic functions, and thus the function (31.23) takes the form

$$\Phi_d(\xi,k) = \sqrt{g/2}\left\{1+\frac{k_1^2}{4}[u+\mathrm{sh}u\mathrm{ch}u]\mathrm{th}u\right\}\mathrm{sch}u, \tag{31.26}$$

$$u \equiv g\xi, \; k = \sqrt{1-k_1^2}.$$

For $k_1^2 \ll 1$, and within the coordinate frame ξ, the function (31.23) describes the bell-like elevations having a width $\approx 1/g$ separated by L from each other. Each elevation characterizes the probability of distribution of paired quasiparticles and deformations of the molecular chain given by the function (31.8). A condition of validity of the continuum approach results in the inequality $g < 1$.

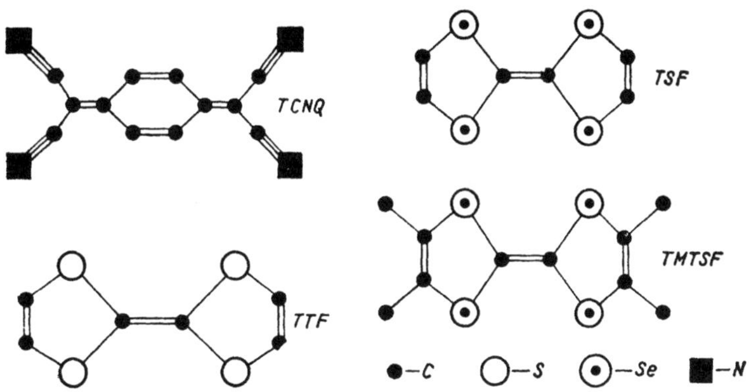

Figure 11. Plane organic molecules.

31.1. Superconductivity of Organic Crystals

As well as α-helix protein molecules, quasi-one-dimensional molecular structures exist in other biological objects, as well as in inanimate nature. A great deal of attention is given currently to the study of organic crystals which are formed by stacks of molecules of $TTF, TSF, TMTSF, TCNQ$ (their structures are plotted in Fig. 11), and other similar molecules, e.g. $(BEDT-TTF)_2J_3$ which are able to accommodate excess electrons or lose electrons with an accompanying transition to the ionic state.

The chains of other atoms, Cs, J, or the molecules, PF_6, ClO_4, etc., being able to lose or bind electrons, are arranged in these crystals parallel to the previously mentioned stacks. A typical arrangement of molecules in a crystal of the $(TMTSF)_2PF_6$ type is represented schematically in Fig. 12 reproduced from the review by Bachgaard and Jerome [139].

As a result of electron transfer from donor to acceptor molecules, the latter have excess electrons. For example, in the crystal $(TMTSF)_2PF_6$ the cesium atoms

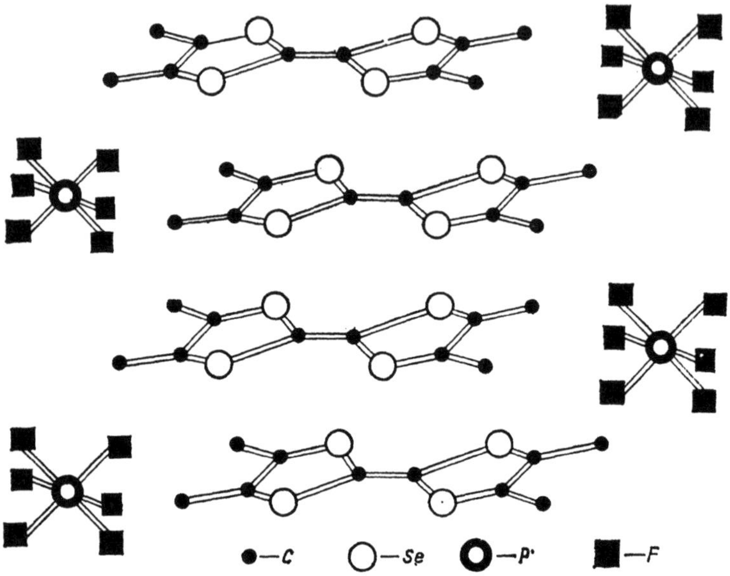

Figure 12. The typical arrangement of the *TMTSF* and PF_6 molecules in a crystal.

attach their electrons to the *TCNQ* molecules having a single excess electron. In novel crystals, $(BEDT-TTF)_2J_3$, electrons transfer from the stacks of the plane molecules of $BEDT-TTF$ type to the iodine atoms (see below). Together with van der Waals interactions, the ionic bonds formed between molecules provide the stability of crystals. In this relation, the total number of electrons transmitted between stacks and atomic chains must depend on the pressure. As the pressure increases, the necessity for forming additional ionic bonds decreases, hence the number of ionized states does too.

Stacks of plane molecules form quasi-one-dimensional structures. An overlap of the wave functions of excess electrons, or holes of neighboring molecules in a stack results in generating quasi-one-dimensional conduction bands for these 'excess' electrons (holes). Due to the collectivization of quasiparticles, these crystals possess a superhigh anisotropy of conductivity, in the ratio of 10^3 to 1.

In 1964, Little [222,223] proposed the idea of realizing high-temperature superconductivity in quasi-one-dimensional molecular systems. This work stimulated wide theoretical and experimental investigations into high-temperature conductivity of quasi-one-dimensional systems. As is well-known, the critical temperature of transition to a superconductivity state in metals is inversely proportional to the square root of the mass of ions constituting the lattice whose displacements from the equilibrium positions as a result of interaction with electrons provide their pairing.

Little suggested that in molecular systems the electrons of side chains of quasi-one-dimensional molecules play the role of ions whose displacements result in a pairing of electrons. The mass of these electrons is many thousand times less than that of ions, hence one can expect that the critical temperature increases by up to a factor of 100. The basic shortcoming of Little's treatment is that the particle displacements which are more inertial, i.e., slower, relative to electrons, can provide a pairing of electrons, and thus, superconductivity. For this reason, the hope of increasing the critical temperature by a large factor is unrealizable. In this context, the search for nonmetallic compounds with an ability of transition to the superconducting state has persistently continued.

Considerable efforts have been directed towards the synthesis of crystals formed by organic molecules. In 1973, crystals consisting of stacks of two types of organic molecules, TTF and $TCNQ$, were synthesized which reveal a sharp increase of conductivity with decreasing temperature. An increase of conductivity with decrease of temperature (up to $50-60K$ only) appeared, however, to be observed in these crystals and similar ones, $TCF-TCNQ$. With further cooling, however, a sharp decrease in conductivity occurs. Such disappointing behavior can be explained by the Peierls transition to the dielectric state, in which the paired electrons form chemical bonds between atoms and thus cannot contribute to conductivity.

Experimental attempts to produce superconducting organic crystals were crowned with success in 1980. In France, it was revealed that the crystals $(TMTSF)_2PF_6$ pass a superconducting state at temperatures below $0.9K$ and at a hydrostatic pressure of $12\bar{k}$ (for review see [139]). The crystals $(TMTSF)_2ClO_4$, passing to a superconducting state at the temperature $1.2K$, even at atmosphere pressure, were soon synthesized in 1981 (see also the review [139]).

Greater successes were achieved in 1984 due to investigations carried out by Schegolev, Yagubskii and their coworkers at the Institute of Chemical Physics, Moscow, who studied monocrystals of the novel organic compound $(BEDT-TTF)_2J_3$ [283-285]. The $BEDT-TTF$ molecules are arranged in two parallel stacks in elementary cells of the triclinic form of these crystals. Filaments of anions J_3^- and neutral atoms J_3 extend along these stacks.

The effect of superconductivity at the record temperatures of $7-8K$ was first discovered at the Institute of Chemical Physics with $(BEDT-TTF)_2J_3$ crystals. This result was reproduced in Japan, at the Institute of Molecular Sciences, Okazaki, and at the Institute of Solid State, Tokyo, in the Laboratory of Organic Metals and Superconductors (Prof. H. Saito) and the Laboratory of Molecular Crystals (Prof. M. Kinoshita). Varying the conditions of synthesis and growth, one can produce crystals with diverse modifications differing from each other by structure and the temperature of superconductivity transition, from $1.5K$ to $7.5K$ at normal pressure. The following two polymeric crystalline modifications have been most studied: the so-called α-phase which is characterized by a sharp transition of the metal dielectric type at $137K$, and the β-phase passing to the superconductivity state at $T_c = 1.5-8K$.

Yagubskii, Pokhodnya, Sheikman, and their coworkers [286] demonstrated that under heating the $(BEDT-TTF)_2J_3$ crystals in the α-phase transform irreversibly into the β-phase, but its novel β-phase differs from the ordinary phase, produced chemically, by the temperature of the critical transition, $6-7K$ at atmospheric pressure. Such a phase is stable over a long period of time.

The nature of the conductivity of quasi-one-dimensional organic compounds has not been elucidated so far. A complete understanding offers a uniquely reasonable way of discovering superconducting materials of a novel nature. The list of papers devoted to the theoretical study of conductivity of quasi-one-dimensional organic compounds possessing a high conductivity which increases with decreasing temperature is rather lengthy (see [8,9,24,82,195,243], and the references therein). Such compounds are usually called Peierls dielectrics. In all the papers mentioned above the description of electron states is given within the model of a one-dimensional metal with a half-occupied conduction band (i.e., the number of electrons is equal to that of atoms in the chain under study). These approaches are also based on linear equations in which the effects of local deformations of the periodic lattice are neglected. The Peierls transition from the metallic to the dielectric state requires that pairs of electrons with opposite spins become localized between adjacent atoms, or molecules at lower temperatures, forming in such a way the additional chemical bond. The lattice constant is hence doubled, and the metal transforms into a dielectric.

The present author suggests that the model of Peierls dielectric is not applicable for an adequate description of organic compounds which pass to a superconducting state at lower temperatures. In fact, the crystal $(TMTSF)_2ClO_4$ possessing the superconductivity transition at $1.2K$, consists of the stacks of plane molecules of tetramethyltetraselenephulvalene $(TMTSF)$. Chains of the perchlorate molecules, ClO_4, are arranged between these stacks. One should note that in this crystal a fraction of $TMTSF$ molecules lose electrons which attach to the perchlorate molecules. The loss of an electron in the molecular electron shell leads to the existence of a hole. Holes formed in this way play the role of positively charged quasiparticles. One should emphasize here that the total number of holes is less than that of molecules. A similar picture takes place in the novel compounds of the $(BEDT-TTF)_2I_3$ type. These quasiparticles can be viewed as the excess ones for neutral molecules. They have the possibility to migrate along a stack due to the exchange effects and are responsible for the high anisotropy of conductivity of the crystal and its transition to the superconductivity state. The total number of sublevels of the hole conduction band conditioned by the exchange interactions is just that of the molecules belonging to a given stack. Since all these molecules do not lose their electrons to the acceptor molecules of the J_3 type, a small part of sublevels of the hole conduction band is occupied. Hence, one cannot refer such a one-dimensional system to the Peierls dielectries.

Assuming that the number of pairs of quasiparticles in the stacks of plane molecules in the superconductivity state is less than the number of molecules, Brizhik and the author proposed a new mechanism of superconductivity of quasi-one-dimensional molecular systems [21] based on the concept of solitons. One should emphasize that the bonds between molecules in stacks of plane molecules refer to weak van der Waals forces. Hence, due to the deformation interaction, a given quasiparticle, hole or electron, causes a local deformation of a stack of molecules which involves intramolecular atomic displacements in molecules, too. The deformation interaction results in the nonlinear equations. Their solution, examined in Subsection 26.1, demonstrates that the excess electrons attached by the molecular chain are paired in the singlet spin state due to indirect interaction through the chain local deformation.

Paired quasiparticles, or bisolitons, are bosons in the singlet spin state. They occupy the bottom of the conduction band of excess quasiparticles of the molecular chain. The motion of paired quasiparticles, electrons or holes, is described by the nonlinear equations (31.4) and (31.5) investigated in the preceding Subsection. We recall that in the state of pairing, the envelope of the wave function of each constituent of a pair obeys the nonlinear Schrödinger equation (31.9) within the coordinate frame $\xi = (x - x_0 - Vt)/a$ moving with a constant velocity V through the chain, or stack of molecules separated by a from each other. Equation (31.9) admits two types of periodic solutions whose spatial periodicity, aL, is determined by the number N of pairs of quasiparticles in the system if two particles refer to a single period.

The constant velocity V takes the following discrete values

$$V_n = 2\pi\hbar n / maL, \quad n = 0, \pm 1, \pm, ... \tag{31.27}$$

where m is the mass of a single quasiparticle which is determined by the energy J of the exchange interaction between neighboring molecules via the relation

$$m = \hbar^2 / 2a^2 J.$$

At small velocities, $V \ll V_0$, the lowest energy periodic solution is given by the Jacobian elliptic function

$$\Phi(\xi, k) = \left[\frac{g}{2E^2(k)}\right]^{1/2} \mathrm{dn}\left[\frac{g\xi}{E(k)}, k\right], \tag{31.28}$$

where a dimensionless parameter g, characterizing the binding of a quasiparticle with a local deformation, is as follows

$$g = \sigma^2 / 2JMV_0^2 = \frac{\sigma^2}{2Ja^2\kappa}, \tag{31.29}$$

in accordance with eqn. (31.10). In eqn. (31.29), M is the mass of a molecule, V_0 is the sound velocity belonging to the acoustic branch of vibrations of the molecules taking part in the formation of the deformation, $E(k)$ is the complete elliptic integral of the second kind depending on the modulus k expressed through the product gL as follows

$$gL = 2E(k)K(k), \tag{31.30}$$

where $K(k)$ is the complete elliptic integral of the first kind.

If the density of excess particles in a given stack is negligible, i.e., $gL > 1$, eqn. (31.30) reduces to the following

$$gL \approx 2\ln[4(1-k^2)^{-1/2}], \quad k \approx 1, \tag{31.31}$$

that is, k slightly differs from unity. Then the function (31.28) is approximated, inside each period, by the hyperbolic function

$$\Phi(\xi,k) = (g/2)^{1/2}\left\{1+\frac{(1-k^2)^{1/2}}{4}[u+\mathrm{sh}u\mathrm{ch}u]\mathrm{th}u\right\}\mathrm{sh}u, \tag{31.32}$$

where $u = g\xi$. Such a function has a bell-like form with the maximum attained at $\xi = 0$. Neighboring maxima are separated by aL from each other. They describe the mutual consistent displacement of all pairs of quasiparticles, each pair of quasiparticles being surrounded by a chain deformation.

The energy of a pair of quasiparticles inside a deformation well created by paired particles is given, in units of the exchange energy J,

$$\epsilon(k) = -g^2(2-k^2), \quad k \approx 1. \tag{31.33}$$

The critical temperature, T_c, at which a pairing is broken (escape of particles out of a well), is proportional to σ^4,

$$k_B T_c \approx \sigma^4/2J^2M^2V_0^4 = \sigma^4/2J^2\kappa^2a^4. \tag{31.34}$$

It is noteworthy that in the context of the given soliton mechanism of superconductivity of complex compounds, the critical temperature T_c of the superconductivity state transition becomes independent of the mass M of heavier particles. Hence, one can suggest that for these compounds the isotopic effect with the peculiar dependence of T_c and M, inherent for superconducting simple metals described by the *BCS* theory, cannot be observed.

For the paired, or bisoliton state described by the function (3.32), each pair of particles carries a double electric charge, energy $\mathcal{E}(V)$, and momentum $\mathcal{P}(V)$ at small velocities together with a stack deformation, and

$$\mathcal{E}(V) = mV^2 - \frac{2\hbar^2g^2}{ma^2}\left[1-\frac{1}{3}\left(1+\frac{2V^2}{V_0^2}\right)\left(1-4gL\exp\left(-\frac{gL}{2}\right)\right)\right], \tag{31.35}$$

$$\mathcal{P}(V) = 2V\left\{m+\frac{2\hbar^2g^2}{3ma^2V_0^2}\left[1-gL\exp\left(-\frac{gL}{2}\right)\right]\right\}. \tag{31.36}$$

At large densities of quasiparticles in a stack, i.e., when the inequality, $gL \geqslant \pi^2/2$, holds, eqn. (31.30) which determines the modulus of the wave function (31.28), takes the form

$$gL = \frac{1}{2}\pi^2(1+k^2/32). \tag{31.37}$$

In this case, the modulus k of the elliptic function is considerably less unity, hence the function (31.32) is approximated by the trigonometric function

$$\Phi(\xi,k) \approx \pi^{-1}\sqrt{2g}\left[1-\frac{k^2}{2}\sin^2(2g\xi/\pi)\right]. \tag{31.38}$$

Therefore, at large densities of pairs of quasiparticles in a stack, their state is characterized by the plane wave given by eqn. (31.6) modulated by the function (31.38). Then the energy and momentum of a quasiparticle pair per period are given approximately by the following formulas,

$$\mathcal{E}(k,V) \approx mV^2 - 4\hbar^2 g^2(ma^2\pi^2)^{-1}\{1-2V^2/V_0^2-k^2/2\},$$

$$\mathcal{P}(k,V) \approx 2V(m+4\hbar^2g^2/ma^2\pi^2V_0^2\}.$$

Hence, if the density of excess pairs of quasiparticles in a stack is small, and the condition,

$$g = \sigma^2/2JMV_0^2 < 3, \tag{31.39}$$

takes place, the crystal passes into the superconducting state as the temperature decreases. The paired quasiparticles are not localized between the molecules of stacks, but, surrounded by deformation, enveloping some molecules of stacks, they (bisolitons) move coherently, as a unique entity along stacks with a single velocity and without resistance.

Anisotropic ceramic materials, which possess rather high temperatures of superconducting transition, have been recently synthesized [288,289]. One should note that in these materials the number of current carriers is considerably smaller than that of molecules in the system. For these compounds, the transition critical temperature $T_c > 40K$, and the isotopic effect is not observed [292,293]. The length of coherency, or the radius of the bound state of an electron pair is of an order of 15Å. According to (26.46a), $g \approx 1.8$ corresponds to this coherency length if $a \approx 4$ Å. For this reason, the BCS theory based on the perturbation treatment relative to the parameter $g \ll 1$, cannot be applicable to these compounds. Small values of g take place for simple metals for which $T_c \leqslant 10K$, $L \sim 10^{-4}$, and the isotopic effect is strikingly manifested.

Naturally, one cannot exclude the point of view that high-temperature superconductivity of novel ceramic compounds is provided by the middle-radius bisolitons. Bisolitons of a small radius corresponding to large $g, g > 10$, localized at an isolated edge of the lattice, possess rather large effective mass, and hence, are really immobile.

32. Autolocalized Periodic Thickening of Particles Described by a Complex Field with Inertialess Self-Interaction

The present Section is devoted to the study of the complex scalar field of spinless quasiparticles with the inertialess self-interaction on the basis of work by Pestryakov and the author [51]. This mathematical model is applicable to describe the energy states of interacting excitons in a one-dimensional molecular system. The problem of the exciton-exciton interaction was studied in the monograph by Agranovich [3]. At small densities of excitons these interactions do not result in generating bound states. For simplicity, we restrict ourselves by treating the case when the interaction between particles is approximated by the term $|\psi|^4$ involved into the energy operator. Therefore, it is assumed that the state of such a field is described by the Hamiltonian density in one-dimensional infinite space,

$$\mathcal{H} = \frac{\hbar^2}{2m}\left|\frac{\partial\psi}{\partial x}\right|^2 \mp G|\psi|^4, \quad \psi = \psi(x,t). \tag{32.1}$$

Unlike the system studied in Sections 29-31, the nonlinearity parameter G is now independent of velocity. It characterizes the inertialess self-interaction of field. The upper sign corresponds to attraction, the lower sign to repulsion. Such a self-interaction can be realized due to the local interaction between quasiparticles corresponding to the field, or to the local interaction of these quasiparticles with the field of inertialess displacements of the density of other particles which are not taken into account explicitly.

Neglecting the self-interaction, $G = 0$, one can characterize the states of noninteracting quasiparticles of mass m by the field $\psi(x,t)$. The motion of these quasiparticles is described by plane waves with wavenumber q and energy $\hbar^2 q^2/2m$. Let us assume that the function $\psi(x,t)$ of the Hamiltonian (32.1) satisfies the periodicity condition,

$$|\psi(x+aL,t)|^2 = |\psi(x,t)|^2 \tag{32.2}$$

and the normalization condition per period,

$$\int_0^{aL} |\psi(x,t)|^2 dx = 1. \tag{32.3}$$

The latter relation represents the conservation law for the number of particles in the region of a single period. The function $|\psi(x,t)|^2$ characterizes the probability density of finding a particle at point x at time t.

Two other integrals of motion corresponding to the Hamiltonian density (32.1) determine the energy,

$$\mathcal{E} = \int_0^{aL} \left\{ \frac{\hbar^2}{2m} \left| \frac{\partial \psi}{\partial x} \right|^2 \pm \frac{1}{2} G |\psi|^4 \right\} dx \tag{32.4}$$

and momentum,

$$\mathcal{P} = \frac{\hbar}{2i} \int_0^{aL} \left(\psi^* \frac{\partial \psi}{\partial x} - \psi \frac{\partial \psi^*}{\partial x} \right) dx \tag{32.5}$$

of the field in a segment aL. The nonlinear Schrödinger equation,

$$\left(i\hbar \frac{\partial}{\partial t} + \frac{\hbar^2}{2m} \frac{\partial^2}{\partial x^2} \pm G |\psi|^2 \right) \psi = 0 \tag{32.6}$$

corresponds to the Hamiltonian density (32.1). This equation admits stationary solutions in the form of modulated plane waves with an inhomogeneous spatial distribution of the density of particles moving with a small constant velocity V along the x-axis. To elucidate the nature of the particle density distribution, one seeks solutions in the following form

$$\psi(x,t) = \frac{1}{\sqrt{a}} \Phi(\xi) \exp\left\{ i \left[qx - \left(\omega + \frac{\hbar q^2}{2m} \right) t + \eta \right] \right\}, \tag{32.7}$$

where ξ is a dimensionless coordinate

$$\xi = (x - x_0 - Vt)/a \tag{32.8}$$

in a reference frame moving with the velocity $V = \hbar q/m$, $\hbar\omega$ is the quasiparticle energy relative to the moving reference frame, η is an arbitrary phase, a is a typical unit of length in the system under study characterized by the parameters m and G,

$$a \equiv \hbar / \sqrt{2mG}. \tag{32.9}$$

The condition of periodicity (32.2) is fulfilled within the moving reference frame $x = x_0 + Vt$, if the real function $\Phi(\xi)$ satisfies the conditions of periodicity

$$\Phi(\xi) = \Phi(\xi + L) \tag{32.10}$$

and normalization

$$\int_0^L \Phi^2(\xi) d\xi = 1 \tag{32.11}$$

for a single quasiparticle localized over the period.

Substituting (32.7) into eqn. (32.6), one obtains the nonlinear Schrödinger equation in terms of dimensionless units

$$\left[\frac{d^2}{d\xi^2} + \epsilon \pm \Phi^2(\xi)\right] \Phi(\xi) = 0, \tag{32.12}$$

where

$$\epsilon = \hbar\omega / \epsilon_0, \quad \epsilon_0 \equiv G. \tag{32.13}$$

The energy $\mathcal{E}$ and momentum $\mathcal{P}$ of the excited states of the field can be represented in terms of the amplitude function $\Phi(\xi)$ within the laboratory coordinate frame

$$\mathcal{E}(V) = \frac{1}{2} mV^2 + \epsilon_0 \int_0^L \left[\left(\frac{d\Phi}{d\xi}\right)^2 \mp \frac{1}{2}\Phi^4\right] d\xi, \tag{32.14}$$

$$\mathcal{P}(V) = mV \int_0^L \Phi^2(\xi) d\xi = mV. \tag{32.15}$$

Multiplying eqn. (32.13) by $\Phi(\xi)$ on the left and integrating it over the interval $(0, L)$, one obtains

$$\epsilon = \int_0^L \left[\left(\frac{d\Phi}{d\xi}\right)^2 \mp \Phi^4\right] d\xi. \tag{32.16}$$

Therefore, formula (32.15) can be rewritten as follows

$$\mathcal{E}(V) = \frac{1}{3}mV^2+\epsilon_0\left[\epsilon\pm\frac{1}{2}\int_0^L\Phi^4(\xi)d\xi\right]. \tag{32.17}$$

32.1. First Type of Autolocalized Solutions: Attractive Forces

We begin by considering the case of attraction. Hence, one should leave the upper sign in all the aforementioned formulas. The problem is to solve the equation,

$$\left[\frac{d^2}{d\xi^2}+\epsilon+\Phi^2(\xi)\right]\Phi(\xi) = 0 \tag{32.18}$$

for the function $\Phi(\xi)$ satisfying the conditions of periodicity (32.11) and normalization (32.12).

Taking into account the equation,

$$\frac{d^2}{du^2}\mathrm{dn}(u,k) = (2-k^2)\mathrm{dn}(u,k)-2\mathrm{dn}^3(u,k), \tag{32.19}$$

for the Jacobian elliptic function $\mathrm{dn}(u,k)$ defined by inverting the elliptic integral,

$$u = \int_1^{\mathrm{dn}(u,k)}\frac{dt}{\sqrt{(1-t^2)(t^2-k_1^2)}} \tag{32.20}$$

for each value of the modulus $k = \sqrt{1-k_1^2}$ varying in the interval (0,1), one can prove that eqn. (32.19) possesses the following solution

$$\Phi(\xi,k) = A\,\mathrm{dn}(\alpha\xi,k) \tag{32.21}$$

where

$$\epsilon = \alpha^2(2-k^2),\ A^2 = 2\alpha^2. \tag{32.22}$$

The constants A and α are determined by means of the conditions of periodicity (32.11) and normalization (32.12). The latter is reduced to the following relation

$$A^2\int_0^{\alpha L}\mathrm{dn}^2(u,k)\mathrm{d}u = \alpha,\quad u\equiv\alpha\xi. \tag{32.23}$$

As is known, the Jacobian function $\mathrm{dn}(u,k)$ is periodic relative to the variable u, with a period equal to the doubled complete elliptic integral $K(k)$ of the first kind. The function (32.21) satisfies the periodicity condition if the relation

$$\alpha L = 2K(k) \tag{32.24}$$

holds. According to eqn. (29.32), one has

$$\int_0^{K(k)} \mathrm{dn}^2(u,k)du = E(k),$$

where $E(k)$ is the complete elliptic integral of the second kind. Thus, from eqn. (32.23) it follows that

$$\alpha = 2A^2E(k). \tag{32.25}$$

Using eqn. (32.22), one finds

$$\alpha = 1/4E(k), \quad A = [2\sqrt{2}E(k)]^{-1} \tag{32.26}$$

and

$$\epsilon = -(2-k^2)/16E^2(k). \tag{32.27}$$

Substituting (32.25) into eqn. (32.24), obtains the equation

$$L = 8K(k)E(k), \tag{32.28}$$

which determines the modulus k for each value of the period L. As k increases, the right-hand side of eqn. (32.28) and hence, the period L increases, too.

Applying eqn. (32.28) for the determination of modulus k of the elliptic function for the period L, one obtains the explicit form of the wave function of the appropriate excitation

$$\Phi_d(\xi,k) = (2\sqrt{2}E(k))^{-1}\mathrm{dn}\left[\frac{\xi}{4E(k)},k\right]. \tag{32.29}$$

The following integral

$$\int_0^L \Phi_d^4(\xi,k)d\xi = Q_d(k)/8E^2(k) \tag{32.30}$$

evolves into the formula for the energy, eqn. (32.18). In eqn. (32.30), $Q_d(k)$ is defined as follows

$$Q_d(k) \equiv \int_0^{K(k)} \mathrm{dn}^4(u,k)\mathrm{d}u = \frac{1}{3}[2-k^2)E(k)-(1-k^2)K(k)]. \tag{32.31}$$

Substituting (32.27) and (32.30) into eqn. (32.18) given with the upper sign, one has the excitation energy

$$\mathcal{E}(V,k) = \frac{1}{2}mV^2 - \frac{\epsilon_0}{16E^2(k)}\left[2-k^2-\frac{Q_d(k)}{E(k)}\right], \tag{32.31a}$$

where, in accordance with (32.9) and (32.10), the velocity V takes discrete values.

For $k_1^2 \ll 1$, one can use the asymptotic expressions for the complete elliptic integrals, (29.35) and (29.36),

$$K(k) \approx \Lambda + \frac{1}{4}(\Lambda-1)k_1^2 + ...,$$

$$E(k) \approx 1 + \frac{1}{2}(\Lambda - \frac{1}{2})k_1^2 + ...,$$

where

$$k_1 = 4\exp(-\Lambda), \quad \text{or} \quad l = \ln(4/k_1).$$

Within this approximation

$$Q_d(k) \approx \frac{2}{3}[1-2\Lambda\exp(-\Lambda)]. \tag{32.32}$$

Then the wave function (32.29) transforms as follows

$$\Phi_d(\xi,k) = \frac{1}{2\sqrt{2}}\text{sch}u\left[1+\frac{k_1^2}{4}(u+\text{sh}u\text{ch}u)\text{th}u\right] \tag{32.33}$$

where $u \equiv \xi/4$. The energy (32.31) takes now the form

$$\mathcal{E}_d(V,k) = \frac{1}{2}mV^2 - \frac{\epsilon_0}{16}\left[\frac{1}{3}+k_1^2\left(1+\frac{1}{3}\Lambda\right)\right]. \tag{32.34}$$

In the limiting case of rather large periods when $k_1^2 \to 0$, the function (32.33) is reduced to a single hyperbolic function

$$\Phi_d(\xi,k) \approx \frac{1}{2\sqrt{2}}\text{sch}(\xi/4), \quad k^2 \to 1. \tag{32.35}$$

This function corresponds to the existence of a single soliton-type excitation with the energy

$$\mathcal{E}_d(V,k\to 1) = \frac{1}{2}mV^2 - \epsilon_0/48 \tag{32.36}$$

in an infinite period.

According to (32.34), a decrease of the period, i.e., an increase of k_1^2, is accompanied by a lowering of energy. Then the region involving one particle is contracted too, relative to the moving reference frame. Hence, this state corresponds to the system with a large average density of particles.

In the opposite limiting case when $k_1^2 \to 1$, the approximate values of the complete elliptic integrals are as follows

$$K(k) \approx \frac{\pi}{2}(1+k^2/4), \quad E(k) \approx \frac{\pi}{2}(1-k^2/4), \quad k^2 \ll 1. \tag{32.37}$$

According to eqn. (32.28), the region of particle localization is determined by the expression

$$L \approx 2\pi^2(1-k^4/16). \tag{32.38}$$

Using (32.38), one can approximate the function (32.29)

$$\Psi_d(\xi,k) \approx \frac{1}{\pi\sqrt{2}}\left[1-\frac{k^2}{4}(1-2\sin^2 u)\right], \quad u \equiv \xi/4. \tag{32.39}$$

Within this approximation, the function (32.31) takes the value

$$Q_d(k) \approx \frac{\pi}{2}(1-k^2/2) \tag{32.40}$$

and the energy of the state given by (32.39) is

$$\mathcal{E}_d(V,k) \approx \frac{1}{2}mV^2 - \frac{2mG}{\pi^2\hbar}(1-k^2/2), \quad k^2 \ll 1. \tag{32.41}$$

For $k \approx 0$, this value corresponds to the smallest possible separation,

$$\mathcal{L}_0 = aL_0 = \pi^2\hbar^2/2mG \tag{32.42}$$

between the particles in the system. The quantity $\mathcal{L}_0$ characterizes the peculiar fundamental length in the system with the inertialess nonlinear interparticle interaction. Therefore, due to the attractive type of self-interaction, the periodic distribution of particles with the spatial modulation $\mathcal{L}_0$ is established in the system. This spatial modulation $\mathcal{L}_0$ depends on the parameters m and G, and the modulation depth k^2. Such modulation vanishes if $k \to 0$.

32.2. Second Type of Autolocalized Solutions: Attractive Forces

As shown in Section 30, eqn. (32.19) has also the second solution under the conditions (32.11) and (32.12),

$$\Phi_c(\xi,k) = B\,\mathrm{cn}(\beta\xi,k). \tag{32.43}$$

Here $\mathrm{cn}(u,k)$ is the second Jacobian elliptic function defined by inverting the elliptic integral

$$\xi\beta = \int_1^{\mathrm{cn}(\beta\xi,k)} \frac{dt}{\sqrt{(1-t^2)(k_1^2+k^2t^2)}}, \quad k_1 = \sqrt{1-k^2}. \tag{32.44}$$

In fact, it is known that the function $c = \mathrm{cn}(u,k)$ satisfies the following differential equation (see Section 30)

$$\frac{d^2c}{du^2} - (2k^2-1)c + 2k^2c^3 = 0. \tag{32.45}$$

Hence the function (32.43) is just the solution of eqn. (32.19) if $u = \beta\xi$ and

$$\epsilon = -\beta^2(2k^2-1), \quad B^2 = 2k^2\beta^2. \tag{32.46}$$

The normalization condition (32.12) for the function (32.43) results in the relation

$$\beta = B^2\int_0^{\beta L} \mathrm{cn}^2(u,k)du, \quad u \equiv \beta\xi. \tag{32.47}$$

The Jacobian function $\mathrm{cn}(u,k)$ is periodic relative to the variable u with the period $4K(k)$ where $K(k)$ is given by the integral (29.11). Hence, the condition of periodicity (32.11) is satisfied if the relation

$$\beta L = 4K(k) \tag{32.48}$$

holds. Using eqn. (32.48) and the equation (see Section 29)

$$\int_0^{4K(k)} \mathrm{cn}^2(u,k)du = 4\delta(k)/k^2, \tag{32.49}$$

where

$$\delta(k) \equiv E(k) - k_1^2K(k), \tag{32.50}$$

one can transform the relation (32.47) into the following $k^2\beta = 4B^2\delta(k)$, from which it follows

$$\beta = [8\delta(k)]^{-1}, \;\; B = k/\delta(k)4\sqrt{2}, \tag{32.51}$$

and

$$\epsilon_c = -(2k^2-1)/64\delta^2(k). \tag{32.52}$$

Substituting β into eqn. (32.48), one finds

$$L = 32\delta(k)K(k), \tag{32.53}$$

which determines the modulus k of the Jacobian function $cn(u,k)$. Therefore, the wave function (32.43) takes the final form

$$\Phi_c(\xi,k) = \frac{k}{4\sqrt{2}\,\delta(k)}\mathrm{cn}\left[\frac{\xi}{8\delta(k)},k\right] \tag{32.54}$$

To evaluate the energy of the state described by the cnoidal function (32.54), one should substitute the integral

$$\int_0^L \Phi_c^4(\xi,k)d\xi = Q_c(k)/32\delta^3(k), \tag{32.55}$$

where Q_c is given by eqn. (28.55), with the upper sign into eqn. (32.18). Applying eqn. (31.52), one obtains

$$\mathcal{E}_c(V,k) = \frac{1}{2}mV^2 - \frac{\epsilon_0}{64\delta^2(k)}\left[2k^2-1-\frac{Q_c(k)}{\delta(k)}\right]. \tag{32.56}$$

Since $Q_c(k) > 0$, from eqn. (32.56) it follows that for $V = 0$ the negative energies correspond only to the states with modulus k satisfying the inequality

$$\frac{1}{2} < k^2 \leqslant 1. \tag{32.57}$$

The latter is valid only at large periods.

In the limiting case of rather small k_1^2, i.e., $k^2 \sim 1$, $\delta(k) \approx E(k) \approx 1$, and for this reason, within the terms of an order of k_1, $Q_c(k)$ is transformed into the following

$$Q_c(k) \approx \frac{2}{3}\left[1-\frac{k_1^2}{2}\Lambda\right] = \frac{2}{3}[1-2\Lambda\exp(-\Lambda)]. \tag{32.58}$$

Within this approximation, the energy (32.56) becomes as follows

$$\mathcal{E}_c(V,k) = \frac{1}{2}mV^2 - \frac{\epsilon_0}{192}[1-26k_1^2], \tag{32.59}$$

and the wave function is

$$\Phi_c(\xi,k) \approx \frac{1}{4\sqrt{2}}\mathrm{sch}u\left[1-\frac{k_1^2}{4}(\mathrm{sh}u\mathrm{ch}u-u)\mathrm{th}u\right] \tag{32.60}$$

with $u = \xi/8$, and $k_1^2 \ll 1$.

As follows from eqn. (32.59), the lowest energy of the states, described by eqn. (32.54), corresponds to $k_1^2 \approx 0$. Unlike the case of the cnoidal waves (32.29), in our case an increase of k_1 within the validity of inequality $k_1^2 \ll 1$ is accompanied by an increase of energy, although the region spanned by excitation and being equal

$$L \approx 32\ln(4/k_1), \;\; k_1^2 \ll 1,$$

in agreement with (32.53), (29.35), and (29.36), is also contracted as for the cnoidal waves.

Within the limit $k_1 = 0$ corresponding to an infinite period, the function (32.60) describes the autolocalized state of the soliton type with the energy

$$\mathcal{E}_c(V,k{\approx}1) = \frac{1}{2}mV^2-\epsilon_0/192.$$

This soliton state differs substantially from that given by the function (32.3) with $k_1 \approx 0$. In the state (32.60), the maximum probability density is four times smaller, and the width of the region where the probability density is appreciably nonvanishing, is two times greater compared with those in the state (32.3).

32.3. Repulsive Forces

Dealing with the inertialess self-interaction with repulsion, one should leave the lower signs in eqns. (32.1) and (32.18). Thus, the problem in question is to solve the following equation

$$\left[\frac{d^2}{d\xi^2}+\epsilon_s-\Phi^2(\xi)\right]\Phi(\xi) = 0 \tag{32.61}$$

for the function $\Phi(\xi)$ satisfying the conditions of periodicity (32.11) and normalization (32.12).

Equation (32.61) has the solution

$$\Phi_s(\xi,k) = C\mathrm{sn}(\mu\xi,k), \tag{32.62}$$

where $\mathrm{sn}(\mu\xi,k)$ is the third Jacobian elliptic function defined by inverting the elliptic integral

$$\mu\xi = \int_0^{\mathrm{sn}(\mu\xi,k)} \frac{dt}{\sqrt{(1-t^2)(1-k^2t^2)}}.$$

In fact, the Jacobian function $\mathrm{sn}(u,k)$ satisfies the differential equation

$$\frac{d^2s}{du^2}+(1+k^2)s-2s^3 = 0, \quad s \equiv \mathrm{sn}(u,k). \tag{32.63}$$

One can prove fairly easy that the function (32.62) satisfies eqn. (32.61) if the relations,

$$\epsilon_s = \mu^2(1+k^2), \quad C^2 = 2\mu^2 \tag{32.64}$$

hold. Then the normalization condition for the function (32.62) becomes

$$\mu = C^2 \int_0^{\mu L} \mathrm{sn}^2(u,k)du, \quad u \equiv \mu\xi. \tag{32.65}$$

The Jacobian elliptic function $\mathrm{sn}(u,k)$ is periodic relative to the variable u with the period $4K(k)$, where $K(k)$ is the complete elliptic integral of the first kind. Hence, one should require the validity of the relation

$$\mu L = 4K(k). \tag{32.66}$$

Using this relation and the integral

$$\int_0^{4K(k)} \mathrm{sn}^2(u,k)du = 4\Gamma(k)/k^2, \tag{32.67}$$

where

$$\Gamma(k) \equiv K(k)-E(k), \tag{32.68}$$

one obtains from eqn. (32.65) that

$$\mu k^2 = 4\Gamma(k)C^2. \tag{32.69}$$

From eqn. (32.64) and (32.69) it follows

$$C = \sqrt{2}k^2/8\Gamma(k), \quad \mu = k^2/8\Gamma(k), \tag{32.70}$$

and

$$\epsilon_s = k^4(1+k^2)/64\Gamma^2(k) > 0. \tag{32.71}$$

Substituting μ into eqn. (32.66), one finds the equation for determining modulus of the Jacobian function $\mathrm{sn}(u,k)$

$$k^2 L = 32\Gamma(k)K(k). \tag{32.72}$$

Positive values of $\mathcal{E}_s$ imply that the states described by the cnoidal function

$$\Phi_s(\xi,k) = \frac{\sqrt{2}k^2}{8\Gamma(k)}\mathrm{sn}\left[\frac{k^2\xi}{8\Gamma(k)},k\right], \tag{32.73}$$

are metastable. The energy minimum is attained at $k = 0$.

Assuming k to be small, one uses the asymptotics (32.37). Then, from eqn. (32.68), it follows that

$$\Gamma(k) \approx \pi k^2/4, \quad k^2 \ll 1. \tag{32.74}$$

Substituting it into eqn. (32.70) and applying the asymptotic value of the function $\mathrm{sn}(u,k)$ for $k^2 \ll 1$, one can transform the function (32.73) into the form

$$\Phi_s(\xi,k) \approx \frac{1}{\pi\sqrt{2}}\left[\sin u + \frac{k^2}{4}\cos u(\sin u\cos u - u)\right], \tag{32.75}$$

where

$$u = \xi/2\pi, \quad k^2 \ll 1, \quad |\xi| \ll L. \tag{32.75a}$$

According to (32.72), (32.74), and (32.38), the period of the function is

$$L = 4\pi^2(1+k^2/4), \quad k^2 \ll 1. \tag{32.76}$$

Taking into account the integral

$$\int_0^L \Phi_s^4(\xi)d\xi = 3(1+k^2)/8\pi^2, \quad k^2 \ll 1, \tag{32.77}$$

and the value

$$\epsilon_s \approx (1+k^2)/4\pi^2 \tag{32.78}$$

one obtains the energy (32.18) of the state (32.75)

$$\mathcal{E}_s(V,k) = \frac{1}{2}mV^2 + \frac{mG^2}{8\pi^2}(1+k^2), \quad k^2 \ll 1 \tag{32.79}$$

in the principal period

$$\mathfrak{L} = La = \pi^2\hbar^2\left[1+\frac{k^2}{4}\right]/mG. \tag{32.80}$$

A periodic distribution of particles with the smallest period (at $k=0$), which is $4\pi^2$ times larger the characteristic length (32.9) is established in the system moving with velocity V.

32.4. Autolocalized Excitations in Closed Molecular Chains

In some cases, one-dimensional molecular systems form closed chains, e.g. such as hydrogen-bonded chains between peptide groups in valinomycin and its derivatives, DNA ring molecules in bacteria, etc. For this reason, it is of interest to study collective excited states in closed, molecular systems consisting of a large number of identical subsystems. Such a problem was first solved by Sergienko [97].

We assume that a given molecular chain consists of identical atomic groups able to pass into excited dipole states with energy $\mathcal{E}_0$. In short, these atomic groups can be called molecules. The latter ones are arranged at equal distances a from each other, forming in such a way a closed ring with N links ($N\gg 1$).

Due to the resonance dipole-dipole interaction between neighboring molecules whose energy is denoted by J, intramolecular excitations inside the ring become collectivized. One can characterize the probability distribution of intramolecular excitations by the function $|\varphi_n(t)|^2$, where the index n designates a number of the molecule occupying the position na in the ring. The transition of the chain into an excited state is accompanied by displacements $\beta_n(t)$ of molecules from their equilibrium positions. The coupling parameter of intramolecular excitations with the displacements $\beta_n(t)$ is denoted by σ.

We restrict matters by treating only those collective excitations which span a large number of molecules. In this case, the functions $\varphi_n(t)$ and $\beta_n(t)$ depend weakly on n, i.e., the application of the continuum approach consisting of the replacement of a discrete set of n by a dimensionless continuous variable ξ, is justified. Within this approach, the collective excitations of the ring can be described by the following set of equations,

$$\left[i\hbar\frac{\partial}{\partial t}+\sigma\rho(\xi,t)+J\frac{\partial^2}{\partial\xi^2}\right]\varphi(\xi,t) = 0, \tag{32.81}$$

$$\left[\frac{\partial^2}{\partial t^2}-\frac{V_0^2}{a^2}\frac{\partial^2}{\partial \xi^2}\right]\rho(\xi,t)+\frac{\sigma}{Ma^2}\frac{\partial^2}{\partial \xi^2}|\varphi(\xi,t)|^2 = 0, \tag{32.82}$$

where

$$\rho(\xi,t) = -\partial\beta/\partial\xi \tag{32.83}$$

is a decrease in the distance between neighboring molecules. In eqn. (32.81), the zeroth energy refers to the energy $\mathcal{E}_0 - J/2$ of the bottom of the exciton band corresponding to the collective excitations with $\sigma = 0$. The quantity V_0 determines the longitudinal sound velocity in the chain with elasticity κ and mass M of molecules. Within the continuum approach, the energy of the resonance dipole-dipole interaction can be expressed in terms of the effective mass m of quasiparticles (electrons) corresponding to a given intramolecular excitation,

$$J = \hbar^2/2ma^2. \tag{32.84}$$

Due to the decrease in intramolecular distances in the excitation region of the molecular chain, its length is contracted by the quantity

$$\int_0^N \rho(\xi)d\xi.$$

Now we study the chains with rather large length na for which the inequality

$$Na \gg \int_0^N \rho(\xi)d\xi \tag{32.85}$$

holds. Then the conditions of normalization and periodicity are simplified,

$$\int_0^N |\varphi(\xi,t)|^2 d\xi = 1, \tag{32.86}$$

$$|\varphi(\xi+N,t)|^2 = |\varphi(\xi,t)|^2, \quad \rho(\xi+N,t) = \rho(\xi,t). \tag{32.87}$$

Further, we deal with the excitations which travel along the chain with a velocity V which is less than the longitudinal sound velocity, V_0. For this purpose, we suggest that

$$\varphi(\xi,t) = \Phi(\zeta)\exp\left\{\frac{i}{\hbar}\left[ma\xi V-\left[\mathcal{E}-\frac{mV^2}{2}\right]t+\eta_0\right]\right\}, \tag{32.88}$$

$$\rho(\xi,t) = \rho(\zeta), \tag{32.89}$$

where $\Phi(\zeta)$ and $\rho(\zeta)$ are the real functions of the variable

$$\zeta = \xi - \xi_0 - Vt. \tag{32.90}$$

According to eqn. (32.86), the function $\Phi(\zeta)$ is normalized by the condition

$$\int_0^N \Phi^2(\zeta)d\zeta = 1. \tag{32.91}$$

In eqn. (32.88), the unknown quantity $\mathfrak{E}$ characterizes the energy of interaction of intramolecular excitation with deformation of the ring, moving with velocity V. The total excitation energy is determined relative to the laboratory frame as follows

$$\ni(V) = \mathfrak{E}(V) + \frac{1}{2}mV^2 + W, \tag{32.92}$$

where

$$W \equiv \frac{1}{2}\int \left[\frac{MV^2}{a^2} + \kappa\right]\rho^2 d\zeta \tag{32.93}$$

is the energy of the mobile local ring deformation.

The condition of periodicity (32.87) is satisfied if the following relations are fulfilled,

$$\Phi(\zeta + N) = \Phi(\zeta), \tag{32.94}$$

and the excitation motion velocity V takes the discrete values

$$\rho(\zeta + N) = \rho(\zeta). \tag{32.95}$$

Taking the condition of periodicity (32.94) into account, one can integrate eqn. (32.82) and obtain

$$\rho(\zeta) = \sigma\Phi^2(\zeta)/MV_0^2(1-s^2), \quad s^2 = (V/V_0)^2. \tag{32.96}$$

Substituting (32.88) and (32.96) into eqn. (32.81), one has the nonlinear equation for determining the function $\Phi(\zeta)$,

$$\left[\frac{d^2}{\partial\zeta^2} + 2g\Phi^2(\zeta)\right]\Phi(\zeta) = 0, \tag{32.97}$$

where

$$\epsilon \equiv 2ma^2\mathcal{E}/\hbar^2, \quad g \equiv \frac{ma^2\sigma^2}{MV_0^2\hbar^2(1-s^2)}. \tag{32.98}$$

The solutions of eqn. (32.97) for negative energies $\mathcal{E}$ are studied in Section 30. They are expressed in terms of the Jacobian elliptic functions (30.8) and (30.18)

$$\Phi_d(\zeta) = \Phi_0 \mathrm{dn}(\sqrt{g}\,\Phi_0\zeta, k), \tag{32.99}$$

$$\Phi_c(\xi) = \varphi_0 \mathrm{cn}\left[\frac{\varphi_0}{k}\sqrt{g}\,\zeta, k\right], \tag{32.100}$$

where the modulus k, and the amplitudes Φ_0 and φ_0 of elliptic functions are determined through the conditions of normalization and periodicity.

Let us begin by dealing with the excitations corresponding to the function (32.99). The normalization condition (32.91) is then reduced to the following one

$$1 = \Phi_0^2 \int_0^N \mathrm{dn}^2(\sqrt{g}\,\Phi_0\xi, k)d\xi = \frac{\Phi_0}{\sqrt{g}} \int_0^{N\Phi_0\sqrt{g}} \mathrm{dn}^2(u,k)du. \tag{32.101}$$

The period of the Jacobian elliptic function $\mathrm{dn}(u,k)$ is equal to $2K(k)$ where $K(k)$ is the complete elliptic integral of the first kind. Assuming that ν periods are packed into the ring, one should write

$$\sqrt{g}N\Phi_0 = 2\nu K(k). \tag{32.102}$$

Using the relation

$$\int_0^{2\nu K(k)} \mathrm{dn}^2(u,k)du = 2\nu E(k),$$

where $E(k)$ is the complete elliptic integral of the second kind, one obtains from eqn. (32.101) that

$$\Phi_0 = \sqrt{g}/2\nu E(k). \tag{32.103}$$

Substituting the latter into eqn. (32.102), one finds the equation for the modulus k of the elliptic function (32.99)

$$gN = 4\nu^2 K(k)E(k), \quad \nu = 1,2,... \tag{32.104}$$

Hence, using eqn. (32.103), one obtains the final form for the excitation function

(32.99)

$$\Phi_d(\xi) = \frac{\sqrt{g}}{2\nu E(k)}\mathrm{dn}\left[\frac{d\xi}{2\nu E(k)},k\right] \tag{32.105}$$

where k is the solution of eqn. (32.104).

Taking into account the value of the second derivative of the Jacobian function (31.99a), with the help of eqn. (32.97) one can evaluate ϵ,

$$\epsilon = -g^2(2-k^2)/4\nu^2E(k), \quad \nu = 1,2,... \tag{32.106}$$

In order to determine the total energy of excitation (32.92) involving the energy of travelling ring deformation, one should evaluate the integral $\int_0^N \Phi_d^4(\zeta)d\zeta$ entering into eqn. (32.93). Using (32.104) and (32.10), one finds

$$\int_0^N \Phi_d^4(\zeta)d\zeta = \frac{g}{8\nu^3E^3(k)}\int_0^{2\nu K(k)} \mathrm{dn}^4(u,k)du = \frac{g\Theta_d(k)}{4\nu^2E^3(k)}. \tag{32.107}$$

Here the function $\Theta_d(k)$ is given by eqn. (29.27) with $N = 1$.

Using eqns. (32.106) and (32.107), one has the total excitation energy

$$\ni(V) = \frac{1}{2}mV^2 + \frac{\hbar^2g^2}{8ma^2\nu^2E(k)}\left\{\frac{(1+s^2)\Theta_d(k)}{E^2(k)} - (2-k^2)\right\}. \tag{32.108}$$

According to (29.14) this excitation carries a momentum,

$$\mathcal{P}(V) = V\left\{m + \frac{\hbar^2g^2\Theta_d(k)}{4mV_0^2\nu^2E^3(k)a^2}\right\}. \tag{32.109}$$

One should recall that the velocity, entering into the formulas (32.108) and (32.109), takes discrete values given by eqn. (32.95).

The formulas obtained are rather cumbersome. However, they are considerably simplified for chains of a great length, $N \gg 1$. Using (29.35) and (29.36), one obtains in this case

$$E(k) \approx 1, \quad K(k) \approx \ln(4/k_1), \quad k_1 = \sqrt{1-k^2}. \tag{32.110}$$

From eqn. (32.104) we obtain the explicit formula for the complementary modulus

$$k_1 = \sqrt{1-k^2} = 4\ln\left[-\frac{gN}{4\nu^2}\right] \ll 1 \tag{32.111}$$

due to which eqn. (32.106) is simplified as follows

$$\epsilon = -\frac{g^2(1+k_1^2)}{2\nu^2} \approx -g^2/2\nu^2, \quad \nu = 1,2,... \tag{43.112}$$

The hydrogen-like dependence of the energy on the integer ν was first noticed by Pestryakov [51,52].

For $k_1^2 \ll 1$, the function Θ_d is approximated as follows

$$\Theta_d(k) \approx \frac{2(2-k^2)}{3}. \tag{32.113}$$

Therefore, the total energy and momentum of the excitation are

$$\ni(V) \approx \frac{1}{2}mV^2 - \frac{\hbar^2 g^2(1+k_1^2)}{24ma^2\nu^2}, \quad k_1^2 \ll 1, \tag{32.114}$$

$$\mathcal{P}(V) = V\left[m + \frac{2g^2\hbar^2(1+k_1^2)}{12mV_0^2\nu^2a^2}\right]. \tag{32.115}$$

The expressions in brackets can be interpreted as the mass carried by the excitation along the ring. Within this approximation, the wave function of the excitation (32.106) is expressed via the hyperbolic function

$$\Phi_d(\zeta) = \frac{\sqrt{g}}{2\nu}\left\{1 + \frac{k_1^2}{4}(u + \text{sh}u\text{ch}u)\text{th}u\right\}\text{sch}u, \tag{32.116}$$

where $u = g\zeta/2\nu$ and $k_1^2 \ll 1$.

Now we investigate the excitations described by the function (32.100). In this case, the normalization condition becomes

$$1 = \varphi_0^2 \int_0^N \text{cn}^2\left[\frac{1}{k}\varphi_0\sqrt{g}\zeta, k\right]d\zeta = \frac{k\varphi_0}{\sqrt{g}} \int_0^{N\varphi_0\sqrt{g}/k} \text{cn}^2(u,k)du. \tag{32.117}$$

The period of the Jacobian elliptic function $\text{cn}(u,k)$ is $4K(k)$. Assuming μ periods to be packed in the ring, one has

$$N\varphi_0\sqrt{g}/k = 4\mu K(k), \quad \mu = 1,2,... \tag{32.118}$$

Using the relation

$$\int_0^{4\mu K(k)} \text{cn}^2(u,k)du = \frac{4\mu}{k^2}\delta(k), \tag{32.119}$$

where

$$\delta(k) \equiv E(k) - k_1^2 K(k), \tag{32.119a}$$

[see eqn. (29.48)], one finds

$$\varphi_0 = k\sqrt{g}\,/\,4\mu\delta(k). \tag{32.120}$$

From the formula (32.118), we obtain the equation

$$gN = 16\mu^2\delta(k)K(k), \tag{32.221}$$

for the modulus k of the elliptic function (32.100). Applying eqn. (32.120), one obtains the final form of this function

$$\Phi_c(\zeta) = \frac{k\sqrt{g}}{4\mu\delta(k)}\,\mathrm{cn}\left[\frac{g\zeta}{4\mu\delta(k)}, k\right]. \tag{32.122}$$

The latter satisfies eqn. (32.97) if

$$\epsilon_c = -\frac{g^2}{12\mu^2\delta(k)}(2k^2 - 1). \tag{32.123}$$

Thus the negative energies correspond to k^2 exceeding $1/2$, or $k_1^2 < 1/2$.

To estimate the total energy and momentum of the excitation, we first calculate the following integral

$$\int_0^N \Phi^4(\zeta)d\zeta = \frac{gk^4}{[4\mu\delta(k)]^3}\int_0^{4\mu K(k)} \mathrm{cn}^4(u,k)du = \frac{g\Theta_c(k)}{16\mu^2\delta^3(k)}, \tag{32.124}$$

where $\Theta_c(k)$ is given by the equations

$$Q_c(k) = k^4 \int_0^{K(k)} \mathrm{cn}^4(u,k)du =$$

$$= \frac{1}{3}\left[2(2k^2-1)E(k) - k_1^2(2-3k^2)K(k)\right].$$

Substituting further this value and eqn. (32.122) into (29.13), one obtains the total excitation energy

$$\ni_c(V) = \frac{1}{2}mV^2 + \frac{\hbar^2 g^2[1 - 2k_1^2 - (1+s^2)\Theta_c(k)/\delta(k)]}{32ma^2\mu^2\delta(k)}. \tag{32.125}$$

Small values of the complementary modulus k_1 correspond to large N. Using the asymptotics of (32.100) and $\delta(k) \approx 1$ following from (32.119a), from eqn. (32.121) one obtains the approximate value for k_1,

$$k_1 \approx 4\exp(\sim gN/12\mu^2), \quad \mu = 1,2,...$$

Then $\Theta_c(k) = 2/3$, and the total excitation energy becomes

$$\ni_c(V) \approx \frac{1}{2}mV^2 - \frac{\hbar^2 g^2}{32ma^2\mu^2}\left[1 - 2k_1^2 - \frac{2}{3}(1+s^2)\right]. \tag{32.126}$$

Therefore, the negative energies, i.e., the energies below the bottom of the band of exciton states, can be realized in a form of the cnoidal wave of the type (32.122) in the rings with large number N of links only. In this case, the function (32.122) has the following asymptotics

$$\Phi_c(\xi) \approx \frac{k\sqrt{g}}{4\mu}\left[1 - \frac{k_1^2}{4}(\text{sh}u\,\text{ch}u - u)\text{th}u\right]\text{sch}, \tag{32.127}$$

where

$$u = g\zeta/4\mu, \quad \mu = 1,2,...,k = \sqrt{1-k_1^2} \approx 1.$$

If the approximation (32.85) becomes invalid, one should include a contraction of the ring length under excitation. This contraction, $\int\rho(\xi)d\xi$, depends on $\rho(\xi)$ which needs to be evaluated. So this problem can only be resolved in an approximate way.

Chapter 9

The Quantum Theory of Quasiparticle Motion in a Molecular Chain Taking into Account Thermal Vibrations

In what has gone before, in our investigation of solitons in molecular systems, we have applied a classical description of displacements of the equilibrium positions of molecules constituting the system in question, or have studied them within the framework of the theory of coherent states. In both these cases, the change of equilibrium positions is determined by classical equations of motion, with quantum fluctuations of equilibrium positions and the thermal motions of molecules relative to the new ones are naturally neglected. We indeed assume that the system under study is in contact with a thermostat at a temperature $\Theta \neq 0$. In this Chapter, we develop a theory which takes all these restrictions into account.

One should note firstly that numerical studies of the dynamics of solitons in a thermalized chain have been carried out in the papers [132,225], where, however, the effect of thermal motions is studied within the framework of the classical model regardless of thermal fluctuations taking place in the chain, even at zero temperature (zero vibrations). This classical description leads to a breakdown of the soliton state even at $4K$. Thermal vibrations affect substantially the dynamics of excited states. A rigorous investigation of the role of thermal vibrations is only possible within a consistent quantum description which involves the energy spectrum of all phonons participating in the excitation. In such a way, one can demonstrate that solitons are not broken at a temperature than higher $300K$ [260,91].

33. Equations for the Stationary States of a Quasiparticle in a One-Dimensional Chain within a Thermostat

The aim of the present Section is the investigation of the energy states of a quasiparticle (exciton or excess electron) in a one-dimensional chain of N neutral molecules. At the start, we assume that this chain is periodically repeated with a large period aN. Taking into account the interaction between the neighboring molecules occupying the fixed nodes $na(n=0,+1,+2,...)$ of the chain, one can write the corresponding Hamiltonian

$$H_c = E_0 \sum_{n=1}^{N} [2B_n^+ B_n - (B_n^+ B_{n+1} + c.c.)], \tag{33.1}$$

$$B_n = B_{n+N}, \quad \sum_{n=1}^{N} |B_n|^2 = 1 \tag{33.1a}$$

where the energy refers to the bottom of the energy band of a free quasiparticle. In eqn. (33.1), E_0 is the energy of exchange (for electrons) or resonance dipole-dipole (for intramolecular vibrations) interaction expressed in terms of the effective mass m of a quasiparticle as follows

$$E_0 = \hbar^2 / 2ma^2. \tag{33.2}$$

Bose operators relating to the presence or absence of a quasiparticle at the na-th node are denoted by B_n^+ and B_n, respectively.

Due to the translational symmetry of the system in question, and the cyclic periodic conditions with the period aN, the states of a quasiparticle determined by the operator (33.1) are characterized by plane waves

$$\psi_k(n) = N^{-1/2} B_n^+ |0\rangle \exp(ikna), \tag{33.3}$$

which depend on the wavenumber k being the eigenvalue of the operator of translation on a segment na and having N discrete values

$$k = 2\pi\mu / aN, \quad \mu = 0, \pm 1, \pm ... N/2. \tag{33.4}$$

In eqn. (33.3), the symbol $|0\rangle$ stands for the vacuum state of quasiparticles. The states, corresponding to the function (33.3), generate the energy band which is described in the long-wave approximation by the following formula

$$E(k) = \hbar^2 k^2 / 2m = (ka)^2 E_0, \quad ka \ll 1. \tag{33.5}$$

Hence, the energy E_0 of the resonance interaction is equal to the energy (33.5) of a quasiparticle with the wavenumber $1/a$. The replacement of the real mass of a particle by its effective mass infers accounting for the interaction of this quasiparticle with molecules arranged at the nodes na of chain. The displacement of molecules from their equilibrium positions results in the additional, so-called deformation interaction which determines behavior of quasiparticle in chain.

To describe displacements in a quantum mechanical way, it is convenient to express the displacement operator u_n of the nth molecule from its equilibrium position na in terms of the operators of creation b_q^+ and annihilation b_q of phonons (see [35], Section 34):

$$\hat{u}_n = \left[\frac{a\hbar}{2NMV_0}\right]^{1/2} \sum_q |\xi|^{-1/2}(b_q+b_{-q}^{+})\exp(in\xi), \quad \xi = aq. \tag{33.6}$$

In eqn. (33.6), M is the mass of molecules,

$$V_0 = a(\kappa/M)^{1/2} \tag{33.7}$$

is the longitudinal sound velocity of the chain having an elasticity coefficient κ. From the definition of V_0, it follows that

$$MV_0^2 = \kappa a^2, \tag{33.8}$$

which characterizes the double deformation energy of the chain assuming a displacement of a molecules on a distance a. Wavenumbers q range over the same N discrete values as the wavenumbers k given by eqn. (33.4).

The operator of the short-range deformation interaction energy of a quasiparticle with the displacements of neutral molecules takes, within the linear approximation, the following form (see [32,35])

$$H_{\text{int}} = N^{-1/2}\sum_{q,n} F(q)B_n^+ B_n(b_q+b_{-q}^{+})\exp(i\xi n), \tag{33.9}$$

where

$$F(q) = -F^*(q) \equiv i\sigma\left[\frac{\epsilon(q)}{2MV_0^2}\right]^{1/2} |\xi|^{-1}\sin\xi. \tag{33.10}$$

σ is the energy of the deformation potential ([35], Section 34) determined via the relation $\sigma = a\chi$ in terms of the parameter χ used in the preceding Chapter; $\epsilon(q)$ is the energy of a phonon with the quantum number q. The operator of the energy of longitudinal deformation and vibrations of molecules in the chain is expressed in terms of $\epsilon(q)$ and b_q through the formula

$$H_{ph} = \sum_q \epsilon(q)b_q^+ b_q. \tag{33.11}$$

Neglecting dispersion, one can define the energy of longitudinal phonons as follows

$$\epsilon(q) = |\xi|\epsilon_0, \quad \epsilon_0 = \epsilon(1/a) = \hbar V_0/a. \tag{33.12}$$

Introducing the amplitude u_q^0 of the zeroth vibration of the qth mode of the chain by using the relation

$$u_q^0 = [\hbar/2M\epsilon_0|\xi|]^{1/2}, \tag{33.12a}$$

one can express the operator u_n of the displacement in terms of the creation and annihilation of qth-mode phonons,

$$\hat{u}_n = N^{-1/2}\sum_q u_q^0(b_q + b_{-q}^+)\exp(\mathrm{i}n\xi). \tag{33.12b}$$

Referring the energy to the bottom of the energy band of a free quasiparticle (33.5), the states of the chain, involving a single particle in the principal period, are determined by the total Hamiltonian

$$H = H_e + H_{\text{int}} + H_{ph}. \tag{33.13}$$

Now one can define the canonical transformation of the operators B_n and b_q to the new bosonic operators A_n and a_q by using the operator

$$S = \exp\left[\sum_n \Omega_n A_n^+ A_n\right], \tag{33.14}$$

where

$$\Omega_n = -\Omega_n^+ = \sum_q(\tilde{\beta}_{qn}a_q - \tilde{\beta}_{qn}a_q^+) \tag{33.15}$$

and

$$\hat{\beta}_{qn} = \beta_{qn}\exp(-i\xi n) - \tag{33.16}$$

are modulated plane waves, In this case, the transformation of the operators are given by the following relations

$$b_q = Sa_qS^+ \equiv a_q + \sum_n A_n^+ A_n \beta_{qn}\exp(-i\xi n), \tag{33.17}$$

$$b_q^+ = Sa_q^+S^+ \equiv a_q + {}^+\sum_n A_n^+ A_n \beta_{qn}^*\exp(i\xi n),$$

and

$$B_n = SA_nS^+ = \exp(-\Omega_n)A_n, \tag{33.18}$$

$$B_n^+ = SA_n^+S^+ = \exp(\Omega_n)A_n^+.$$

Using eqns. (33.17) and (33.18), one represents the energy operator (33.13) in terms of the new operators

$$\tilde{H} = SHS^+ \equiv \sum_n E_0\{2A_n^+A_n - [A_{n+1}^+A_n\exp(-W_n) + c.c.]\} + \tag{33.19}$$

$$+\sum_{n,q} A_n^+A_n e^{i\xi n}[\epsilon(q)\beta_{qn}^* - N^{-1/2}F(q)]a_q + c.c.\} -$$

$$-\sum_{n,m,q}\{N^{-1/2}F(q)\beta_{qm}A_n^+A_nA_m^+A_m\exp(i\xi(n-m)) + c.c.\} +$$

$$+\sum_{n,m,q}\{\epsilon(q)\beta_{qn}^*\beta_{qm}A_n^+A_nA_m^+A_m\exp[i\xi(n-m)]\} + \sum_q \epsilon(q)a_q^+a_q$$

where the operator function $\exp(-\tilde{W}_n)$ is

$$\exp(-W_n) = \exp(\Omega_n)\exp(-\Omega_{n+1}). \tag{33.20}$$

From eqn. (33.17), it follows that the operators a_q^+ and a_q describe molecular vibrations relative to the new equilibrium positions of molecules in the chain determined by the functions β_{qm}.

Let us introduce the eigenfunctions $|\nu>$ of the operators $a_q^+a_q$ characterizing the states of the chain with the definite number ν_q of real phonons of the qth mode,

$$(a_q^+a_q - \nu_q)|\nu_q> = 0, \quad \nu_q = 0, 1, ...$$

These functions are determined by the formula

$$|\nu_q> = (\nu_q)^{-1/2}(a^+)^{\nu_q}|0), \tag{33.21}$$

where $|0)$ is the function of a vacuum of real phonons. The functions of single intramolecular excitations have the form

$$A_n^+|0> \tag{33.21a}$$

where $|0>$ is the function of the state of the chain without intramolecular excitations.

The excited states of the whole chain are described by the functions

$$|\Psi\rangle = \sum_n \varphi_n A_n^+ |0\rangle \prod_q |\nu_q\rangle, \tag{33.22}$$

where the unknown functions φ_n, which characterize a distribution of intramolecular excitations along the chain, are normalized by the condition

$$\sum_n |\varphi_n|^2 = 1. \tag{33.23}$$

The energy states of the chain in thermodynamics equilibrium with a thermostat at temperature Θ are determined by the functional $\mathcal{H}$ which is derived from the operator (33.19) under the statistical average over the wave functions (33.22) with the following density matrix

$$\rho = \frac{\exp[-\Sigma_q \epsilon(q) a_q^+ a_q / \Theta]}{\mathrm{sp}\{\exp[-\Sigma_q \epsilon(q) a_a^+ a_q / \Theta]\}}.$$

Therefore, one obtains the energy functional

$$\mathcal{H} = E_0 \left\{ 2 - \sum_n [\varphi_n^* \varphi_{n+1} \exp(-\tilde{W}_n) + \varphi_n \varphi_{n+1}^* \exp(-\tilde{W}_{n+1})] \right\} - \tag{33.24}$$

$$-N^{-1/2} \sum_{q,n} F(q) |\varphi_n|^2 (\beta_{q,n} + \beta_{-q,n}^*) +$$

$$+ \sum_q \epsilon(q) \left[\langle\nu_q\rangle + \sum_n |\beta_{qn}|^2 |\varphi_n|^2 \right],$$

where

$$\langle\nu_q\rangle = \{\exp[\epsilon(q)/\theta] - 1\}^{-1} \tag{33.25}$$

is the average number of real phonons of the qth mode, and the function W_n is

$$W_n = \sum_q \left\{ \left(\langle\nu_q\rangle + \frac{1}{2} \right) (|\beta_{q,n}|^2 + |\beta_{q,n+1}|^2) - (1 + \langle\nu_q\rangle) \beta_{qn}^* \beta_{q,n+1} e^{-i\xi} - \right. \tag{33.26}$$

$$\left. - \langle\nu_q\rangle \beta_{q,n+1}^* \beta_{qn} e^{i\xi} \right\}.$$

Let us start by considering only those excited states whose amplitudes β_{qn} of the plane waves (33.16) depend weakly on n, i.e., which satisfy the following inequalities

$$|\beta_{qn} - \beta_{q,n+1}| \ll \beta_{qn}. \tag{33.27}$$

Then the formula (33.26) becomes simplified as follows

$$W_n = \sum_q \{(1+2<\nu_q>)[1-\cos\xi]+i\sin\xi\} \mid \beta_{qn} \mid^2.$$

Taking into account the fact that the summation in this expression is carried out over positive and negative $\xi = qn$, one obtains finally

$$W_n = \sum_q (1+2<\nu_q>)[1-\cos\xi] \mid \beta_{qn} \mid^2 \approx \frac{1}{2}\sum_q (1+2<\nu_q>)\xi^2 \mid \beta_{qn} \mid^2. \qquad (33.28)$$

The term $\exp(-W_n)$ is the energy functional (33.24) is called the *Debye-Waller factor*. It characterizes a weakness of the resonance (exchange) interaction of the nth molecule with the neighboring ones due to quantum fluctuations of intermolecular distances.

Thus, with the help of the given transformation, in the Hamiltonian (33.13) we perform the transformation characterizing coherent states of displaced molecules, to the probability amplitudes φ of the distribution of a quasiparticle over the chain nodes, and to the mean values of the quantum numbers $<\nu_q>$ of real phonons which determine the vibrations of molecules relative to their new equilibrium positions. Imposing periodic boundary conditions, the function (33.22) remains translational invariant. Hence, the states of a quasiparticle in a given chain are characterized by the quantum number (33.4) which takes discrete values.

The energy functional (33.24) describes the nonlocalized (band) states as well as the states which are not so strongly autolocalized and satisfy the inequality (33.27). We recall that the autolocalized state is the stationary state in which the intermolecular separations or orientations are changed over some finite region of the chain, i.e., there occurs a local breakdown of translational symmetry. This localization region can be anywhere in the chain. Thus, the total translational symmetry is conserved, and the stationary states are characterized by energy and total momentum provided by the translocation of the localized region along the chain with a constant velocity depending on the wavenumber k.

Stationary nonlocalized states are also characterized by a definite value of the wavenumber k; nevertheless, in these states the probability of distribution of a quasiparticle is independent of n, and hence, motion of the quasiparticle is absent. By means of the superposition of stationary nonlocalized states one can generate the nonstationary ones which do not have a strictly definite energy and in which the probability of finding a particle in a finite region is nonvanishing, this region being translated along the chain with a group velocity. However, the dimension of the localized region in this state increases continuously in the course of time, i.e., the wave packet is 'running'.

Autolocalized states which arise under the short-range interaction (33.9), are described via nonlinear differential equations. We denote these states *solitons* to distinguish them from the autolocalized states introduced firstly by Landau [20] in 1933 and investigated in detail by Pekar [98,99] for describing electron movement in ionic crystals. Such states are called *polarons* since they arise from the long-range (Coulomb) interaction of electrons with the field of electric polarization of a crystal described by longitudinal optical phonons. The properties of polarons are determined by integro-differential equations.

In the following Sections, we study separately the nonlocalized and autolocalized states taking into account the short-range interaction of a quasiparticle with longitudinal acoustic phonons only.

34. Nonlocalized Stationary States of a Quasiparticle

Imposing periodic boundary conditions with large periods aN, one can describe the nonlocalized (band) states of a quasiparticle by plane waves,

$$\varphi_n(t) = N^{-1/2}\exp[i(kna - \omega t), \quad \sum_{n=1}^{N} |\varphi_n|^2 = 1. \tag{34.1}$$

The function $\varphi_n(t)$ depends on given wavenumbers k taking discrete values (33.4) which are the eigenvalues of the operator of translation of the chain on the segment na (see Section 4). Finally, they determine the quasi-momentum $\hbar k$ of a particle and its phase velocity $V_p = \omega/k$.

For fixed value of k, the function $\varphi_n(t)$ describes the idealized stationary state of a quasiparticle with a uniform probability distribution, $|\varphi_n|^2 = \text{const}$. In this state, a real translocation of the particle is absent.

The real motion of the particle is described by a wave packet; that is, by the superposition of plane waves having wavenumbers near some fixed value k_0. The average velocity of this particle coincides with the group velocity of a wave packet

$$V(k_0) = (\partial\omega/\partial k)_{k=k_0}. \tag{34.2}$$

The uniform distribution of a quasiparticle over a chain takes place in the strongly stationary states of the type (34.1). For this reason, the functions (33.16) are plane waves having amplitudes $\beta_{qn} = \beta_q$ independent of the number of the node of chain. Hence, the function (33.28) is independent of n too, and transforms into a real, temperature-dependent function of the following type

$$W = \frac{1}{2}\sum_q \eta(q)\xi^2 |\beta_q|^2, \tag{34.3}$$

$$\eta(q) \equiv 1+2<\nu_q> = \text{ctgh}[\epsilon(q)/2\Theta] \tag{34.4}$$

where Θ is the temperature. Using the normalization condition for the function (34.1) and neglecting the energy of real phonons one can transform the functional (33.24) for a fixed wavenumber k as follows

$$\mathcal{H}_k(\beta) = 2E_0 - \epsilon_0\gamma_k e^{-W} - N^{-1/2}\sum_q F(q)(\beta_q + \beta^*_{-q}) + \sum_q |\xi|\epsilon_0|\beta_q|^2. \tag{34.5}$$

Minimizing the expression (34.5), one finds

$$\beta_q^0 = N^{-1/2}F^*(q)[|\xi|\epsilon_0(1+\delta k\Xi(\xi)]^{-1} \tag{34.6}$$

where the following abbreviated notations have been introduced,

$$\Xi(\xi) \equiv |\xi|\eta(q), \tag{34.7}$$

$$\delta_k \equiv \gamma_k \exp(-W), \tag{34.8}$$

and

$$\epsilon_0\gamma_k = \left[1 - \frac{1}{2}a^2k^2\right]E_0. \tag{34.8a}$$

For $k = 0$, the parameter γ_0,

$$\gamma_0 = E_0/\epsilon_0 \tag{34.9}$$

determines the ratio of the energy of a free quasiparticle to that of phonons for the wavenumber $k = q = 1/a$. Hence, a small value of γ_0 corresponds to rigid chains having large ϵ_0 and narrow energy bands of free quasiparticles. It is also convenient to introduce the quantity G_0 which determines the effective energy of the interaction of a quasiparticle with the chain deformation,

$$G_0 = \sigma^2/MV_0^2 = \sigma^2/\kappa a^2. \tag{34.10}$$

Substituting the minimizing value (34.6) into the energy functional (34.5), one can obtain the energy of a dressed quasiparticle,

$$\mathcal{E}(k) = \mathcal{H}_k(\beta^0). \tag{34.11}$$

Taking into account the identities

$$\sum_q F(q)(\beta_q^0 + \beta_{-q}^0) = N^{-1} G_0 \sum_q [1 + \delta_k \Xi(\xi)]^{-1}, \tag{34.12}$$

$$\sum_q \epsilon(q) |\beta_q^0|^2 = (2N)^{-1} G_0 \sum_q [1 + \delta_k \Xi(\xi)]^{-1} \tag{34.13}$$

and eqn. (34.6), and imposing the condition $\delta_k \Xi(\xi) \ll 1$, one can transform the energy (34.11) as follows

$$\mathcal{E}(k) = \frac{\hbar^2}{ma^2}\left[1 - \frac{m}{\tilde{m}}\right] + \frac{\hbar^2 k^2}{2\tilde{m}} - \frac{1}{2} G_0 \mathcal{K}(k). \tag{34.14}$$

Here, $\tilde{m}$ is the effective mass of a dressed quasiparticle given by

$$\tilde{m} \equiv m \exp(W)/(1 + W), \tag{34.15}$$

and

$$\mathcal{K}(k) = N^{-1} \sum_q [1 + \delta_k \Xi(\xi)]^{-1}. \tag{34.16}$$

The equality (34.12) characterizes the gain of energy resulting from the binding of a quasiparticle with deformation, whereas the equality (34.13) gives the energy required for the deformation generation.

The first term in eqn. (34.14) raises the bottom of the energy band of a dressed quasiparticle compared with that of a free quasiparticle. This increase is conditioned by a weakening of the resonance (exchange) interaction manifested in the increase of the effective mass $\tilde{m}$, given by eqn. (34.15), due to the Debye-Waller factor. The term W in the denominator of eqn. (34.15) partially compensates this effect. The second term in eqn. (34.14) characterizes the kinetic energy of a dressed quasiparticle with effective mass $\tilde{m}$. The last term in eqn. (34.14) gives the gain of energy resulting from the dressing of a quasiparticle.

Substituting eqn. (34.6) into (34.3), one obtains the following transcendental equation

$$W = (2\epsilon_0)^{-1} G_0 \mathcal{K}_1(k), \tag{34.17}$$

which determines the Debye-Waller factor, $\exp(-W)$, at the value

$$\mathcal{K}_1(k) = N^{-1} \sum_q \Xi(\xi)[1+\delta_k \Xi(\xi)]^{-2} \tag{34.18}$$

together with eqn. (34.8). One can easily check that the functions $\mathcal{K}(k)$ and $\mathcal{K}_1(k)$ are linked by the relation

$$\mathcal{K}_1(k) = -\frac{\partial}{\partial \delta_k} \mathcal{K}(k). \tag{34.19}$$

For that reason, it is sufficient only to evaluate the sum (34.16).

The theory developed is justified for small values of k (the long-wave approximation, $ka \ll 1$). According to eqn. (34.8a), the function γk depends therefore weakly on k. Hence evaluating the sums (34.16) and (34.18), one can assume $\delta_k \approx \delta_0$ where

$$\delta_0 = \gamma_0 \exp(-W). \tag{34.20}$$

To simplify calculations, one replaces the summation over q by integration with the help of the relation

$$N^{-1} \sum_q f(q) = \pi^{-1} \int_0^{\pi} f(\xi) d\xi. \tag{34.21}$$

The sum (34.16) transforms into the integral

$$\mathcal{K}(0) = \pi^{-1} \int_0^{\pi} \frac{d\xi}{1+\delta_0 \Xi(\xi)}. \tag{34.22}$$

According to eqn. (34.19), one obtains the value of $\mathcal{K}_1(0)$,

$$\mathcal{K}_1(0) = -\frac{\partial}{\partial \delta_0} \mathcal{K}(0). \tag{34.23}$$

Even with the simplifications given, the explicit evaluation of the integral (34.22) is rather difficult in the general case since in the integrand there appears the complex and temperature-dependent function $\Xi(\xi)$. Hence, one should consider this integral separately for high ($\epsilon_0 \ll 2\Theta$) and low ($\epsilon_0 \gg 2\Theta$) temperatures. In both limiting cases, the function $\Xi(\xi)$ can be approximated as follows

$$\Xi(\xi) \approx \begin{cases} 2\Theta/\epsilon_0 + \epsilon_0\xi^2/6\Theta, & 2\Theta > \epsilon_0; \\ |\xi|\{1+2\exp[-\epsilon_0|\xi|/\Theta]\}, & 2\Theta < \epsilon_0. \end{cases} \tag{34.24}$$

The quantity $\Theta_p \approx \epsilon_0$ can be conditionally named the Debye temperature.

(A) *High temperatures* (which exceed the Debye temperature).

In this case, one obtains

$$\Xi(\xi) \approx 2\Theta/\epsilon_0, \tag{34.25}$$

and

$$\mathcal{K}(0) = [1+4\Theta E_0\epsilon_0^{-2}/\exp W]^{-1}, \tag{34.26}$$

$$\mathcal{K}_1(0) = 2\Theta\epsilon_0^{-1}[1+4\Theta E_0\epsilon_0^{-2}/\exp W]^{-2}.$$

Therefore, eqn. (34.17) which determines W, takes the form

$$[1+4\Theta E_0/\epsilon_0^2\exp W]^2 W = \Theta G_0/2\epsilon_0^2. \tag{34.27}$$

If W_0 is the root of eqn. (34.27), the energy of a dressed quasiparticle is given for small k by the formula

$$\mathcal{E}(k) = \hbar^2(ma^2)^{-1}[1+(1+W_0)/\exp W_0]+ \tag{34.28}$$

$$+\frac{\hbar(1+W_0)k^2}{2m\exp W_0} - \frac{1}{2}G_0[1+4\theta E_0/\epsilon_0^2\exp W_0].$$

Besides the explicit dependence on temperature, entering into the last term, the dependence is on W_0, too.

(B) *Low temperatures*

In this case, the function (34.19) can be replaced by the approximate value

$$\Xi(\xi) \approx |\xi|. \tag{34.29}$$

Within this approximation, the integral (34.18) takes the value

$$\mathcal{K}_1(0) = \pi^{-1}\int_0^\pi \frac{\xi d\xi}{(1+\delta_0\xi)^2} = \pi\mathcal{P}(\delta_0), \tag{34.30}$$

$$\mathcal{P}(\delta_0) = (\pi\delta_0^{-2})[\ln(1+\pi\delta_0) - \pi\delta_0/(1+\pi\delta_0)]. \tag{34.30a}$$

Hence, eqn. (34.17) which determines the Debye-Waller factor, becomes

$$W = A\mathcal{P}(\delta_0), \tag{34.31}$$

where

$$A \equiv \pi G_0 / 4\epsilon_0 = \pi\beta_{\max}^2 / 8 < x_0^2 >. \tag{34.31a}$$

$<x_0^2> = \hbar / 2\sqrt{\kappa M}$ is the square of the amplitude of the zeroth vibrations of molecules relative to their equilibrium positions; $\beta_{\max}^2 = \sigma^2 / \kappa^2 d^2$ is the square of the largest displacement of molecules from the equilibrium positions [see eqn. (5.5)]. Thus, the quantity A is proportional to the square of the ratio of the deformation displacement to the amplitude of the zeroth vibrations.

Together with eqn. (34.20), eqn. (34.31) enables the function W to be obtained for any A depending on the ratio G_0/ϵ_0 and the parameter Ω involving in the relation

$$\pi\delta_0 = \Omega / \exp W, \quad \Omega \equiv \pi E_0 / \epsilon_0 = \pi\gamma_0. \tag{34.32}$$

In accordance with eqn. (34.32), $\pi\delta_0$ is small at small values of Ω (a narrow energy band of free quasiparticles and high sound velocity). Then $\mathcal{P}(\delta_0) \approx (1+\pi\delta_0)^{-1} \approx 1$, and eqn. (34.31) is reduced to the following

$$W = A \equiv \pi G_0 / A\epsilon_0. \tag{34.33}$$

Within the approximation (34.29), the integral (34.22) entering the formula (34.14) takes the following value for a dressed quasiparticle

$$\mathcal{K}(0) = (\pi\delta_0)^{-1}\ln(1+\pi\delta_0). \tag{34.34}$$

If $\pi\delta_0 \ll 1$, this is equal to unity. In this case and for a small ratio $\pi E_0/\epsilon_0$, the factor W becomes small too, and the effective mass of a dressed quasiparticle is just that of a free quasiparticle, i.e., $\tilde{m} = m$. Hence, the first term in eqn. (34.14) vanishes, and a lowering of the bottom of the energy band of a quasiparticle under its dressing is given by $G_0/2$.

If A increases, according to eqns. (34.31) and (34.15), then W and $\tilde{m}$ increase also. Thus the first term in eqn. (34.14) has the same tendency, and the energy level of a dressed quasiparticle may appear to be higher than that of a free quasiparticle at small G_0, and so the 'dressing' process becomes less profitable in an energy sense.

Let us now consider the case when both quantities A and Ω are larger than

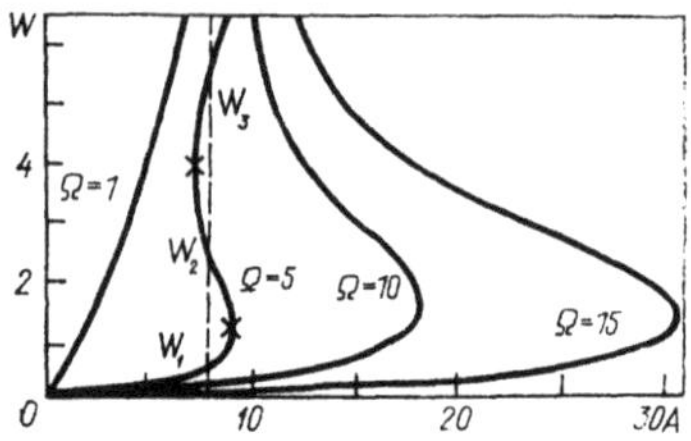

Figure 13. Dependence of the function W on A proportional to the square of the coupling coefficient for four values of Ω.

unity. The dependence of W on A is given by eqns. (34.17) and (34.20), and is depicted in Fig. 13 for $\Omega = 1.5$, 10, and 15. The condition of applicability of the adiabatic approach, for which a quasiparticle motion is adapted to the instant positions of molecules, is better satisfied as Ω becomes larger. It follows from eqns. (34.17) and (34.20) that for $\Omega < 2$ each value of A corresponds to a unique value of W. However, if $\Omega > 2$, the function $W(A)$ becomes a three-valued one over some domain of A depending on Ω. (This fact was firstly found by Toyozawa [272,273] in his study of the strong interaction of an electron with deformation vibrations of a crystal). The corresponding values W_1, W_2, and W_3 can considerably differ from each other. For example, as follows from Fig. 13, one has the roots $W_1 = 0.2$, $W_2 = 2.2$; and $W_3 = 5.6$ at $\Omega = 5$ and $A = 8$.

The stationary states of a dressed quasiparticle with wavenumber k corresponding to the lowest root $W_1 = 0.2$ are characterized by an effective mass which exceeds that of a free quasiparticle by a factor of 1.02 only. Taking into account eqns. (34.32) and (34.4), the energy band is given by the function

$$\mathcal{E}_1(k) = 0.02\hbar^2/ma^2 - 0.08G_0 + \hbar^2k^2/2.04m \tag{34.35}$$

in the region of small k. A quasiparticle is immobile in this strong stationary state. Real motion corresponds to the quasistationary state described by a wave packet with wavenumbers in the neighborhood of k_0. In this state, a particle moves with group velocity

$$V_1 = 0.98\hbar k_0/m \tag{34.36}$$

which accounts for 0.98 of that of a free quasiparticle having mass $\tilde{m}$ and quasimomentum $\hbar k_0$. The localization region of such a particle is smaller the greater is the deviation of wavenumbers k in the wave packet.

In the state corresponding to the third root $W_3 = 5.6$, the dressed quasiparticle possesses an effective mass which exceeds 41 times the mass of a free quasiparticle. In this case, the energy band for small k is described by the following formula

$$\mathcal{E}_3(k) = 0.996\frac{\hbar^2}{ma^2} - 0.964G_0 + \hbar^2 k^2 / 82m. \tag{34.37}$$

The larger effective mass of the dressed quasiparticle, and hence, its narrower energy band means that the wave packets consisting of the corresponding plane waves with wavenumbers in neighborhood of k_0 have a group velocity $V_3 = 0.024\hbar k_0 / m$ which is considerably smaller than that of a free quasiparticle with effective mass m and the same quasi-momentum $\hbar k_0$. The energy band $\mathcal{E}_3(k)$ lies considerably higher than the band $\mathcal{E}_1(k)$. For $\Omega = 5$ and $A = 9$, the states corresponding to the second root $W_2 = 2.2$ are unstable since an increase of A in the domain of $A = 8$ is accompanied by a decrease in W. If $\Omega = 5$, all other states which correspond to those values of W which lie between the points designated by crosses in Fig. 13, become unstable.

If a quasiparticle corresponds to the intramolecular dipole excitation, the aforementioned nonlocalized states are called *excitons*. In the standard theories of excitons only the resonance interaction between molecules fixed rigidly at the nodes of a chain are taken into account. As shown before, the 'dressing' effect for the states, corresponding to the low energy band, is negligible. Thus, the interaction with displacements of molecules from equilibrium positions can be estimated by means of the perturbation treatment. This interaction results in dispersion and damping of the excitons of the zeroth-order treatment and the extension of the absorption band due to the participation of the group of states with different wavenumbers in an excitation. The approximate treatment is also justified by the fact that in the interaction with light an excitation occurs without a change in the equilibrium positions of molecules due to the Franck-Condon principle.

The nonlocal nature of exciton excitations results in a very important phenomenon known as spatial dispersion which arises when the permittivity of a system depends on the frequency as well as on the wavelength of exciting radiation. For a detailed discussion of this phenomenon and its consequences, see the book by Pekar [100].

35. Smooth Autolocalized State - Single Soliton

We mentioned in Section 33, the energy functional (33.24) describes localized states as well as nonlocalized ones. Let us consider those cases when such autolocalization occurs over a finite region of a large chain. In other words, we shall assume that a quasiparticle state in the chain is described by the function

$$\varphi_n(t) = \Phi(n)\exp[i(kna-\omega t)] \tag{35.1}$$

with a fixed wavenumber (33.4) and real amplitude $\Phi(n)$. The latter is nonvanishing only within a finite (but not sufficiently small) domain of chain, i.e., for those n which satisfy the following condition

$$|\tilde{n}-n| \leqslant N_0 \ll N \gg 1. \tag{35.2}$$

In this case, the function $|\beta_{qn}|^2$ for the modulated wave (33.16) vanishes also beyond the domain (35.2) and weakly depends on n inside it. Due to the second property of the function $|\beta_{qn}|^2$, we can put it equal to $|\beta_{q\tilde{n}}|^2$. Thus, we may reserve this value in the sum $\Sigma_n |\varphi_n|^2|\beta_{qn}|^2$ which appears in the energy functional (33.24), beyond the domain (35.2) too, since $|\varphi_n|^2 = 0$ there. Thus, we can use the approximate equality

$$\sum_n |\varphi_n|^2|\beta_{qn}|^2 \approx |\beta_{q\tilde{n}}|^2\sum_n |\varphi_n|^2 = |\beta_{q\tilde{n}}|^2. \tag{35.3}$$

If the relation (35.3) is valid in the neighborhood of the nth node where the quasiparticle is mainly distributed, one can rewrite the energy functional (33.24) as follows

$$\mathcal{H}_m = E_0[2|\varphi_n|^2-(\varphi_n^*\varphi_{n+1}+c.c.)\exp(-W_n)+ \tag{35.4}$$

$$+\sum_q \epsilon(q)[<\nu_q>+|\beta_{qn}|^2]-N^{-1/2}\sum_q F(q)|\varphi_n|^2(\beta_{qn}+\beta^*_{-q,n}),$$

with $n = \tilde{n}$. Since the $\tilde{n}$th node can be anywhere in the chain, in what follows we omit the 'tilde' sign. The amplitudes β_{qn}, appearing in the function (33.26), weakly depend on n in the excitation region. Hence, one can use the approximate relation

$$|\beta_{q,n}\beta^*_{q,n+1}| \approx |\beta^*_{q,n+1}| \approx |\beta_{q,n+1}|^2 \approx |\beta_{qn}|^2,$$

to transform the function (33.26) into the following form

$$W_n \approx \frac{1}{2}\sum_q \eta_a\xi^2|\beta_{q,n}|^2. \tag{35.5}$$

Using eqns. (35.4) and (35.5), one obtains the difference-differential equations

$$i\hbar\frac{\partial\varphi_n}{\partial t} = \frac{\partial\mathcal{H}_n}{\partial\varphi_n^*} \equiv E_0[2\varphi_n-(\varphi_{n+1}+\varphi_{n-1})\exp(-W_n)- \tag{35.6}$$

$$-N^{-1/2}\sum_q F(q)|\varphi_n|^2(\beta_{qn}+\beta_{-q,n}),$$

$$i\hbar\frac{\partial\beta_{qn}}{\partial t} = \frac{\partial\mathcal{H}_n}{\partial\beta_{qn}^*} \equiv \epsilon(q)\beta_{qn}-N^{-1/2}F^*(q)|\varphi_n|^2+ \tag{35.7}$$

$$+E_0\eta_q\xi^2|\varphi_n|^2\beta_{qn}\exp(-W_n).$$

Real phonons with wavenumbers q correspond to vibrations of molecules relative to equilibrium positions with the frequencies $\omega q = |q|V_0$. The time-dependence of the amplitudes of the coherent displacements β_{qn} of the equilibrium positions is determined by the velocity V of a particle moving along a chain,

$$i\hbar\frac{\partial\beta_{qn}}{\partial t} = |q|V\beta_{qn}. \tag{35.8}$$

Taking this equation into account and using eqn. (35.7), and its complex conjugate, we find

$$\beta_{qn} = F^*(q)|\varphi_n|^2/N^{-1/2}|\xi|\epsilon_0\{1+\pi^{-1}\delta_n\Xi(\xi)-s\}, \tag{35.9}$$

and

$$\beta_{qn}^* = F^*(q)|\varphi_n|^2/N^{-1/2}|\xi|\epsilon_0\{1+\pi^{-1}\delta_n\Xi(\xi)+s\}. \tag{35.10}$$

Here we use the abbreviated notations

$$s = V/V_0,\ \ \delta_n \equiv \frac{1}{2}\pi\gamma_0|\varphi_n|^2\exp(-W_n). \tag{35.11}$$

The quantities $\Xi(\xi)$ and γ_0 are given by eqns. (34.7) and (34.9), respectively.

With the help of eqns. (35.9) and (35.10), one can transform eqn. (35.6) into the nonlinear difference-differential equation

$$i\hbar\frac{\partial\varphi_n}{\partial t} = E_0[2\varphi_n-(\varphi_{n+1}+\varphi_{n-1})\exp(-W_n)]+G(n)|\varphi_n|^2\varphi_n, \tag{35.12}$$

where the energy of nonlinearity $G(n)$ is

$$G(n) \equiv G_0I_{1s}(n), \tag{35.13}$$

and

$$I_{1s}(n) \equiv N^{-1/2}\sum_q (1+\pi^{-1}\delta_n)\Xi(\xi)\{[1+\pi^{-1}\delta_n \Xi(\xi)]^2 - s^2\}^{-1}. \tag{35.14}$$

By virtue of the formula (35.9), eqn. (35.5), which determines the function W_n, takes the form

$$W_n = G_0(2\epsilon_0)^{-1}|\varphi_n|^4 I_{2s}(n), \tag{35.15}$$

where

$$I_{2s}(n) \equiv N^{-1}\sum_q \Xi(\xi)\{(1+\pi^{-1}\delta_n \Xi(\xi))^2 - s^2\}^{-1}. \tag{35.16}$$

The energy $\mathcal{E}_{\text{int}}(n)$ of the interaction of a quasiparticle, located in the region of the nth node of a chain deformation, and the energy $\mathcal{E}_{\text{def}}(n)$ of the chain deformation in the neighborhood of this node are given, respectively, by the following expressions

$$\mathcal{E}_{\text{int}}(n) = -|\varphi_n|^2 G_0(n), \tag{35.17}$$

$$\mathcal{E}_{\text{def}}(n) = \sum_q |\xi|\epsilon_0|\beta_{q,n}|^2 = \frac{1}{2}G_0|\varphi_n|^4 I_{3s}(n), \tag{35.18}$$

where

$$I_{3s}(n) \equiv N^{-1}\sum_q \{[1+\pi^{-1}\delta_n \Xi(\xi)]^2 - s^2\} - 1. \tag{35.19}$$

The total momentum of the quasiparticle in a state with fixed k moving together with virtual phonons, is

$$\mathcal{P} = \hbar k + \sum_{n,q>0} \hbar q(|\beta_{qn}|^2 - |\beta_{-q,n}|^2).$$

Substituting eqns. (35.9) and (35.10) into this formula, one obtains

$$\mathcal{P} = \hbar\left[k + s\sum_n G_0|\varphi_n|^2(2a\epsilon_0)^{-1} I_{3s}(n)\right]. \tag{35.20}$$

Within the domain of a quasiparticle localization there occurs a decrease of intermolecular separations which is

$$\rho_n = \langle\psi|(u_n - u_{n-1})|\psi\rangle.$$

Using eqns. (33.6) and (33.9), one has

$$\rho_n = a\sigma(MV_0^2)^{-1}|\varphi_n|^2 I_{1s}(n).$$

Replacing summation over q by integration in accordance with eqn. (34.21), at low temperatures (where $\Xi(\xi) \approx |\xi|$) one obtains the following values of the integrals involved in the foregoing formulas:

$$I_{1s}(n) = (2\delta_n)^{-1}\ln[1+\delta_n(2+\delta_n)/(1-s^2)], \tag{35.21}$$

$$I_{2s}(n) = (2\delta_n)^{-1}\left\{\ln\frac{(1+\delta_n)^2-s^2}{1-s^2} - \frac{1}{s}\ln\frac{(1+s)(1-s+\delta_n)}{(1-s)(1+s+\delta_n)}\right\},$$

$$I_{3s}(n) = (2s\delta_n)^{-1}\ln\left[1+\frac{2\delta_{n'}}{(1-s)\delta_n+(1-s^2)}\right].$$

It is of interest to deal with the states in which the localization region of a quasiparticle exceeds considerably the intermolecular distance. In this case, $|\varphi_n|^2 \ll 1$, and $\delta_n \ll 1$ at low temperatures. For small δ_n, the integrals (35.21) are simplified:

$$2(3\delta_n)^{-1}I_{2s} = I_{1s} = I_{3s} = (1-s^2)^{-1}, \tag{35.22}$$

and the energy of nonlinearity (35.13) takes the value

$$G(s) = G_0/(1-s^2), \tag{35.23}$$

which is independent of n. As the soliton velocity increases, so does the nonlinearity parameter. Within this approximation, according to eqn. (35.22), eqn. (35.15) for determining W_n has the form

$$W_n = B_n\exp(-W_n), \tag{35.24}$$

$$B_n \equiv 3\pi\gamma_0|\varphi_0|^6 G_0/16\epsilon_0(1-s^2). \tag{35.24a}$$

The energy of chain deformation in the region of the nth molecule becomes

$$\mathcal{E}_{\text{def}}(n) \equiv \sum_q |\xi|\epsilon_0|\beta_{qn}|^2 = G_0|\varphi_n|^4/2(1-s^2). \tag{35.25}$$

Within the same approximation, the total momentum $\mathcal{P}$ and the decrease of intermolecular separations are given, respectively, by the following formulas

$$\mathcal{P} = \hbar\left[k + sG_0(2a\epsilon_0)^{-1}\sum_n|\varphi_n|^2/(1-s^2)\right], \tag{35.26}$$

$$\rho_n = a\sigma|\varphi_n|^2/MV_0^2(1-s^2). \tag{35.26a}$$

At not so low temperatures, when, in agreement with (34.24), one can use the approximate expression

$$\Xi(\xi) = |\xi|\{1+2\exp[-|\xi|\epsilon_0/\Theta]\},$$

and if the inequality $\delta_n \ll 1$ is valid, one can replace eqn. (35.16) by its approximate one

$$I_{2s}(\Theta) \approx \frac{1}{\pi}\int_0^{\pi}\frac{\xi[1+2\exp(-\xi\epsilon_0/\Theta)]d\xi}{(1+\pi^{-1}\delta_n\xi)^2-s^2} \approx [1+2\exp(-\pi\epsilon_0/\Theta)]I_{2s},$$

where I_{2s} is determined by eqn. (35.22). In this case, eqn. (35.4) transforms as follows

$$W_n = B_n[1+2\exp(-\pi\epsilon_0/\Theta)]\exp(-W_n), \tag{35.27}$$

where B_n is given by eqn. (35.24a).

Therefore, with a relatively small increase of temperature there occurs an increase of W, in accordance with eqn. (35.27). In this case, the expressions (35.25) and (35.26) remain unchanged. Their temperature dependence is determined through the temperature dependence of the function φ_n only. The nonlinear equation (35.12) for determining φ_n contains W_n which in turn depends on φ_n, in accordance with eqn. (35.24). For this reason it is difficult to obtain its exact solution, however, this can be found via the self-consistent procedure.

Now we deal with the smooth autolocalized states. Applying the continuum approach, one transforms eqn. (35.12) as

$$\left[i\hbar\frac{\partial}{\partial t}-2E_0(1-e^{-W})+\frac{\hbar^2 e^{-W}}{2m}\frac{\partial^2}{\partial x^2}+G(s)|\varphi(x,t)|^2\right]\varphi(x,t) = 0. \tag{35.28}$$

According to eqn. (35.23), the energy of nonlinearity is independent of x, and due to eqn. (35.24) the function W depends on $\varphi(x,t)$. Let us study the solutions corresponding to small value of W. Within the zeroth-order treatment, one finds the solutions of eqn. (35.28) with vanishing W. The solutions of the equation for an infinite chain correspond to the equation

$$\left[i\hbar\frac{\partial}{\partial t}+\frac{\hbar}{2m}\frac{\partial^2}{\partial x^2}+G(s)|\varphi_0(x,t)|^2\right]\varphi_0(x,t) = 0, \tag{35.29}$$

investigated in Section 3. Within the long-wave approach, one can seek the solutions of eqn. (35.29) in the following form

$$\varphi_0(x,t) = \Phi_0(z)\exp[i(kx-\omega t)], \tag{35.30}$$

where $\Phi(z)$ is a smooth real function relative to the coordinate frame

$$z = x - x_0 - Vt, \tag{35.31}$$

moving with a constant velocity V. The wavenumber k in eqn. (35.30) is related to the velocity V and effective mass m of a free quasiparticle according to the relation

$$V = \hbar k/m. \tag{35.32}$$

The velocity V must be smaller than the maximum velocity $V_g = \hbar/ma$ of a free quasiparticle. This constraint is always fulfilled within the long-wave approach, $ka \ll 1$.

Substituting the function (35.30) into eqn. (35.29), one obtains the equation for the real envelope

$$\left[\Lambda_0 + \frac{\hbar^2}{2m}\frac{d^2}{dz^2} + G(s)\Phi^2(z)\right]\Phi_0(z) = 0. \tag{35.33}$$

For fixed wavenumber k, an eigenvalue determines the energy Λ_0 of a quasiparticle in the deformation field created by the quasiparticle

$$\hbar\omega = \Lambda_0 + \hbar^2 k^2/2m. \tag{35.34}$$

The particular solution of eqn. (35.33), normalized for an infinite chain by the condition

$$a^{-1}\int \Phi^2(z)dz = 1 \tag{35.35}$$

is given by the function

$$\Phi_0(z) = \left[\frac{1}{2}aQ_0\right]^{1/2} \operatorname{sech}(Q_0 z), \tag{35.36}$$

with the following parameters

$$Q_0 = G(s)/4aE_0, \tag{35.37}$$

$$\Lambda_0 = -G^2(s)/16E_0. \tag{35.38}$$

The autolocalized stationary states with wavenumber $k = mV/\hbar$, described by the wave function (35.30) with a bell-like envelope (35.36) which satisfies the non-linear Schrödinger equation (35.33), are called solitons.

The amplitude of the wave function (35.36) has a maximum at $z = 0$. This function approaches zero far from the soliton centre as

$$\left[\frac{1}{2}aQ_0\right]^{1/2} \exp(-Q_0|z|), \quad Q_0 \neq 0.$$

We define the localization region of soliton, l_0, by those values $\pm z$ at which $\Phi^2(z) \leqslant 0.1\Phi^2(0)$. Then $l_0 = 2\pi/Q_0$. Therefore, the average value is

$$<|\varphi(x,t)|^2> \approx a/l_0 = G(s)/8\pi E_0. \tag{35.39}$$

If the condition of applicability of the continuum approach is satisfied, i.e., $a \ll l_0$, the average value (35.39) is rather small. Substituting this into eqn. (35.27), one can estimate W_0 in the region of soliton with the help of the following equation

$$W(\theta) = B(\theta)/\exp W(\Theta). \tag{35.40}$$

In eqn. (35.40), the velocity- and temperature-dependent factor $B_0(T)$ is given by the expression

$$B(T) = B_0\left[1+2\exp\left(-\frac{\pi\epsilon_0}{\Theta}\right)\right], \tag{35.41}$$

where

$$B_0 \approx 3.71\cdot 10^{-5}\gamma_0 G_0/\epsilon_0 E_0^3(1-s^2). \tag{35.41a}$$

So, at low temperatures, $\Theta < \epsilon_0$, and small velocities, $s^2 \ll 1$, $B(T)$ is small. It increases weakly with an increase of temperature and a small increase of velocity when the condition of applicability of the harmonic approximation for describing the displacements of the equilibrium positions of molecules remains so far to be valid. The $B(\Theta)$-dependence of $W(\Theta)$ is plotted in Fig. 14. Stable soliton states correspond only to the smallest root which increases as $B(\Theta)$ does. Evaluating $W(\Theta)$, corresponding to the chain with the parameters γ_0, E_0, and $G(s)$ for a given velocity V, via eqn. (35.40), one obtains the first-order approximation to eqn. (35.28)

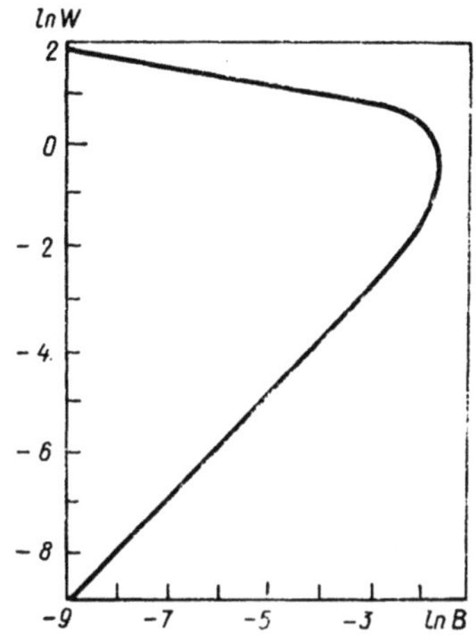

Figure 14. Plot of the function W of the parameter B on a logarithmic scale.

$$\left\{i\hbar\frac{\partial}{\partial t}-2E_0(1-e^{-W(\theta)})-\frac{\hbar^2 e^{-W(\Theta)}}{2m}\frac{\partial^2}{\partial x^2}+G(s)|\varphi_1(x,t)|^2\right\}\times \tag{35.42}$$

$$\times\,\varphi_1(x,t) = 0.$$

Given the velocity V, one can seek the solution of eqn. (35.42) in the form

$$\varphi_1(x,t) = \Phi_1(z)\exp[i(kz-\omega t)], \tag{35.43}$$

where a variable z is given by eqn. (35.31). The envelope $\Phi_1(z)$ satisfies then the equation

$$\left[\Lambda_1+\frac{\hbar^2}{2\tilde{m}}\frac{d^2}{dz^2}+G(s)\Phi_1^2(z)\right]\Phi_1(z) = 0, \tag{35.44}$$

which involves the effective mass $\tilde{m}$,

$$\tilde{m} = m\exp W(\Theta), \tag{35.45}$$

of a dressed quasiparticle. The velocity V determines the value of the wavenumber via the relation

$$k = \tilde{m}V/\hbar. \tag{35.46}$$

The energy $\hbar\omega$ of a dressed quasiparticle in the deformation field is expressed through Λ_1 as

$$\hbar\omega = \Lambda_1 + 2E_0(1 - e^{-W(\Theta)}) + (ka)^2 E_0 \exp[-W(\Theta)]. \tag{35.47}$$

The solution of eqn. (35.44), normalized for an infinite chain, corresponds to a single soliton

$$\Phi_1(z) = \left[\frac{1}{2}aQ_1\right]^{1/2} \operatorname{sech}(Q_1 z) \tag{35.48}$$

which moves with velocity $V = sV_0$ with $s^2 \ll 1$. The quantities Q_1 and Λ_1 take into account a weakening of the resonance interaction due to quantum fluctuations,

$$Q_1 = G(s)\exp W(\Theta)/4aE_0, \tag{35.49}$$

$$\Lambda_1 = -G^2(s)\exp W(\Theta)/16E_0. \tag{35.50}$$

Therefore, under a weakening of the resonance interaction, the localization region of a soliton contracts as the temperature increases. The quantity x_0, appearing in the variable z, eqn. (35.31), reflects a translational invariance of the system in question, i.e., an arbitrary position of a soliton centre in the chain, under local breakdown of the translational symmetry described by the function (35.48).

The total soliton function (35.43) differs from the plane waves (34.1) which describe stationary nonlocalized states by the fact that its amplitude in the neighborhood of a soliton centre is not negligibly small as the length of the chain approaches infinity.

One can consider formally eqn. (35.44) as the Schrödinger equation for a single quasiparticle with an effective mass $\tilde{m}$ which moves in a potential well $U(z)$. The latter is generated by the displacements of the equilibrium positions of molecules in the chain,

$$U(z) = -G(s)\Phi^2(z) = -\frac{Q_0^2 \exp W(\Theta)}{8E_0(1-s^2)^2 \operatorname{ch}^2(Q_1 z)}. \tag{35.51}$$

The energy of a quasiparticle in this well moving with a velocity V is given by the expression

$$\hbar\omega = 2E_0(1 - e^{-W(\Theta)}) + a^2(k^2 - Q_1^2)E_0/\exp W(\Theta). \tag{35.52}$$

This energy attains its minimum,

$$(\hbar\omega)_{\min} = 2E_0(1-e^{-W(\Theta)}) - G_0^2 \exp W(\Theta)/16E_0 \tag{35.53}$$

at $V = 0$. The first term in eqn. (35.53) is determined by decreasing the resonance interaction due to fluctuations of intermolecular distances. The second term characterizes a gain in energy due to the binding of a quasiparticle in the field (35.51). As the temperature increases, $W(\Theta)$ tends to increase also, so the first term increases the energy while the second one reduces it.

To evaluate the total energy $\mathcal{E}(V)$ carried by the mobile soliton one should add to eqn. (35.52) the energy involved in producing a deformation. According to eqns. (35.25) and (35.48), this energy is

$$\mathcal{E}_{\text{def}} = \frac{G_0}{2a(1-s^2)} \int \Phi^4(z)dz = Q_0^2 \exp W(\Theta)/24E_0(1-s^2)^2. \tag{35.54}$$

Hence, using eqn. (35.46) one obtains the total soliton energy

$$\begin{aligned} \mathcal{E}(\Theta, V) &= \hbar\omega + \mathcal{E}_{\text{def}} \\ &= 2E_0(1-e^{-W(\Theta)}) + \frac{1}{2} mV^2 e^{W(\Theta)} - \frac{G_0 \exp W(\Theta)}{48E_0(1-s^2)^2}. \end{aligned} \tag{35.55}$$

For small velocities $s^2 \ll 1$, this can be written as follows

$$\mathcal{E}(\Theta, V) = \mathcal{E}_{\text{sol}}(\Theta) - \frac{G_0^2 \exp W(\Theta)}{48E_0} + \frac{1}{2} M_{\text{sol}}(\Theta) V^2, \tag{35.56}$$

where

$$M_{\text{sol}}(\Theta) = m\left[1 + \frac{a^2\sigma^4}{12\hbar^2 M^2 V_0^6}\right] \exp W(\Theta), \tag{35.57}$$

is the effective mass of a slow soliton. The energy $\mathcal{E}_{\text{sol}}(\Theta)$ of an immobile soliton, relative to the bottom of the energy band of a free quasiparticle, is given by the expression

$$\mathcal{E}_{\text{sol}}(\Theta) = 2E_0(1-e^{-W(\Theta)}) - \frac{\sigma^4 \exp W(\Theta)}{48E_0 M^2 V_0^4}. \tag{35.58}$$

According to eqn. (35.26), the momentum carried by the soliton becomes

$$\mathcal{P}_{\text{sol}}(\Theta) = mV\left[1 + \frac{a^2\sigma^4}{12\hbar^2 M^2 V_0^6 (1-s^2)^2}\right] \exp W(\Theta). \tag{35.59}$$

At small velocities, this takes the form

$$\mathcal{P}_{\text{sol}}(\Theta) = M_{\text{sol}}(\Theta)V. \tag{35.60}$$

where $M_{\text{sol}}(\Theta)$ is given by eqn. (35.57).

In the study of solitons for $\Theta = 0$ within the classical theory [49,169], we only take into account long-wave displacements with the maximum wavenumber of virtual phonons participating in the formation of a soliton; this wavenumber being inversely proportional to the soliton dimension: $\sim Q^{-1}$. In this case, $W = 0$, and all the foregoing expressions are classical. At high temperatures $\mathcal{E}_0 < 2\Theta$, when $\Xi(\xi) \approx 2\Theta/\mathcal{E}_0$, and for small velocities $s^2 \ll 1$, the sums (35.14) and (35.16) can be evaluated by replacing summation by integration

$$I_{1s}(n) \approx \frac{1}{\pi}\int_0^{\pi} \frac{d\xi}{(1+2\delta_n\Theta/\pi\epsilon_n)} \approx 1-2\Theta\delta_n/\pi\epsilon_0, \tag{35.61}$$

$$I_{2s}(n) \approx 2\Theta\epsilon_0^{-1}[1-2\Theta\delta_n/\pi\epsilon_0]. \tag{35.62}$$

In this case, the energy of nonlinearity G, involved in eqn. (35.28), takes the value

$$G(\Theta) \approx G_0[1-2\Theta\delta_n/\pi\epsilon_0]. \tag{35.63}$$

This linearity decreases with increasing temperature. Within the same approximation, using eqn. (35.39) one obtains the value of the factor $W(\Theta)$

$$W(\Theta) = \frac{\Theta G_0^3[1-2\Theta\delta_n/\pi\epsilon_0]}{8\pi^2\epsilon_0^2 E_0^2}, \tag{35.64}$$

which increases with increasing temperature. The quantity δ_n, appearing in eqns. (35.63) and (35.64), is given by formula (35.11).

One should take into account that high temperatures are bounded by Θ_{cr} at which there occurs the thermal destruction of the condition of equality of the mean heat energy to the difference in the energies of a free exciton and soliton,

$$24E_0M^2V_0^4\Theta_{\text{kp}} \approx \sigma^4 \exp W(\Theta_{\text{kp}}), \quad MV_0^2 = a^2\kappa.$$

The essential peculiarity of the autolocalized states examined here is their stationary at small velocities and not so high temperatures. Due to this property, they differ from the nonstationary localized states described by wave packets consisting of nonlocalized states investigated in Section 34. The localization region of a quasiparticle in the soliton state remains unchanged under motion with any small (compared with V_0) constant velocity. The wavenumber which characterizes the phase velocity of a quasiparticle in the stationary nonlocalized state takes discrete values, as mentioned in Section 33.

36. Solutions Describing the Movement of Periodically Arranged Quasiparticles

Besides the solution (35.48) describing the motion of a single soliton, the nonlinear equation (35.44) for the envelope function (35.46) possesses periodic solutions. They characterize the motion of periodically arranged quasiparticle in the chain. Let us now study these solutions. In dimensionless variables, eqn. (35.44) takes the form

$$\left[\Lambda_1 + \hbar^2(2\tilde{m}a^2)^{-1}\frac{d^2}{d\zeta^2} + G(s)\Phi_1(\zeta)\right]\Phi_1(\zeta) = 0, \tag{36.1}$$

$$\tilde{m} = m\exp W(\Theta), \quad \zeta = (x - x_0 - Vt)/a.$$

At low temperatures $W(\Theta)$ is evaluated with the help of eqn. (35.40), and at high temperatures by means of eqn. (35.64). Equation (36.1) determines the real amplitude of the wave function

$$\varphi(x,t) = a^{-1/2}\Phi(\zeta)\exp\{i[k(x-x_0)-\omega t]\}. \tag{36.2}$$

We are searching for the solutions of eqn. (36.1) which satisfy the condition of periodicity

$$|\varphi(x,t)|^2 = |\varphi(x+aN,t)|^2. \tag{36.3}$$

Here we assume that within each period there exists a single quasiparticle, i.e., the normalization condition

$$a^{-1}\int_0^{Na} |\varphi(x,t)|^2 dx = 1, \tag{36.4}$$

is obeyed. The periodic condition for the function (36.4) results in that for the envelope

$$\Phi(\zeta + N) = \Phi(\zeta). \tag{36.5}$$

The envelope satisfies the condition of normalization per single period,

$$\int_0^N \Phi^2(\zeta)d\zeta = 1. \tag{36.6}$$

To obtain the periodic solutions of eqn. (36.1), we begin by multiplying this by $2\tilde{m}a^2\hbar^{-2}\, d\Phi/d\zeta$. Integrating further over ζ, we obtain the equation

$$(d\Phi/d\zeta)^2 = g[2\tau\Phi^2 - \Phi^4 + C], \tag{36.7}$$

where $C > 0$ is an integration constant, and q and τ the dimensionless parameters,

$$g \equiv G\exp W/2E_0, \quad \tau = -\Lambda_1/G. \tag{36.8}$$

Introducing the following quantities

$$\Phi_0^2 \equiv \tau + [\tau^2 - C]^{1/2}, \quad \Phi_1^2 \equiv \tau - [\tau^2 - C]^{1/2}, \tag{36.9}$$

extracting the root and integrating over the interval $(0, \Phi(\zeta)/\Phi_0)$, we obtain

$$\pm\zeta g^{1/2}\Phi_0 = \int_0^{\Phi(\zeta)/\Phi_0} \frac{dl}{[(1-l^2)(l^2-\lambda_1^2)]^{1/2}}. \tag{36.10}$$

Here

$$l \equiv \Phi(\zeta)/\Phi_0, \quad \lambda_1^2 = (\Phi_1/\Phi_0)^2 < 1. \tag{36.11}$$

The inversion of the integral (36.10) is expressed in terms of the even Jacobian elliptic function

$$\Phi(\zeta,\lambda) = \Phi_0 \mathrm{dn}(u,\lambda), \quad u = g^{1/2}\Phi_0\zeta. \tag{36.12}$$

The elliptic function $\mathrm{dn}(u,\lambda)$ is determined by the modulus λ taking continuous values in the interval (0,1), or by the complementary modulus $\lambda_1 = (1-\lambda)^{1/2}$. At $\zeta = 0$, the function (36.12) attains its maximum Φ_0 which is defined by the normalization condition (36.4),

$$1 - \int_0^N \Phi^2(\zeta,\lambda)d\zeta = g^{1/2}\Phi_0 \int_0^{g^{1/2}N\Phi_0} \mathrm{dn}^2(u,\lambda)du. \tag{36.13}$$

The period of $\mathrm{dn}(u,\lambda)$ relative to the variable u is equal to $2\tilde{K}(\lambda)$, where $\tilde{K}(\lambda)$ is the complete elliptic integral of the first kind (see Section 30). Hence, taking the normalization (36.13) per period into account, one can write

$$\Phi_0 g^{1/2} N = 2\tilde{K}(\lambda). \tag{36.14}$$

If eqn. (36.14) holds, it follows from eqn. (36.13) that

$$\Phi_0 = g^{1/2}/2\tilde{E}(\lambda), \tag{36.15}$$

where

$$\tilde{E}(\lambda) \equiv \int\limits_0^{K(\lambda)} \mathrm{dn}^2(u,\lambda)du \tag{36.16}$$

is the complete elliptic integral of the second kind. Using eqns. (36.14) and (35.15), one obtains

$$gN = 4K(\lambda)\tilde{E}(\lambda). \tag{36.17}$$

This formula enables us to determine the modulus of the elliptic function for a given spatial period N and nonlinearity parameter g. Evaluating the modulus λ and Φ_0 from eqn. (36.15), one finds the explicit form of the envelope

$$\Phi(\zeta,\lambda) = \left[\frac{\sqrt{g}}{2\tilde{E}(\lambda)}\right]\mathrm{dn}\left[\frac{g\zeta}{2\tilde{E}(\lambda)},\lambda\right]. \tag{36.18}$$

For nonvanishing g and large period, the complete elliptic integrals have the following asymptotes

$$\tilde{E}(\lambda) \approx 1, \quad \tilde{K}(\lambda) \approx \ln(4/\lambda_1). \tag{36.19}$$

In this case, eqn. (36.17) takes the form

$$g(N) = 4\ln(4/\lambda_1),$$

from which it follows that

$$\lambda_1 = 4^{-1}\exp(-g(N)/4) \ll 1. \tag{36.20}$$

If $\lambda_1^2 \ll 1/2$, the elliptic function (36.18) is approximated by the hyperbolic functions

$$\Phi(\zeta,\lambda) = \frac{1}{2}g^{1/2}\left[1+\frac{1}{4}\lambda_1^2(u+\mathrm{sh}u\mathrm{ch}u)\mathrm{th}u\right]\mathrm{sech}u, \quad u \equiv \frac{1}{2}g\zeta. \tag{36.21}$$

Relative to the coordinate frame ζ moving with velocity V, this function describes bell-like elevations of height $1/2\sqrt{g}$ and width $\sim 2/g$ which are separated by a distance N from each other. Given $\lambda_1 < 1/2$, each soliton occupies a region with length $N_0 = 4\ln(4/\lambda_1)/g$. If

$$gN \gg 1 \tag{36.21a}$$

it follows from eqn. (36.20) that $\lambda_1 \ll 1$. The function (36.21) then results in that of a single soliton which has been studied before for rather lengthy chains. The inequality (36.21a) gives the condition under which a chain can be considered to be rather lengthy. If $gN \approx \pi$, the complete elliptic integrals have the following asymptotes

$$\tilde{E}(\lambda) \approx \frac{\pi}{2}\left[1-\frac{1}{2}\lambda^2-\frac{3}{64}\lambda^4\right] \tag{36.22}$$

$$\tilde{K}(\lambda) \approx \frac{\pi}{2}\left[1+\frac{1}{4}\lambda^2+\frac{9}{64}\lambda^4\right].$$

Hence,

$$gN \approx \pi\left[1+\frac{\lambda^4}{32}\right], \quad \lambda^2 \ll 1. \tag{36.23}$$

In this case, the Jacobian elliptic function is approximated by trigonometric functions, and the envelope (36.18) takes the following form

$$\Phi(\zeta,\lambda) = \frac{\sqrt{g}}{\pi}\left[1-\frac{1}{2}\lambda^2\sin^2\left[\frac{g\zeta}{\pi},\lambda\right]\right], \quad \lambda^2 \ll 1. \tag{36.24}$$

Therefore, the total wave function (36.2) becomes a modulated plane wave. The modulation depth is determined by λ^2, and the modulation period accounts for $4\pi/g$.

According to eqn. (36.8), g is given by the formula

$$g = G_0 \exp W(\Theta)/2E_0(1-s^2). \tag{36.25}$$

Each quasiparticle occupies a region of chain having a length π/g. All these particles move as a unique whole with velocity V. As the temperature and velocity (in limits of $s^2 \ll 1$) increase, g does too. In the limiting case $\lambda \ll 1$,

$$N = \pi/g = 2\pi E_0(1-s^2)/G_0 \exp W(\Theta).$$

We now evaluate the energy and momentum carried within each period by the quasiparticle in the state described by the function (36.2) with the envelope (36.18). Bearing in mind eqns. (36.8) - (36.11), one obtains the value of Λ_1 which in agreement with (35.47), determines the energy $\hbar\omega$ of the quasiparticle,

$$\Lambda_1 = -\tau G = -G\exp W(\theta)(1+\lambda_1^2)/16E_0\tilde{E}^2(\lambda). \tag{36.26}$$

If $N \to \infty$ and $\lambda_1 \to 0$, when $\tilde{E}^2(\lambda) \to 1$, Λ_1 coincides with that of a single soliton given by eqn. (35.50). To evaluate the total energy, one should add the deformation energy,

$$\mathcal{E}_{\rm def} = \frac{1}{2}G\int_0^N \Phi^4(\zeta)d\zeta = \frac{G^2\exp W(\theta)T(\lambda)}{16E_0\tilde{E}^3(\lambda)} \tag{36.27}$$

to $\hbar\omega$. In eqn. (36.27), we introduce the following function

$$T(\lambda) \equiv \int_0^{K(\lambda)} {\rm dn}^4(u,\lambda){\rm dn} = \frac{1}{3}[2(2-\lambda^2)\tilde{E}(k)-(1-\lambda^2)\tilde{K}(\lambda)] \tag{36.28}$$

which possesses the asymptotes

$$T(\lambda) = \begin{cases} \frac{1}{3}[2-\lambda_1^2\ln(4/\lambda_1),] \ \lambda^2 \ll 1, \\ \frac{\pi}{2}[1-3\lambda^2/4], \ \lambda^2 \ll 1. \end{cases}$$

So, for each period, the energy carried by the excitation moving with velocity V and characterized by the complementary modulus λ_1 becomes

$$\mathcal{E}(V) = \frac{\hbar^2}{ma^2}[1-\exp W(\theta)]+\frac{1}{2}mV^2\exp W(\Theta)- \tag{36.29}$$

$$-\frac{G^2\exp W(\Theta)[1+\lambda_1-T(\lambda)/\tilde{E}(\lambda)]}{16E_0\tilde{E}^2(\lambda)}.$$

This excitation carries the following momentum

$$\mathcal{P} = mV\exp W(\Theta)\left[1+\frac{a^2\sigma^2T(\lambda)}{8M^2V_0^3\tilde{E}^3(1-s^2)^2}\right]. \tag{36.30}$$

37. Strong Quasiparticle Autolocalization

To describe the stationary states of a quasiparticle, such as an exciton or an excess electron, in molecular systems, we usually apply the continuum approximation and the method of effective mass of a free quasiparticle. The effective mass methods takes into account the periodic field of molecules arranged rigidly at the nodes of a quasi-one-dimensional lattice. The local displacement of molecules from the lattice nodes causes the additional, or deformational interaction (σ) of a quasiparticle with the lattice. The aforementioned approximation is justified if the inequality

$$2g = \sigma^2 / E_0 \kappa d^2 < 2\pi \tag{37.1}$$

holds. In this case, the linear dimension of the soliton, proportional to d/g (see Section 26) exceeds the intrapeptide separation d.

However, this inequality loses its validity for some polypeptide chains. In this case, quasiparticle localization in the region, comparable with the distance d between neighboring monomers (peptide groups), is quite possible. A study of these states with an account of zeroth vibrations and heat motion of monomers requires a complete quantum description which is given in this Section.

The total energy functional (33.24) introduced before and given as the function W_n, determined by eqn. (33.26), enables the investigation of strong localized states, too. However, in this case on should take into account that the inequality (33.27) becomes invalid for strong quasiparticle autolocalization.

For strong autolocalization, the effective mass of a 'dressed' quasiparticle is large, while its group velocity is rather small. For this reason, within the first-order treatment one can consider the particle to be immobile. According to eqn. (35.1), the state of an immobile autolocalized quasiparticle with energy $\hbar\omega_0$ in a molecular chain is described by the following wave function

$$\varphi_n(t) = \Phi_n \exp(-i\omega_0 t) \tag{37.2}$$

with the real amplitude Φ_n. Hence, the functional (33.24) depends on the function Φ_n only.

Let us assume that autolocalization occurs in the neighborhood of the node $n = 0$. The functions Φ_n, which determine the amplitudes of the probability distributions of a quasiparticle on monomers, are rapidly decreasing functions of $|n|$

and obey the condition $\Phi_n = \Phi_{-n}$. Neglecting the energy of real phonons, the functional (33.24) takes the form

$$\mathcal{H}(\Phi,\beta) = E_0\left\{2-\sum_n \Phi_n\Phi_{n+1}(e^{-W_n}+e^{-W_{n+1}})\right\}+ \tag{37.3}$$

$$+\sum_{q,n}\left\{\epsilon(q)\,|\beta_{qn}|^2-\frac{1}{\sqrt{N}}F(q)(\beta_{qn}+\beta^*_{-q,n})\right\}\Phi^2_n.$$

If a given chain involves a large number, $N+1$, of molecules, summation over n is carried out over all integers $n = 0,\pm 1,+...,\pm N/2$, and that over q is carried out over N discrete values of wavenumber

$$q = 2\pi j/Nd,\ \ j = \pm 1,\pm ...\pm N/2.$$

Since the function β_{qn} is proportional to Φ^2_n in the states of strongly localized particles, this decreases with increasing $|n|$ faster than Φ_n. So the expression (33.26) can be replaced by the following approximate one

$$W_n \approx \frac{1}{2}\sum_q \eta_q(|\beta_n|^2+|\beta_{n-1}|^2), \tag{37.4}$$

where the abbreviated notation,

$$\eta_a \equiv 2\langle\nu_q\rangle+1 = \mathrm{ctgh}\left[\frac{\epsilon(q)}{\Theta}\right], \tag{37.4a}$$

is used.

The energy band $\epsilon(q)$ of free longitudinal acoustic phonons in the Hamiltonian (37.2) is defined as follows

$$\epsilon(q) = |\xi|\epsilon_0,\ \xi = qd,\ \ -\frac{\pi}{2}\leqslant\xi\leqslant\frac{\pi}{2}. \tag{37.5}$$

The quantity

$$\epsilon_0 \equiv \hbar(\kappa/M)^{1/2} = \hbar V_0/d \tag{37.6}$$

characterizes the width of the energy band, V_0 is the velocity of acoustic phonons, and M is the mass of a monomer. The interaction of a quasiparticle with acoustic phonons is described by the formula (33.10). In the limiting case of the excitation localization at a single node when there is nonvanishing $\Phi_n = \delta_{no}$ only for the functions Φ_n, the sum of the energies of the quasiparticle and lattice deformation is determined by minimizing the functional

$$\mathcal{H}(\beta) = 2E_0+\sum_q[\epsilon(q)|\beta_{q0}|^2-N^{-1/2}F(q)(\beta_{q0}+\beta^*_{-q,0})]. \tag{37.7}$$

Minimizing the functional (37.7), one finds

$$\beta^0_{q0} = N^{-1/2}F^*(q)/|\xi|\epsilon_0.$$

Substituting this value into eqn. (37.7), one obtains the total energy

$$\mathcal{E}\equiv\mathcal{H}(\beta^0) = 2E_0\left\{1-\frac{1}{2}g\right\}, \tag{37.8}$$

where the dimensionless parameter g is

$$g = \frac{\sigma^2}{2E_0\kappa d^2} = \frac{\sigma^2}{2E_0MV_0^2}. \tag{37.9}$$

Expression (37.8) characterizes the energy of a quasiparticle at the node $n = 0$ in the deformation field regardless of the resonance (exchange) interaction with neighboring monomers. This energy is just that obtained by the author (see [32], Ch. V, §4) via the exact diagonalization of the Hamiltonian of a system with $E_0 = 0$. From eqn. (37.8), it follows that a quasiparticle localization on a single molecule takes place only if the inequality

$$g>2 \tag{37.10}$$

holds ($E_0\neq0$).

In the general case, when the resonance interaction $E_0 \neq 0$ is particularly taken into account, there occurs a partial autolocalization of a quasiparticle. In this case, one can minimize the total functional (37.5) by using the normalized trial function

$$\Phi_n = (\text{th}\lambda)^{1/2}\exp(-\lambda|n|),\ \ \lambda>0, \tag{37.11}$$

which decays exponentially with increasing $|n|$ and which corresponds to the excitation localization in the region of the node $n = 0$. With an account of eqn. (37.3) and fixed λ, the functional (37.2) is minimized at the value

$$\beta_{qn} = F^*(q)\Phi_n^2/N^{1/2}\epsilon_0[|\xi|+\pi B_n], \tag{37.12}$$

where

$$B_n = (2\pi)^{-1}(1+\delta_{n0})\gamma_0\eta_q e^{-W_n}\Phi_n^2\operatorname{sech}\lambda, \tag{37.13}$$

$$\gamma_0 \equiv E_0/\epsilon_0. \tag{37.14}$$

Substituting eqn. (37.12) into the expression (37.2) and taking into account the approximate formula

$$\sum_n \Phi_n\Phi_{n+1}(e^{-W_n}+e^{-W_{n+1}}) \approx 2\sum_n \Phi_n^2\exp(-\lambda-W_n),$$

where

$$W_n \approx \frac{1}{2}\eta_q|\beta_{qn}|^2, \tag{37.15}$$

one finds the functional containing a single unknown quantity

$$H(\lambda) = 2E_0\left\{1-\sum_n\Phi_n^2 e^{-W_n}\exp(-\lambda)-g\sum_n\Phi_n^4\left[D_1(n)-\frac{1}{2}D_2(n)\right]\right\}. \tag{37.16}$$

In eqn. (37.16), the following notations

$$D_1(n) \equiv N^{-1}\sum_q|\xi|(|\xi|+\pi B_n)^{-1}, \tag{37.17}$$

$$D_2(\eta) \equiv N^{-1}\sum_q\xi^2(\xi|+\pi B_n)^{-2}, \tag{37.18}$$

are used.

By virtue of the formulas (37.12) and (37.18), one can derive the equation which determines the function

$$W_n = \frac{1}{8}\mu\Phi_n^4 D_3(n), \tag{37.19}$$

where

$$D_3(n) \equiv N^{-1}\sum_q\eta_q|\xi|(|\xi|+\pi B_n)^{-2}. \tag{37.20}$$

Equation (37.19) involves the dimensionless parameter

$$\mu \equiv 4g\gamma_0 = \beta_{\max}^2/\langle z^2\rangle. \tag{37.21}$$

which characterizes the contribution of zeroth vibrations since this is expressed in terms of the ratio of the square of the maximum displacement of molecules of a chain from the equilibrium positions, $\beta_{\max}^2 \equiv \sigma^2/\kappa^2 d^2$, to the average value of the square of the zeroth vibrations $<z^2> \equiv \hbar/2\sqrt{\kappa M}$; the maximum displacement being conditioned by the interaction of virtual phonons with a quasiparticle.

In the limiting cases of low and high temperatures when the approximate formulas,

$$\eta_q = \begin{cases} \eta_0 \equiv 1+2\exp(-\pi\epsilon_0/\Theta), & \Theta < \epsilon_0\pi; \\ \eta_\xi \equiv 2\Theta/\epsilon_0|\xi|, & \Theta > \pi\epsilon_0, \end{cases} \tag{37.22}$$

can be applied, the aforementioned sums can be evaluated by conversion to integration over the q-space via the relation

$$N^{-1}\sum_q f(q) = \pi^{-1}\int_0^\pi f(\xi)d\xi, \quad \xi = qd.$$

Using the relation, in the case of low temperatures one has

$$D_1(n) = 1-B_n\ln(1+B_n^{-1}), \tag{37.23}$$

$$D_2(n) = 1+(1+B_n^{-1})-2B_n\ln(1+B_n^{-1}), \tag{37.23a}$$

$$D_3(n) = \pi^{-1}\eta_0[\ln(1+B_n^{-1})-(1+B_n)^{-1}]. \tag{37.23b}$$

At high temperatures, these functions take the values

$$D_1(n) = 1-A_n\mathrm{arctg}(A_n^{-1}), \tag{37.24}$$

$$D_2(n) = 1+\frac{1}{2}\left[\frac{A_n}{1+A_n}\right]^2-\frac{3}{2}A_n\mathrm{arctg}(A_n^{-1}), \tag{37.24a}$$

$$D_3(n) = 2\Theta(\pi^2\epsilon_0)^{-1}[(2A_n)^{-1}\mathrm{arctg}(A_n^{-1})+(1+A_n^2)^{-1}] \tag{37.24b}$$

$$A_n^2 \equiv \pi^{-1}|\xi|B_n = 2(1+\delta_{n0})\gamma_0\Theta(\pi^2\epsilon_0)^{-1}e^{-W_n}\Phi_n^2\exp(-\lambda). \tag{37.25}$$

These formulas are analyzed in the next Subsections separately, for chains having narrow ($E_0 \ll \epsilon_0$) and broad ($E_0 > \epsilon_0$) energy bands of quasiparticles.

37.1. Chains with Narrow Energy Bands

For chains having narrow energy bands of quasiparticles we consider those for which the inequalities

$$E \ll \epsilon_0, \text{ and } \gamma_0 \equiv E_0/\epsilon_0 \ll 1 \tag{37.26}$$

are satisfied. Let us introduce the following notations: $\omega_p = E_0/\hbar$ for the maximum frequency of a free quasiparticle in its energy band; and $\omega_{ph} = \epsilon_0/\hbar$ for the maximum velocity of acoustic phonons. The inequality (37.26) can then be rewritten as follows

$$\omega_p \ll \omega_{ph}. \tag{37.27}$$

Therefore, in this case the phonon system becomes faster compared with that of quasiparticles. For this reason, quantum fluctuations, or the zeroth vibrations of molecules relative to their equilibrium positions, play an important role for excited states, even at absolute zero. Now we study low and high temperatures separately.

Low temperatures, $\Theta < \pi\epsilon_0$

If the inequalities (37.26) are fulfilled, the quantity B_0 defined by eqn. (37.13) is considerably less than unity,

$$B_0 = \pi^{-1}\gamma_0 e^{-W_0}\,\mathrm{sech}\lambda \ll 1. \tag{37.28}$$

Hence, eqn. (37.19) which determine the function $W_0(\Theta)$, transforms into the following

$$W_0(\Theta) = \frac{\mu\eta}{8\pi}\ln B_0\,\mathrm{th}^2\lambda. \tag{37.29}$$

By virtue of eqn. (37.28), one obtains

$$W_0(\Theta) = -L\ln[\pi^{-1}\gamma_0\eta_0\,\mathrm{sech}\lambda], \tag{37.30}$$

where

$$L \equiv L(\Theta) = \eta_0\mu\,\mathrm{th}^2\lambda/[8\pi - \eta_0\mu\,\mathrm{th}^2\lambda]. \tag{37.31}$$

In the case of strong autolocalized $\mathrm{th}^2\lambda \approx 1$, one can replace eqn. (37.31) by the simpler equation,

$$L = \mu\eta_0(8\mu - \mu\eta_0)^{-1}. \tag{37.32}$$

The function L increases weakly with growth of temperature.

It follows from eqn. (37.30) that the Debye-Waller factor is given by the formula

$$\exp(-W_0) = (\pi^{-1}\gamma_0\eta_0\,\mathrm{sech}\lambda)^{L}. \tag{37.33}$$

Positivity of W_0 and L imposes the following constraint upon λ:

$$1 \approx \mathrm{th}^2\lambda < 8\pi/\mu\eta_0. \tag{37.34}$$

Therefore, the localization ($\lambda \ll 1$) of a quasiparticle in chains with narrow energy bands takes place if the inequality

$$\mu < 8\pi/\eta_0, \tag{37.35}$$

holds, i.e., if the zeroth vibrations are larger than the displacements. Taking eqn. (37.21) into account, from inequality (37.35) there follows a constraint on the upper bound of the dimensionless nonlinearity parameter

$$g < \pi\epsilon_0/E_0\eta_0 \equiv \pi/\gamma_0\eta_0. \tag{37.36}$$

For complete localization, the lower bound of g is given by the inequality (37.10) at the zeroth temperature. If the inequality $\mu \ll 8\pi$ holds, the function (37.32) takes the simplest form,

$$L = \mu\eta_0/8\pi. \tag{37.37}$$

The value of the localization parameter λ is determined by minimizing the functional (37.16) with $W_n \approx W_0$ and taking account of eqn. (37.12). If $B_n \ll 1$, the functions $D_1(n)$ and $D_2(n)$ are equal to unity, and the functional (37.16) becomes

$$\mathcal{H}(\lambda) = 2E_0\left\{1-(\pi^{-1}\gamma_0\eta_0)^L(\mathrm{sech}\lambda)^{L+1}-\frac{1}{4}g\,\mathrm{th}^2\lambda(1+\mathrm{th}^2\lambda)\right\}. \tag{37.38}$$

Differentiating eqn. (37.38) relative to $x = \mathrm{th}\lambda$, one obtains

$$(1+3x^2)(1-x^2)^{1/2(1-L)} = 4xJ, \tag{37.39}$$

where

$$J = g^{-1}(1+L)(\pi^{-1}\gamma_0\eta_0)^L. \tag{37.40}$$

If $\mu < 8\pi$, or $g < \pi/2\gamma$, eqn. (37.39) has a solution within the interval $0 \leqslant x \leqslant 1$. The complete localization corresponds to $x = 1$ which refers to $J \approx 0$, i.e., the large g and small γ_0. A localization becomes incomplete at small and nonvanishing J. To obtain the appropriate solution x_0, one can substitute $x = 1-\delta$, which slightly differs from unity, into eqn. (37.39). Within first-order terms relative to δ, from eqn. (37.39) one has

$$\delta \approx 1+\frac{1}{2}J^{2/(1-L)}.$$

Therefore, the minimizing value λ_0 of the localization parameter is

$$\lambda_0 = \operatorname{arcth}(1-\delta) \approx \ln 2+(1-L)^{-1}\ln\left\{\frac{g}{1+L}\left[\frac{\pi}{\gamma_0\eta_0}\right]^2\right\}. \tag{37.41}$$

In the case of rather large zeroth vibrations and $\mu \ll 1$, and $\gamma_0 \ll 1$, this expression is simplified

$$\lambda_0 \approx \ln\{2g(\pi\epsilon_0 / E_0\eta_0)^L\}. \tag{37.42}$$

In the limit $\mu \to 0$ (L tends to zero, too), λ_0 approaches its limiting value

$$\lambda_0 \to \tilde{\lambda}_0 \equiv \ln(2g) \approx \ln(\sigma^2 / E_0 M V^2).$$

If the function L is given by eqn. (37.37), the Debye-Waller factor takes the following value

$$\exp(-W_0) \approx (\gamma_0\eta_0 / \pi g)^{\mu\eta_0/8\pi}.$$

Substituting (37.42) into eqn. (37.38) and taking into account that strong localization is realized if $\operatorname{th}\lambda_0 \approx 1$ and $\operatorname{sech}\lambda_0 = 1/g$, one obtains the energy of a dressed quasiparticle

$$\mathcal{E} = H(\lambda_0) = 2E_0 - gE_0 - 2\pi\epsilon_0\eta_0^{-1}(\gamma_0\eta_0 / \pi g)^{1+L}. \tag{37.43}$$

From the expression (37.43), it follows that the strong autolocalization becomes profitable ($\epsilon<0$) if $g>2$. Then the third term in eqn. (37.43) is immaterial. It weakly depends on temperature through the exponent $L \approx \mu\eta_0 / 8\pi$.

The envelope of the wave function (37.12) decreases via a power law with increasing $|n|$ and has the form

$$\Phi_n = \Phi_0\{2g(\pi / \gamma_0\eta_0)^L\}^{-|n|}, \tag{37.44}$$

where

$$\Phi_0 = \frac{(\pi\epsilon_0 / \gamma_0\eta_0)^{2L} - (2g)^{-2}}{(\pi\epsilon_0 / \gamma_0\eta_0)^{2L} + (2g)^{-2}}.$$

Due to (37.42) and (37.43), it follows that as the temperature increases in the range of $\Theta < \pi\epsilon_0$, the localization of a quasiparticle becomes stronger, and its energy reduces.

It is useful to compare the results given for strongly localized states, in which the discreteness of the molecular chain becomes important, with those for weak localization, when the continuum approach is justified. In the latter case $g<2$. Let us consider for simplicity, the case of lower temperatures and $L \approx 0$. Then the functional (37.38) is as follows

$$H_0(\lambda) = 2E_0\left\{1-\mathrm{sech}\lambda-\frac{1}{4}g\,\mathrm{th}\lambda\right\}. \tag{37.45}$$

Minimizing this, one obtains

$$\left(\mathrm{sh}\lambda-\frac{1}{4}g\right)\mathrm{sech}^2\lambda = 0.$$

Hence, the minimizing value λ is given by the expression

$$\mathrm{sh}\lambda = \frac{1}{4}g<1 \ \text{ and } \ \lambda_0 \approx \frac{1}{4}g. \tag{37.46}$$

Substituting this into eqn. (37.45), one finds the energy

$$\mathcal{E} = H_0(\lambda_0) = 2E_0\{1-(1+g^2/16)^{1/2}\}. \tag{37.47}$$

For a weak localization, $g^2 \ll 1$, one has

$$\mathcal{E} \approx -E_0 g^2/16. \tag{37.48}$$

Inserting (37.46) into the expression (37.11), one determines the envelope of the wave function in this state,

$$\Phi_n \approx \frac{1}{2}\sqrt{g}\exp(-|n|g/4). \tag{37.49}$$

The latter expressions are just those given by (11.4) and (11.5) and obtained in Chapter 3 within the continuum approach for the rest solitons. So, in the case of a weak quasiparticle localization ($g<2$), the energy gain under localization is proportional to the square of the parameter g. In the opposite case of strong autolocalization, a linear dependence on g dominates.

Let us consider now a crystal having narrow energy bands of quasiparticles, $\gamma_0 \ll 1$, at high temperatures. We suggest that at $\Theta > \pi\epsilon_0$ the inequality

$$\Xi \equiv \epsilon_0^{-1}\gamma_0\Theta < 1 \tag{37.50}$$

is also fulfilled. We also expand the function (37.24b) in powers of this small parameter and substitute this into eqn. (37.19). Within the lowest-order treatment relative to Ξ, one obtains the equation for the factor W_0

$$W_0 \exp\left[-\frac{1}{2}W_0\right] = G \tag{37.51}$$

where

$$G \equiv \frac{1}{4}g\,\mathrm{th}^2\lambda\left[\frac{1}{2}\Xi\,\mathrm{ch}\lambda\right]^{1/2}. \tag{37.52}$$

If particle localization is strengthened, the function G increases. However, the left-hand side of eqn. (37.51) has a maximum at $W = 2$ which is equal to $2\exp(-1) \approx 0.74$. Hence, this equation possesses only solutions for $G < G_0 = 0.74$. The greatest localization then corresponds to $W_0 \approx 2$. Evaluating the sums D_1 and D_2 by means of eqns. (37.24) and (37.24a) at $W \approx 2$ within the terms proportional to $\Xi^{1/2}$ and substituting the values obtained into the functional (37.16), one finds its approximate magnitude

$$\mathcal{H}(\lambda) = 2E_0\{1-(2,7)^{-2}\mathrm{sech}\lambda - \frac{1}{2}g\,\mathrm{th}^2\lambda + (126)^{-1}g\,\mathrm{th}^2\lambda \times \tag{37.53}$$

$$\times\, (2\Xi\,\mathrm{sech}\lambda)^{1/2}\}.$$

Assuming $\mathrm{th}^2\lambda \approx 1$ and differentiating with respect to $x = (\mathrm{sech}\lambda)^{1/2}$, one determines the minimizing value

$$x_0^2 = \mathrm{sech}\lambda_0 = \frac{1}{16}g^2\Xi. \tag{37.54}$$

Substituting the latter quantity into eqns. (37.53) and (37.11), one obtains the energy of the system,

$$\mathcal{E}(\Xi) \approx 2E_0\left[1-\frac{1}{2}g+\frac{1}{128}g^2\Xi\right] \tag{37.55}$$

and the amplitude of the excitation probability distribution,

$$\Phi_n = \left[1-\left(\frac{1}{16}\Xi g^2\right)^2\right]^{1/4}\left(\frac{1}{8}\Xi g\right)^{-|n|}. \tag{37.56}$$

Therefore, if an excitation is localized ($\lambda > 1$), the dimensionless nonlinearity parameter g must satisfy the inequality

$$g < 2(3/\Xi)^{1/2} = 2(3\epsilon_0/\gamma_0\Theta)^{1/2}. \tag{37.57}$$

The exponent W_0 in the Debye-Waller factor is close to two. One obtains its temperature dependence by expanding the left-hand side of eqn. (37.51) near the maximum. Substituting further the expressions (37.54) and (37.52), one obtains

$$W_0 \approx 2\left[1+\frac{\sqrt{2}}{16}g^2\Xi\right]. \tag{37.58}$$

So, at a temperature $\Theta > \pi\epsilon_0$, $\gamma\Theta/\epsilon_0 < 1$, there occurs an increase in excitation energy and its spatial extension over a chain. We recall that if $\Theta < \pi\epsilon_0$, a temperature increase results in a lowering of the quasiparticle energy and an increase in its localization.

The aforementioned localization conditioned (37.57) at the temperature $\Theta > \pi\epsilon_0$ is profitable only at temperatures for which the mean heat energy of phonons is smaller than the energy of the binding of a quasiparticle with a local deformation, i.e., when the inequalities

$$\mathcal{E}(\Xi) < 0 \quad \text{and} \quad |(\Xi)\mathcal{E}| > \frac{1}{2}\Theta \tag{37.59}$$

and satisfied. A simultaneous fulfillment of the conditions (37.57) and (37.59) imposes an essential constraint on the possibility of quasiparticle localization at high temperatures.

37.2. Chains with Broad Energy Bands

Chains for which the inequality

$$E_0 \gg \epsilon_0, \quad \text{i.e.,} \quad \gamma_0 \gg 1, \tag{37.60}$$

is satisfied are referred to as chains with broad energy bands for quasiparticles. In such chains, quasiparticles play the role of a fast subsystem, while the local displacements of the equilibrium positions of molecules play the role of a slow subsystem. In states having a strong localization of quasiparticles, which are of present interest, the translational motion, accompanied by a local deformation, is characterized by a considerably large effective mass. This is a reason to consider the strongly localized quasiparticle as being immobile, regardless of the large resonance interaction between molecules by virtue of the condition (37.60). In these localized states, the quasiparticles are described by the functional (37.2) involving the amplitude function Φ_n. The latter characterizes the envelope of the total wave function (37.1) describing fast vibrations of a quasiparticle with frequencies $E_0/\hbar$ within the potential well generated by slow inertial displacements of molecules in the deformation region.

Together with the inequality (37.60), we also assume the validity of the following equation:

$$\mu = 4g\gamma_0 \gg 1. \tag{37.61}$$

This inequality indicates the insignificant contribution of zeroth vibrations to the state in question. In this case, B_0 becomes rather large and is determined by eqn. (37.13):

$$B_0 = \pi^{-1}\gamma_0\eta_0 e^{-W}\text{sech}\lambda \gg 1. \tag{37.62}$$

Hence, at the temperatures defined by the relation

$$\eta_0 = 1+2\exp(-\pi\epsilon_0/\Theta), \quad \Theta < \pi\epsilon_0,$$

one can approximate the function (37.23b) as follows:

$$D_3(0) = \pi^{-1}\eta_0 B^{-2}(0). \tag{37.63}$$

So, eqn. (37.19), which determines the function W_0, now takes the form

$$W_0\exp(-2W_0) = (2\eta_0\gamma_0)^{-1}\pi g\,\text{sh}^2\lambda. \tag{37.64}$$

This equation possesses a solution if its right-hand side does not exceed the maximum value of the left-hand side which occurs at $W_0 = 1/2$.

Evaluating the energy functional (37.16), within the terms proportional to $1/\gamma$, one obtains

$$\mathcal{H}(\lambda) = 2E_0\{1-(\exp 1)^{-1/2}\text{sech}\lambda - 0{,}8g(\eta_0\gamma_0)^{-1}\text{th}^2\lambda\,\text{ch}\lambda\}. \tag{37.65}$$

For a strong localization, $\text{th}^2\lambda \approx 1$. Within the approach accepted, differentiating $\mathcal{H}(\lambda)$, one obtains the minimizing value λ_0,

$$\text{ch}\lambda_0 = (0{,}7\gamma_0\eta_0/\pi g)^{1/2}. \tag{37.66}$$

In this state, the energy of the system and the wave-function envelope take the form, respectively,

$$\mathcal{E}_0 = \mathcal{H}(\lambda_0) = 2E_0[1-(2\Xi)^{1/2}], \tag{37.67}$$

where

$$\Xi = \pi g/\gamma_0\eta_0, \tag{37.67a}$$

$$\Phi_n = (1-1{,}4\Xi)^{1/4}(\Xi)^{-1/2|n|}. \tag{37.68}$$

Therefore, the autolocalized state of the given type is profitable energy-wise if the energy (37.67) is lower than the bottom of the energy band of a free quasiparticle which is determined by the zeroth value. This condition is equivalent to the requirement of validity of the following inequality

$$\Xi > 1/2. \tag{37.69}$$

The second necessary condition of stability of the localized excitation requires that the decrease of energy exceeds the mean heat energy of vibrations of molecules in the chain, i.e., fulfillment of the inequality

$$(2\Xi)^{1/2} - 1 > \pi/2E_0. \tag{37.70}$$

In order to evaluate W_0, one can approximate the left-hand side of eqn. (37.64) by a parabola near $W_0 = 1/2$. Substituting λ_0 given by eqn. (37.66) into this, one obtains

$$W_0 \approx \frac{1}{2} - (0.6\pi g/\gamma_0\eta_0)^{1/2}. \tag{37.71}$$

The necessity of satisfying eqns. (37.69) and (37.70) imposes an essential constraint on the possibility of autolocalization of a quasiparticle in a chain having broad energy bands.

38. Radiation Excitation of Optical Solitons in Crystalline Acetanilide

Crystals of acetanilide $(CH_3CONHC_6H_5)_n$, or, for short, *ACN*, are generated by lengthy chains of molecules consisting of repeating plane peptide groups which are held over the chains by hydrogen bonding. These chains are held together transversely by van der Waals forces. This compound was utilized as an analgesic in the nineteenth century. Present interest in it is due to its remarkable similarity to the chain structure of peptide groups in α-helical proteins.

In 1973, Careri, from the University of Rome, demonstrated [148] that as the temperature is lowered, there arises a novel line in the infrared absorption spectrum of crystalline *ACN* in the region of $1650\, cm^{-1}$, which is shifted relative to the maximum of the main Amide-I band ($1665\, cm^{-1}$) by $15\, cm^{-1}$ (see Fig. 15).

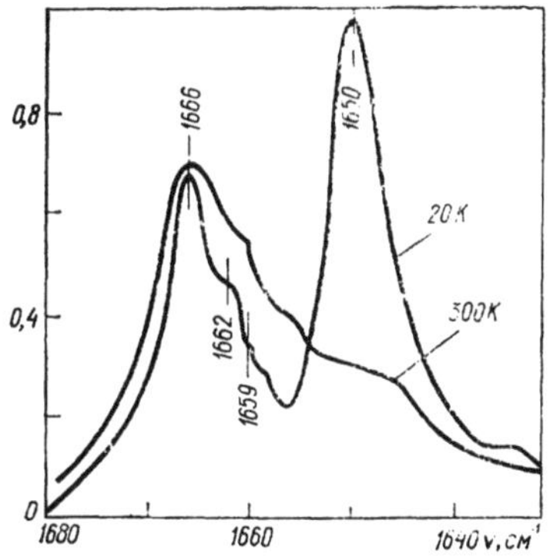

Figure 15. Absorption infrared band of an *ACN* crystal. The bands with maxima 1666, 1662, and 1659 cm^{-1} correspond to the formation of excitons (Davydov splitting), and the 1650 cm^{-1} band corresponds to the formation of an optical soliton.

The main absorption band has the three peaks, $1666\, cm^{-1}$, $1662\, cm^{-1}$, and $1659\, cm^{-1}$, at low temperatures. This structure is due to the splitting of the exciton states (Davydov splitting) conditioned by the resonance dipole-dipole interaction of Amide-I vibrations of eight peptide groups involved in the unit cell of the *ACN* crystal. This splitting is well understood on the basis of the theory of excitons developed by Davydov [32].

The intensity of this new absorption band having a width of about $5\, cm^{-1}$ is low at room temperature, but increases considerably as the temperature is lowered (see Fig. 15). This new absorption band is absent in amorphous samples. The appearance of this band upon cooling the crystalline *ACN* cannot be explained on the basis of traditional molecular spectroscopy (crystal defects, Fermi resonance, etc.). This is a reason to refer to it as the *anomalous Amide-I absorption band.*

In 1983, Careri et al. [149] suggested that the anomalous line of Amide-I vibrations in crystalline *ACN* is due to a soliton-like excitation which is similar to Davydov's solitons in α-helical proteins resulting from the coupling of the

Amide-I vibrations to out-of-plane acoustic displacements of peptide groups. The authors [149-150] suggested that the appearance of an anomalous absorption band in ACN crystals is due to the coupling of an $N-H$ bond inside the peptide groups. Such solitons are localized over a small number of peptide groups and are really immobile.

As mentioned in Section 5, Davydov's solitons in α-helical proteins cannot be excited directly by electromagnetic radiation since the heavier peptide groups move too slowly about their equilibrium positions. The new type of solitons, corresponding to the anomalous absorption band, results from the coupling of Amide-I vibrations to comparatively high-frequency intrapeptide optical vibrations of $N-H$ bonds ($\sim 700-1000\,cm^{-1}$). For this reason, these solitons can be directly excited by radiation. To emphasize this fact, we describe these as *optical solitons*.

In order to describe the solitons arising from the absorption of radiation by crystalline ACN, the authors modified Davydov's theory and studied a simple model which takes into account the interaction between neighboring peptide groups belonging to the single chain. In 1985, the authors of the paper [259] attempted to involve the additional bonds between neighboring unit cells of the ACN crystal. It was theoretically suggested that solitons are localized on few peptide groups and are, in fact, immobile. In the next Subsections we present the simplest theory of optical solitons by considering a single chain of peptide groups and their generation by infrared radiation.

38.1. Optical Solitons in Crystalline Acetanilide

As the simplest model of the ACN crystal, we treat a single chain of repeating peptide groups coupled together by hydrogen bonds. We assume that each peptide group of this chain is characterized by the following parameters: $\mathcal{E}_0 = 1665\,cm^{-1}$ is the energy of the Amide-I vibration; $J = 3-5.8\,cm^{-1}$ is the energy of dipole-dipole interaction; $d = 4.9$ Å is the intrapeptide distance; $\Omega \approx 700-1000\,cm^{-1}$ is the frequency of the deformation vibration of the $N-H$ bonds which is taken only into account; $q_D = 6.8\,cm^{-1}$ is the maximum wavenumber of the optical mode; $m \sim 1.67 \times 10^{-24} g$ is the reduced mass of an optical vibration; $\chi = (2-5) \times 10^{-10}$ newton is the parameter of coupling of the Amide-I vibration to the optical mode. Due to the small value of J compared with $\mathcal{E}_0$ and Ω, one cannot take into account a displacement of excitation over the chain of peptide groups to a first-order treatment. We also assume that the excitation is localized near a single, e.g. nth peptide group. Putting $J \approx 0$, the energy operator of the chain of N peptide groups can be written in the following form

$$H = B_n^+ B_n(\epsilon_0 + \chi u_0) + \hbar \sum_q \Omega b_q^+ b_q, \tag{38.1}$$

where B_n^+ and B_n are Bose operators of the presence and absence of excitation on the peptide groups, respectively, and u_n is a displacement of a proton of mass m in the $N-H$ bond in the presence of an Amide-I vibration. Taking heat motion into account, one can express this displacement in terms of creation, b_q^+, and annihilation, b_q, operators of a phonon of the optical mode with wavenumber q by using the relation

$$u_n = \beta N^{-1/2} \sum_q (b_q + b_{-q}^+). \tag{38.2}$$

Here

$$\beta \equiv (\hbar / 2m\Omega)^{1/2} \tag{38.3}$$

is the amplitude of the zeroth vibration with frequency Ω of a particle with mass m. Using the cyclic boundary conditions, one can carry out the summation over N discrete and equidistant values of wavenumber q defined by the following relation

$$q = \pi j / Nd, \quad j = 0 \pm 1, \quad \pm 2, \quad \pm ... \pm N/2. \tag{38.4}$$

In eqn. (38.1), the parameter χ characterizes a change of the energy of the Amide-I vibration as a result of a proton displacement.

The energy operator (38.1) admits a complete diagonalization under the transformation from the operators B_n and b_q into the new operators, A_n and a_q, by the use of the unitary transformation

$$B_n = S A_n^+ S^+, \quad b_q = S a_q S^+,$$

where the unitary operator S is given by the expression

$$S = \exp(\sigma A_n^+ A_n), \tag{38.5}$$

$$\sigma = \zeta N^{-1/2} \sum_q (a_0^+ - a_q), \tag{38.6}$$

$$\zeta \equiv \chi\beta / \hbar\Omega = \chi / (2m\hbar\Omega^3)^{1/2}. \tag{38.7}$$

Using the technique of unitary transformations of operators, one can find the relation between the operators B_n and b_q and the new operators, A_n and a_q:

$$B_n = A_n \exp(-\sigma), \ b_q = a_q - \zeta N^{-1/2} A_n^+ A_n. \tag{38.8}$$

Substituting (38.8) into eqn. (38.1), one obtains the transformed Hamiltonian in diagonal form,

$$H = \mathcal{E}_0 A_n^+ A_n - \hbar\Omega\zeta^2 (A_n^+ A_n)^2 + \sum_q \hbar\Omega a_q^+ a_q. \tag{38.9}$$

Let us define the eigenfunctions ν_q of the operators $a_q^+ a_q$ which characterize the states of a chain with a given number ν_q of real phonons of the qth mode. These functions are defined through the following relation

$$a_q^+ a_q |\nu_q\rangle = \nu_q |\nu_q\rangle. \tag{38.10}$$

If $A_n^+ |0\rangle$ is the function of the Amide-I vibrational excitation, the states of the given chain with fixed numbers ν_q of real phonons are described by the functions

$$|\Psi\rangle = A_n^+ \,||0\rangle |\nu_q\rangle. \tag{38.11}$$

According to eqn. (38.8), these states correspond to the energy

$$\langle\Psi|H|\Psi\rangle = \mathcal{E}_0 - \hbar\Omega\zeta^2 + \sum_q \hbar\Omega\nu_q. \tag{38.12}$$

The energy states of a chain in thermodynamic equilibrium at temperature Θ (in units of energy) are defined by the statistical average of the mean value (38.12) obtained over the functions (38.7) with the density matrix

$$\rho = \frac{\exp(-\Sigma_q \hbar\Omega\nu_q/\Theta)}{Sp\exp(-\Sigma_q \hbar\Omega\nu_q/\Theta)}$$

This double average of any operator F is briefly designated by the symbol $\langle\langle F\rangle\rangle = \mathrm{Sp}\{\rho\langle\Psi|F|\Psi\rangle\}$. Under the statistical average (38.12), the quantum numbers ν_q, are replaced by the mean values of real phonons in the chain,

$$\langle\nu_q\rangle = [\exp(-\hbar\Omega/\Theta - 1)]^{-1}. \tag{38.13}$$

Therefore, the energy of a given chain with a single quanta of the Amide-I vibration, interacting with optical vibrations of frequency Ω, which is in thermodynamic equilibrium at the temperature Θ, is given by the formula

$$\langle\langle H\rangle\rangle = \mathcal{E}_0 - \hbar\Omega\zeta^2 + \sum_q \hbar\Omega\langle\nu_q\rangle. \tag{38.14}$$

At zero temperature, $<\nu_q> = 0$, and the energy of the chain takes the value

$$\hbar\omega_0 = \mathcal{E}_0 - \hbar\Omega\zeta^2 = \mathcal{E}_0 - \chi^2/2m\Omega^2. \tag{38.15}$$

Hence, the quantity

$$\Delta = -\chi^2/2m\Omega \tag{38.16}$$

determines the decrease of excitation energy under autolocalization of an Amide-I vibration by optical phonons.

38.2. Absorption of Radiation Resulting from the Creation of Optical Solitons

The absorption coefficient $\kappa(\omega)$ and refractive index $n(\omega)$ of an electromagnetic wave of frequency ω are determined by the dielectric permittivity $\epsilon(\omega)$ of a crystal via the following relations

$$2n(\omega)\kappa(\omega) = \mathrm{Im}\epsilon(\omega), \quad n^2(\omega) - \kappa^2(\omega) = \mathrm{Re}\epsilon(\omega). \tag{38.17}$$

According to the general theory of interaction of an electromagnetic wave of frequency ω with dipole (d) excitations of frequency ω_0 in solids ([32], Ch. IV, §2; [35], §46), the dielectric permittivity is expressed in terms of the Fourier transform $G(\omega)$ of the retarding Green's function $G(t)$ as follows

$$\epsilon(\omega) = \epsilon_0 - \Lambda G(\omega), \quad \Lambda \equiv 4\pi d^2\omega_0/\mathrm{v}. \tag{38.18}$$

In eqn. (38.18), ϵ_0 is the dielectric permittivity provided by other excitations in a crystal, and v is the volume per molecule. In our case, Green's function $G(t)$ is determined through the operators B_n and B_n^+ of the intrapeptide Amide-I vibration,

$$G(t) = i\delta(t) \ll [B_n(t), B_n^+] \gg, \tag{38.19}$$

where $[a,b] \equiv ab - ba$,

$$\delta(t) = \begin{cases} 1 \text{ if } t \geqslant 0; \\ 0 \text{ if } t < 0, \end{cases} \tag{38.20}$$

and

$$B_n(t) = \exp(iHt)B_n\exp(-iHt) \tag{38.21}$$

is the operator excitation annihilation within the framework of the Heisenberg formalism, and H is given by eqn. (38.9).

To evaluate (38.19), one needs to transform the operator B_n into the new operator A_n which enters into the Hamiltonian (38.9). Substituting (38.8) into eqn. (38.21), one finds

$$<\Psi|B_n(t)|\Psi> = A_n e^{-\sigma(t)}\exp(-i\omega_0 t), \tag{38.22}$$

where

$$\sigma(t) \equiv \zeta N^{-1/2}\sum_q (a_q^+ e^{i\Omega t} - a_q e^{-i\Omega t}). \tag{38.23}$$

Therefore, Green's function (38.19) takes the form

$$G(t) = -i\delta(t)\ll e^{-\sigma(t)}e^{\sigma(0)}\gg \exp-(i\omega_0 t). \tag{38.24}$$

By virtue of eqn. (38.23), one can write

$$\exp[g(t)] = \ll e^{\sigma(t)}e^{\sigma(0)}\gg = \prod_q \ll M_q \gg.$$

Here

$$M_q = \exp[\zeta^2 N^{-1}(e^{-i\Omega t}a_q^+ - e^{i\Omega t}a_q)(a_q - a_q^+)].$$

Carrying out the double average, one obtains finally

$$g(t) = \frac{1}{N}\zeta^2\sum_q\{(<\nu_q>+1)e^{-i\Omega t} + <\nu_q>e^{i\Omega t} - (2\nu_q+1)\}. \tag{38.25}$$

Using eqn. (38.24), we can determine the Fourier transform of the retarding Green's function

$$G(\omega) = \int_{-\infty}^{\infty} \exp(i\omega t - \eta t)G(t)dt \tag{38.26}$$

$$= -i\int_0^{\infty} \exp\{i(\omega - \omega_0 + i\eta)t + g(t)\}dt.$$

A small real η, appearing in this integral determines a rule of by-passing a pole $\omega = \omega_0$ under its evaluation. Formally, this quantity takes into account the relaxation processes occurring in the system.

Neglecting the interaction of the Amide-I vibration with the deformation ($\chi=0$), one can easily evaluate the function $g(t)$ and the integral (38.26):

$$\underset{\chi=0}{G(\omega)} = [\omega-\omega_0+i\eta]^{-1}.$$

Then the relations (38.17) take the form

$$n^2(\omega)-\kappa^2(\omega) = \epsilon_0+\frac{\Lambda(\omega_0-\omega)}{(\omega_0-\omega)^2+\eta^2},$$

$$2n(\omega)\kappa(\omega) = \Lambda\eta[(\omega_0-\omega)^2+\eta^2]^{-1}.$$

Taking the interaction with the deformation into account ($\chi\neq 0$), one can transform eqn. (38.25) as follows

$$g(t) = -W+\frac{1}{N}\sum_q(D_q e^{-i\omega_0 t}+C_q e^{i\omega_0 t}), \tag{38.27}$$

where

$$D_q = \zeta^2(<\nu_q+1), \quad C_q = \zeta^2<\nu_q>, \quad W = \frac{1}{N}\sum_q(D_q+C_q). \tag{38.28}$$

The Fourier transform of the Green's function becomes

$$G(\omega) = -ie^{-W}\int_0^\infty \exp[i(\omega-\omega_0+i\eta)t]\sum_{n,m}I_{n,m}(t)dt. \tag{38.29}$$

Here n and m are positive integers (including zero), and

$$I_{n,m}(t) = (n!m!)^{-1}D^n C^m \exp[i\Omega(m-n)t], \tag{38.30}$$

where

$$D = \frac{1}{N}\sum_q D_q = \zeta^2 N^{-1}\sum_q(<\nu_q>+1), \tag{38.30a}$$

$$C = \frac{1}{N}\sum_q C_q = \zeta^2\frac{1}{N}\sum_q<\nu_q>, \tag{38.30b}$$

$$W = \zeta^2\frac{1}{N}\sum_q(2<\nu_q>+1). \tag{38.30c}$$

Integrating eqn. (38.29), one obtains

$$G(\omega) = -ie^{-W}\sum_{n,m}\frac{D^n C^m\{\eta+i[\omega-\omega_0-\Omega(n-m)]\}}{n!m!\{[\omega-\omega_0-(n-m)\Omega|^2+\eta^2\}}.$$

According to eqn. (38.17), the absorption coefficient is determined by the following expression

$$\kappa(\omega) = \frac{\Lambda\eta}{2\sqrt{\epsilon_n}} e^{-W} \sum_{n,m} \frac{D^n C^m}{n!m!\{[\omega-\omega_0-(n-m)\Omega]^2+\eta^2\}}. \tag{38.31}$$

Therefore, the absorption curve $\kappa(\omega)$ is represented by a set of equidistant bands separated from the frequency ω_0 by the distances $(n-m)\Omega$. Each term corresponds to the excitation of a single autolocalized vibration of the Amide-I type when n real phonons are generated simultaneously with the annihilation of m ones.

At low temperatures, when $C \approx 0$, a single term with $m = 0$ is preserved in the sums over m in eqn. (38.31), i.e.,

$$\kappa(\omega) = \frac{\Lambda\eta}{2\sqrt{\epsilon_0}} e^{-W} \sum_n (n!)^{-1} D^n [(\omega-\omega_0-n\Omega)^2+\eta^2]^{-1}. \tag{38.32}$$

The ratio of the maxima of two neighboring bands, separated by Ω from each other, is

$$\frac{D^n(n+1)!}{n!D^{n+1}} = (n+1)/D.$$

If $D < 1$, each following maximum is smaller than the previous one. Maximum absorption corresponds to the transition into the states without real phonons. In crystalline ACN, such a transition is the creation of an optical soliton without real phonons. The absorption band is given by the following curve

$$\kappa(\omega) = \frac{\eta\Lambda}{2\sqrt{\epsilon_n}} e^{-W}/[(\omega-\omega_0)^2+\eta^2] \tag{38.33}$$

whose width is η. As has been determined experimentally, this is approximately $5\,cm^{-1}$ at low temperatures. This width is conditioned by the lifetime of an optical soliton relative to transitions into other excited states of the crystal, which are not taken into account here.

According to eqns. (38.7) and (38.30c), the absorption maximum,

$$\kappa(\omega_0) = \Lambda \exp(-W)/2\eta\sqrt{\epsilon_0}, \tag{38.34}$$

becomes proportional to

$$\eta^{-1}\exp(-W) = \eta^{-1}\exp[-\chi^2/2m\hbar\Omega^3] \tag{38.35}$$

at low temperatures. Therefore, appreciable absorption takes place if there is an interaction with an optical mode of rather large frequency Ω and small damping η. The absorption maximum decreases with increasing temperature, since W increases with temperatures, where

$$W = \zeta^2 \begin{cases} 1+2e^{\hbar\Omega/\Theta}, & \text{if } \hbar\Omega \gg \Theta, \\ 2\Theta/\hbar\Omega, & \text{if } \hbar\Omega \ll \Theta, \end{cases}$$

in agreement with eqn. (38.30c). However, a decrease of the experimentally observed absorption maximum occurs more rapidly with increase of temperature. One can assume that a temperature lowering of the absorption maximum, which corresponds to generating an optical soliton in the crystal, is provided by an increase of the phenomenological parameter η.

Careri et al. [149] note that together with optical solitons, radiation results in excitations of acoustic-soliton type in crystalline *ACN*. These solitons do not contribute to the absorption spectrum. One cannot exclude the possibility that the optical solitons, arising under absorption, transform further into long-lived Davydov solitons under the radiation of acoustic phonons. The comparatively large width ($\sim 5\, cm^{-1}$) of the anomalous absorption band and its fast decrease with temperature increase indicates indirectly such transformation. A theory of this transformation has not yet been developed.

In the next Subsection we investigate the properties of acoustic solitons localized at a single peptide group.

38.3. Acoustic Solitons in an ACN Crystal

At the very start, we assume that together with optical solitons, strongly autolocalized states can also be generated in *ACN* crystals due to the interaction of Amide-I vibrations with the acoustic due to the interaction of Amide-I vibrations with the acoustic branch of longitudinal vibrations. One can label these solitons 'acoustic'. Since the acoustic branch of vibrations in a given crystal arises due to the coupling between peptide groups arranged not only in a single chain, one should consider the three-dimensional field of vibrations. For simplicity, we consider the isotropic case.

So, we assume that phonons in a crystal are characterized by a wavevector $\vec{q}$ and a frequency $\Omega_{\vec{q}}$,

$$\Omega_q = 2\Omega_0 \sin(qa/2), \quad \Omega_0 = (\kappa/M)^{1/2}. \tag{38.36}$$

Here M is the mass of a peptide group, κ is the elasticity coefficient of the crystal, and a is the mean value of the lattice constant. The excited states of the crystal with a single quanta of Amide-I vibrations of energy $\mathcal{E}_0$ on the peptide group located at the zeroth node are described by the following Hamiltonian operator

$$H_a = B_0^+ B_0 + \sum_q \hbar\Omega_q b_{\vec{q}}^+ b_{\vec{q}} + H_{\text{int}}. \tag{38.37}$$

One can express the relative displacement u_d of neighboring peptide groups separated by a distance $d = 4.74\,\text{Å}$ from each other, in terms of the creation, $b_{\vec{q}}^+$, and annihilation, $b_{\vec{q}}$, operators of the longitudinal phonon $(\vec{q}, \Omega_{\vec{q}})$,

$$\vec{u}_q = N^{-1/2} \sum_{\vec{q}} \beta_{\vec{q}} (b_{\vec{q}}^+ + b_{-\vec{q}}) \frac{\vec{q}}{|q|} e^{i\vec{q}\vec{d}}, \tag{38.38}$$

where

$$\beta_{\vec{q}} = (\hbar / 2M\Omega_{\vec{q}})^{1/2} \tag{38.38a}$$

is the amplitude of the zeroth vibrations. The operator of interaction of the intra-peptide vibration of Amide-I type with the longitudinal acoustic vibrations has the form,

$$H_{\text{int}} = -\frac{\chi}{2}(\vec{u}_d - \vec{u}_{-d}) \frac{\vec{d}}{|d|} B_0^+ B_0 = -\frac{1}{\sqrt{N}} \sum_{\vec{q}} \zeta_{\vec{q}} (b_{\vec{q}} + b_{-\vec{q}}^+) B_0^+ B_0 \tag{38.39}$$

where

$$\zeta_{\vec{q}} = i\chi\mu \sin(\vec{q}\vec{d})(\hbar / 2M\Omega_q)^{1/2}, \quad \mu = \frac{\vec{q}\vec{d}}{|q||d|}, \tag{38.40}$$

and $\vec{d}/|d|$ is a unit vector directed along the chain of peptide groups. One can further diagonalize the operator (38.37) by means of the canonical transformation to the new operators A_0, $a_{\vec{q}}$ by using the unitary operator

$$S = \exp\{A_0^+ A_0 \sigma\}, \tag{38.41}$$

where

$$\sigma = \frac{1}{\sqrt{N}} \sum_{\vec{q}} \zeta_{\vec{q}}^* (a_{\vec{q}}^+ - a_{\vec{q}})(\hbar\Omega_{\vec{q}})^{-1}.$$

Using the standard technique of unitary transformations, one finds

$$B_0 = SA_0S^{-1} = A_0\exp(-\sigma),\ b_{\vec{q}} = Sa_{\vec{q}}S^{-1} = \tag{38.42}$$

$$= a_{\vec{q}} - \zeta^*_{\vec{q}}(\hbar\Omega_{\vec{q}})^{-1}A_0^+A_0.$$

Substituting these expressions into the operator (38.37), we can transform the latter into the diagonal form relative to the new operators

$$H_a = \mathcal{E}_0A_0^+A_0 - \frac{1}{N}\sum_{\vec{q}}(\hbar\Omega_{\vec{q}})^{-1}|\Xi_{\vec{q}}|^2(A_0^+A_0)^2 + \sum_{\vec{q}}\hbar\Omega_{\vec{q}}a^+_{\vec{q}}a_{\vec{q}}.$$

Averaging this operator over the functions

$$|\Psi> = A_0^+|0>\prod_{\vec{q}}|\nu_{\vec{q}}>$$

and carrying out a statistical average over the state of thermodynamical equilibrium at the temperature Θ, one obtains the energy of excitation in the given chain:

$$\ll H_a \gg = \mathcal{E}_0 - \frac{1}{N}\sum_{\vec{q}}|\zeta_{\vec{q}}|^2/\hbar\Omega_{\vec{q}}. \tag{38.43}$$

Neglecting the contribution of real acoustic phonons, the excitation energy is given by the relation

$$\hbar\omega_a = \mathcal{E}_0 - \frac{1}{N}\sum_{\vec{q}}|\zeta_{\vec{q}}|^2/\hbar\Omega_{\vec{q}}. \tag{38.44}$$

Hence, a decrease of energy of the Amide-I vibration due to the creation of an acoustic soliton, as a result of the interaction with virtual phonons, is determined by the following quantity

$$\Delta = -\chi^2[8NM\Omega_0^2]^{-1}\sum_{\vec{q}}\mu^2\sin^2(q\mu d)/\sin^2(qa/2).$$

Replacing here the summation by integration over $\vec{q}$-space,

$$\sum_{\vec{q}}\ldots = \frac{3}{2q^3}\int_{-1}^{1}d\mu\int_0^{q_D}\ldots q^2dq,$$

where $q_D = (6\pi^2n)^{1/2}$ is the Debye wavenumber in the crystal with n unit cells per unit volume, and also the sine-functions by their arguments, one obtains

$$\Delta = -3\chi^2d^2/10Ma^2\Omega_0^2. \tag{38.45}$$

Green's function, which determines the interaction of an acoustic soliton with radiation of frequency ω, can be rewritten now according to eqn. (38.18) as follows

$$G(t) = -i\delta(t)\exp\{g_a(t)-i\omega_a t\}, \tag{38.46}$$

where

$$g_a(t) = -W_a+\frac{1}{N}\sum_{\vec{q}}(\hbar\Omega_{\vec{q}})^{-2}|\zeta_{\vec{q}}|^2\{(<\nu_{\vec{q}}>+1)e^{-\Omega_{\vec{q}}t}+<\nu_{\vec{q}}>e^{i\omega_{\vec{q}}t}\} \tag{38.47}$$

and

$$W_a = \frac{1}{N}\sum_{\vec{q}}(\hbar\Omega_{\vec{q}})^{-2}|\zeta_{\vec{q}}|^2(2<\nu_q>+1). \tag{38.48}$$

The absorption coefficient $\kappa(\omega)$ takes the form

$$\kappa(\omega) = -\Lambda \mathrm{Im} G(\omega), \quad \Lambda\equiv 1/2n(\omega_a). \tag{38.49}$$

Here $G(\omega)$ is the Fourier transform of the Green's function $G(t)$:

$$G(\omega) = \int_{-\infty}^{\infty} G(t)\exp[i\omega t-\eta t]dt = -ie^{-W_a}\int_0^{\infty}\exp\left\{i(\omega-\omega_a+i\eta)t+\right. \tag{38.50}$$

$$\left.+\frac{1}{N}\sum_{\vec{q}}(\hbar\Omega_{\vec{q}})^{-2}|\zeta_{\vec{q}}|^2[(<\nu_{\vec{q}}>+1)e^{-i\Omega_{\vec{q}}t}+<\nu_{\vec{q}}>e^{i\Omega_q t}]\right\}.$$

At low temperatures, when $<\nu_q>\ll 1$, the terms appearing in the sum over $\vec{q}$ do not contribute to the integral due to their fast oscillations. Hence,

$$G(\omega) = \frac{\exp(-W_a)}{\omega-\omega_a+i\eta}, \quad <\nu_{\vec{q}}>\ll 1. \tag{38.51}$$

Substituting this expression into eqn. (38.49), one finds explicitly the absorption coefficient

$$\kappa(\omega) = \frac{\eta\Lambda\exp(-W_a)}{(\omega-\omega_a)^2+\eta^2}, \quad <\nu_{\vec{q}}>\ll 1. \tag{38.52}$$

This is therefore characterized by a single Lorentz curve with the maximum

$$\kappa(\omega_a) = \Lambda\eta^{-1}\exp(-W_a) \tag{38.52a}$$

at $\omega=\omega_a$. In order to evaluate W_a, we substitute $<\nu_q> = [\exp(\hbar\Omega_q/\Theta)-1]^{-1}$ into eqn. (38.48). By virtue of (38.48) and (38.40) and replacing the sine-functions by their arguments and summation by integration over $\vec{q}$-space, one obtains

$$W = A + B\Theta^2 \int_0^{x_0} \frac{x\,dx}{e^x - 1}, \tag{38.53}$$

where

$$x_0 = a\hbar\Omega_0 q_D / 2\Theta, \quad A \equiv \frac{3\chi^2 d^2}{20M\hbar_q D a^3 \Omega_0^3}, \quad B \equiv \frac{3\chi^2 d^2}{2M\hbar^3 q_D^3 a^3 \Omega_0^5}.$$

At low temperatures, when $x_0 \gg 1$, the integral in eqn. (38.53) is equal to $\pi^2/6$, and therefore

$$W = A + \frac{\pi^2}{6} B\Theta^2. \tag{38.54}$$

Thus the maximum of the absorption band (38.52) decreases rapidly with increasing temperature according to the law

$$\exp\left[-\frac{B\pi^2}{6}\Theta^2\right]. \tag{38.55}$$

At high temperatures, when the mean value $<\nu_q>$ is considerably larger than unity,

$$<\nu_q> = \Theta / \hbar\Omega_d \gg 1, \tag{38.56}$$

the integration region in (38.50) can be partitioned into two parts, from zero to t_0, and from t_0 to ∞. Here $t_0 = (\max\Omega_q)^{-1}$. Due to large oscillations in the second subregion, from t_0 to ∞, the terms appearing in the square bracket do not contribute to the integral. Therefore, the Fourier transform of Green's function now takes the form

$$G_1(\omega) = \frac{\exp(-\tilde{W}_a)}{\omega - \omega_a + i\eta}, \tag{38.57}$$

where

$$\tilde{W}_a = \frac{1}{N}\sum_{\vec{q}} \left[1 + \frac{2\Theta}{\hbar\Omega_q}\right] |\zeta_q|^2 (\hbar\Omega_q)^{-2}.$$

The imaginary part in eqn. (38.57) determines the absorption coefficient

$$\kappa(\omega) = \frac{\eta\Lambda\exp(-\tilde{W}_a)}{(\omega-\omega_a)^2+\eta^2}, \tag{38.58}$$

which corresponds to the narrow phononless spectral line with half-width η and a maximum near the frequency ω_a which is proportional to $\eta^{-1}\exp(-\tilde{W}_a)$. At high temperatures, the intensity of the phononless line is negligible.

Investigating the second part of the integral (38.50) in the subregion 0, t_0 where $t<t_0$, one can utilize the approximate relation

$$(<\nu_q>+1)e^{-i\Omega_q t}+<\nu_q>e^{i\Omega_q t} \approx 2<\nu_q>+1-<\nu_q>\Omega_q^2 t^2-i\Omega_q t. \tag{38.59}$$

In this case, one can analyze the factor $\exp(-\tilde{W}_a)$. The Fourier transform is given by the following integral

$$G_2(\omega) = -i\int_0^{t_0}\exp[izt+D^2t^2]dt, \tag{38.60}$$

where

$$z \equiv \omega-\omega_a+i\eta-A, \tag{38.61}$$

$$D^2 = \frac{\Theta}{\hbar^2 N}\sum_q \Omega_q^{-2}|\zeta_q|^2 = \frac{\Theta\chi^2}{2N\hbar M}\sum_{\vec{q}}\Omega_q^{-3}\frac{\vec{q}\vec{d}}{|q||d|}\sin^2(\vec{q}\vec{d}),$$

and

$$A = \frac{1}{N}\sum_q(\hbar\Omega_q^{-1})|\zeta_q|^2 = \frac{\chi^2}{2MN}\sum_{\vec{q}}\Omega_a^{-2}\frac{\vec{d}\vec{q}}{|d||q|}\sin^2(\vec{q}\vec{d}). \tag{38.62}$$

If the parameter D is rather large, so that the inequalities

$$Dt_0 = \frac{D}{\max\Omega_q} \gg 1 \text{ and } D \gg \eta$$

are satisfied, the upper limit of integration in eqn. (38.60) can be extended to infinity. In this case, one obtains

$$G_2(\omega) = \begin{cases} -i\sqrt{\pi}(2D)^{-1}\exp(-|z|^2/4D^2), & \text{if } |z|<D; \\ 1/z, & \text{if } |z|>D. \end{cases}$$

Therefore, the absorption coefficient is given by the expression

$$\kappa(\omega) = -\Lambda \mathrm{Im} G_2 =$$

$$= \begin{cases} (\sqrt{\pi}\Lambda/2D)\exp\{-\dfrac{(\omega-\omega_a-A)^2+\eta^2}{4D^2}\}, & \text{if } |z|<D; \\ \dfrac{\Lambda\eta}{(\omega-\omega_a-A)^2+\eta^2}, & \text{if } |z|>D, \end{cases}$$

and the resonance frequency $\omega = \omega_a + A$ is shifted from the resonance frequency ω_a of a weak phononless line appearing in eqn. (38.58) to larger frequencies by the quantity A. In the neighborhood of the resonance frequency ($|z| < D$) the dispersion law of the absorption coefficient takes a Gaussian form. In the region far from resonance, $|z| > D$, the absorption curve possesses the Lorentz form.

The quantity D, defined by eqn. (38.61), characterizes the width of the Gaussian curve and its maximum D^{-1}. As the temperature increases, its width increases too, but its maximum decreases. The quantity A, given by eqn. (38.62), determines a shift of the frequency ω_r of the absorption maximum relative to the frequency ω_a of a weak phononless line given by (38.52), i.e.,

$$\omega_r = \omega_a + A.$$

Virtual phonons only participate in this shift.

Chapter 10

The Nonlinear Sine-Gordon Equation

39. Canonic Form of the Sine-Gordon Equation

The nonlinear Gordon equation describes diverse nonlinear phenomena, such as the propagation of ferromagnetic waves coupled with a rotation of the magnetization vector (motion of domain walls), the rotational phase transitions in molecular crystals, the motion of dislocations in crystals, the self-induced transparency of crystals under the propagation of ultrashort intense optical pulses, the properties of extended Josephson junctions, etc. This equation describes a real one-component field $\varphi(x,t)$ which depends on time t and a single variable x. Such an equation determines the difference in angular rotational phases characterizing some nonlinear processes as described in the following Sections.

Let Λ_0 be the characteristic length and ω_0 the characteristic frequency of a given phenomenon. In the dimensionless variables of length, $\zeta = x/\Lambda_0$, and time, $\tau = t\omega_0$, the canonical form of the nonlinear equation involving the sine of the function $\varphi(\zeta,\tau)$ which characterizes a rotational phase is

$$\varphi_{\xi\xi}(\xi,\tau)-\varphi_{\tau\tau}(\xi,\tau) = \sin\varphi(\xi,\tau). \tag{39.1}$$

Henceforth, we use subindices of a function to designate the appropriate derivatives. The first derivatives of a scalar function $\varphi(\xi,\tau)$ define the so-called electric field

$$\mathrm{v}(\xi,\tau) = \varphi_\tau(\xi,\tau), \tag{39.2}$$

and the magnetic field

$$h(\xi,\tau) = \varphi_\xi(\xi,\tau). \tag{39.3}$$

Equation (39.1) is usually called the sinusoidal Gordon equation. The term 'sine-Gordon Equation', or *SGE* for short was first proposed in 1970 by Kruskal [246] and used by Rubinstein [244]. Now it appears widely in the scientific literature.

The Sine-Gordon equation can be derived from the variational principle for the following Lagrange function density

$$\mathfrak{L} = \frac{1}{2}[\varphi_\xi^2 - \varphi_\tau^2 - 2(\cos\varphi - 1)]. \tag{39.4}$$

Really, using the Euler-Lagrange equation,

$$\frac{\partial}{\partial\tau}\left[\frac{\partial\mathfrak{L}}{\partial\varphi_\tau}\right] + \frac{\partial}{\partial\xi}\left[\frac{\partial\mathfrak{L}}{\partial\varphi_\xi}\right] - \frac{\partial\mathfrak{L}}{\partial\varphi} = 0,$$

one obtains directly eqn. (39.1). A momentum canonically conjugated to φ is

$$\varphi_\tau = -\partial\mathfrak{L}/\partial\varphi_\tau. \tag{39.5}$$

The Hamiltonian function is given by the formula

$$\mathcal{E} = \frac{1}{2}\int_{-\infty}^{\infty} d\xi[\varphi_\tau^2 + \varphi_\xi^2 + 2(1-\cos\varphi)]. \tag{39.6}$$

It defines the total energy of the system in question. The first term in eqn. (39.6) can be interpreted as the kinetic energy, the second as the elastic energy, and the third as the oscillating external potential. Together with the energy (39.6), the momentum,

$$\mathcal{P} = -\int_{-\infty}^{\infty} d\xi\varphi_\tau\varphi_\xi \tag{39.7}$$

is an integral of motion, too. The explicit form of other integrals of motion are beyond our present interest.

Let us define the density of rotation via the formula

$$\rho(\xi) = \frac{\partial\mathfrak{L}}{\partial\varphi_\xi} = \varphi_\xi.$$

If $\varphi(\xi,\tau)$ is an arbitrary solution of the sine-Gordon equation satisfying the boundary conditions

$$\lim_{\xi\to\infty}\varphi(\xi,\tau) = 2\pi m, \quad \lim_{\xi\to-\infty}\varphi(\xi,\tau) = 2\pi n,$$

the total rotation defined as

$$R \equiv \int_{-\infty}^{\infty} \rho(\xi) d\xi = 2\pi(m-n)$$

becomes the integral of motion.

Equation (39.1) is translational invariant for an infinite extended system. This means that eqn. (39.1) is invariant relative to the Lorentz transformation

$$\begin{pmatrix} \tau' \\ \xi' \end{pmatrix} = \gamma \begin{pmatrix} 1, & -W \\ -W, & 1 \end{pmatrix} \begin{pmatrix} \tau \\ \xi \end{pmatrix}, \tag{39.8}$$

where

$$\gamma = (1-W^2)^{-1/2} \tag{39.9}$$

is the so-called Lorentz contraction, and W is the velocity of relative motion measured in units of the characteristic velocity $c_0 = \omega_0 \Lambda_0$. It is clear that $W \leqslant 1$. The energy and momentum are thus transformed as the covariant bivector

$$\begin{pmatrix} \mathcal{E}' \\ \mathcal{P}' \end{pmatrix} = \gamma \begin{pmatrix} 1, & W \\ W, & 1 \end{pmatrix} \begin{pmatrix} \mathcal{E} \\ \mathcal{P} \end{pmatrix}. \tag{39.10}$$

The sine-Gordon equation can be viewed as a continuum approximation of the equation

$$\frac{d^2\varphi_m}{dt^2} - \varphi_{n+1} - \varphi_{n-1} + 2\varphi_n = \sin\varphi_n, \tag{39.11}$$

proposed by Frenkel and Kontorova [124] in 1939 to describe the motion of dislocations in solids (see Section 43). The right-hand side of eqn. (39.11) involves a periodic potential of the lattice. Equation (39.11) originated the mechanical model, the so-called 'Scott's pendulum', consisting of pendulums suspended by elastic spring at equal distances in a gravitation field. Such a model gives a clear illustration of different solutions of the sine-Gordon equation. Diverse applications of this equation have been reviewed by Scott et al. [248].

40. Single-Phase Solutions of the Sine-Gordon Equation

The nonlinear sine-Gordon equation refers to the class of equations which possess the remarkable property of complete integrability. For this reason, it has been extensively studied in the field of nonlinear equations [89,130,133,151,180,218].

A simple class of solutions of eqn. (39.1) for an infinite extended system can be obtained in the form of running stationary waves with a fixed shape propagating with a constant velocity W. To describe such waves, one can assume that we function φ depends on the variables ξ and τ as follows

$$\varphi(\xi,\tau) = \varphi(\theta), \tag{40.1}$$

$$\theta \equiv \xi - \xi_0 - W\tau, \tag{40.2}$$

where ξ_0 is an arbitrary constant. The solutions of the type (40.1) depending on a single variable θ, are usually *single-phase solutions*. The velocity W appearing in eqn. (40.2) is measured in units of the characteristic velocity, $c_0 = \omega_0\Lambda_0$, which is too much less that in a vacuum.

Using the variable θ, one can transform eqn. (39.1) to the following

$$(1 - W^2)\varphi_{\theta\theta}(\theta) = \sin\varphi(\theta). \tag{40.3}$$

Integrating this, one obtains

$$(1 - W^2)\varphi_\theta^2 = 2(E - \cos\varphi), \tag{40.4}$$

where E is constant of integration. The solutions of eqn. (40.4) may be divided into two large classes: the solutions of a magnetic type, of M-type, moving with a velocity less than the characteristic velocity $c_0(W^2<1)$, and the solutions of an electric type, or E-type, whose velocity of propagation exceeds the characteristic velocity ($W^2>1$). The term 'magnetic-type waves' arises as a result of the fact that, assuming $W^2 < 1$ and applying the Lorentz transformation (39.8), one can perform a transition to a reference frame where a single-phase wave depends only on the spatial coordinate and hence, creates a magnetic field. Alternatively $W^2 > 1$, with the help of the Lorentz transformation one can pass to a reference frame where a single-phase wave depends only on time and creates thus an electric field. For this reason, single-phase waves with $W^2 > 1$ are called electric, or E-type waves.

40.1. Magnetic-Type Wave Solutions

We begin by considering the case of velocities less than the characteristic velocity c_0, i.e., $W^2 < 1$. Extracting the square root of both sides of eqn. (40.4) and integrating further, one obtains

$$\int_{\varphi_0}^{\varphi(\theta)} \frac{d\varphi}{\sqrt{E - \cos\varphi}} = \eta\sqrt{2}\gamma\theta, \quad \eta = \pm 1, \tag{40.5}$$

where γ is given by eqn. (39.9). Evaluating the phase difference $\varphi(\theta)$ by means of eqn. (40.5), one should consider the following three cases: $E = 1$, and $|E| < 1$. Let us consider these separately.

Case 1: $E = 1 (W^2 < 1)$. Assuming $\varphi_0 = \pi$, one transforms eqn. (40.5) with $E = 1$ as follows:

$$\frac{1}{2}\int_{\pi}^{\varphi(\theta)} \frac{d\varphi}{\sin(\varphi/2)} = \eta\gamma\theta.$$

Evaluating this integral, one obtains

$$\eta\gamma\theta = \ln \mathrm{tg}(\varphi/4)$$

or

$$\varphi(\theta) = 4\mathrm{arctg}[\exp(\eta\gamma\theta)]. \tag{40.6}$$

If $\eta = 1$ then, according to eqn. (40.6), a variation of the coordinate θ over the interval $-\infty, \infty$ is accompanied by a change (increase) of the phase difference from 0 to 2π (see Fig. 16). Unlike the solitons whose wave functions vanish as $|\theta| \to \infty$, excitations of the type (40.6) have different asymptotes under $\theta \to \pm\infty$. Such excitations are called topological solitons, travelling *extended domain walls, kink-solitons*, or simply, *kinks*. For brevity, we use henceforth the latter term. To emphasize that under the variation of θ from $-\infty$ to ∞, the phase φ varies from zero to 2π, one can sometimes refer to such excitations as *2π-pulses*.

Assuming $\eta = -1$ in eqn. (40.6), one has that a variation of θ from $-\infty$ to ∞ is accompanied by a decrease of the function (40.6) from 2π to 0. Such excitations are called *antikink-solitons*, or *antikinks*. Unlike the functions of usual solitons, the wave functions (40.6) are not normalized.

Using the relation

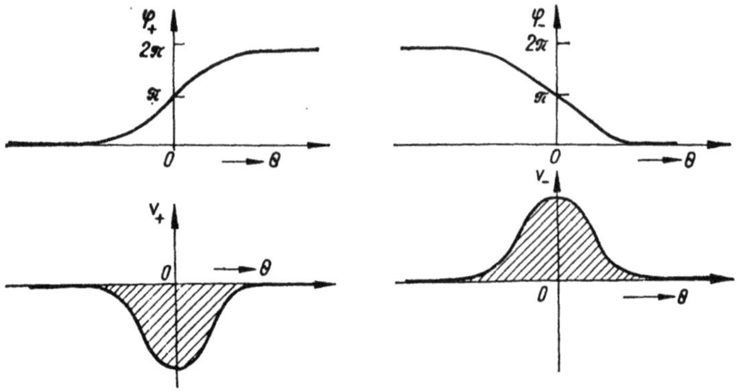

Figure 16. The single-phase solution of the sine-Gordon equations: $\eta = +1$ (kink, or fluxon); $\eta = -1$ (antikink, or antifluxon), and ϕ_η is the phase, v_η is the reduced potential difference.

$$\varphi_\epsilon(\theta) = -W_{\varphi\theta}(\theta), \tag{40.7}$$

one can transform eqn. (39.7) which determines momentum, into the following one

$$\mathcal{P} = W\int_{-\infty}^{\infty} \varphi_\theta^2(\theta)d\theta. \tag{40.8}$$

For $E = 1$ from eqn. (40.4) it follows

$$\frac{1}{2}(1-W^2)\varphi_\theta^2 = 1-\cos\varphi.$$

Using this relation and eqn. (40.7), one transform the expression (39.6) for energy as follows

$$\mathcal{E} = \int_{-\infty}^{\infty} \varphi_\theta^2(\theta)d\theta. \tag{40.9}$$

From eqn. (40.6), it follows that the function

$$\varphi_\theta(\theta) = 2\eta\gamma\mathrm{sch}(\gamma\theta)$$

possesses the bell-like form inherent for a common soliton. Using the latter formula and eqns. (40.8), one can evaluate the energy and momentum

$$\mathcal{E} = 8\gamma, \ \ \mathcal{P} = 8W\gamma = W\mathcal{E}. \tag{40.10}$$

Singling out the explicit dependence on ξ and τ in (40.6), one obtains

$$\varphi(\xi,\tau) = 4\mathrm{arctg}\{\exp[\eta\gamma(\xi-\xi_0-W\tau)]\}.$$

Therefore, the immobile excitation is described by the function

$$\varphi(\xi) = 4\mathrm{arctg}\{\exp[\eta(\xi-\xi_0)]\}. \tag{40.11}$$

So, the transition from an immobile excitation (40.11) to that moving with velocity W is carried out by means of the Lorentz transformation

$$\xi \rightarrow \gamma(\xi - W_\tau). \tag{40.12}$$

Case 2: $E > 1$.

If the integration constant $E > 1$, eqn. (40.5) is reduced to

$$\int_\pi^{\varphi(\theta)} \frac{d\varphi}{\sqrt{A+\sin^2(\varphi/2)}} = 2\eta\gamma\theta \tag{40.13}$$

where

$$A \equiv (E-1)/2 > 0, \quad 0 \leqslant A \leqslant 1. \tag{40.13a}$$

Introducing a new variable, $t = \cos(\varphi/2)$, and a parameter k via the relation

$$A = k_1^2/k^2, \quad k_1^2 = 1-k^2 \tag{40.14}$$

which varies in the interval (0,1), one can transform eqn. (40.13):

$$\int_0^{\cos(\varphi/2)} \frac{dt}{\sqrt{(1-t^2)(1-k^2t^2)}} = -\eta\theta\gamma/k. \tag{40.15}$$

Taking into account the definition of the Jacobian elliptic function $\mathrm{sn}(u,k)$ of modulus k (see Section 29),

$$\int_0^{\mathrm{sn}(u,k)} \frac{dt}{\sqrt{(1-t^2)(1-k^2t^2)}} = u, \tag{40.16}$$

eqn. (40.15) can be rewritten as follows

$$\cos(\varphi/2) = -\eta\mathrm{sh}(u,k), \quad u \equiv \gamma \tag{40.17}$$

or

$$\sin\left(\frac{\varphi-\pi}{2}\right) = \eta\mathrm{sn}(u,k), \tag{40.17a}$$

$$\varphi(\theta) = 2\arcsin[\eta \mathrm{sn}(u,k)]+\pi. \tag{40.17b}$$

For a fixed modulus k, the Jacobian elliptic function $\mathrm{sn}(u,k)$ is an odd periodic function of the variable u with period $4K(k)$, where $K(k)$ is the complete elliptic integral of the first kind. The spatial period l of the function (40.17) relative to θ is defined by the formula

$$l = 4kK(k)/\gamma. \tag{40.18}$$

As k varies from zero to unity, the complete elliptic integral $K(k)$ varies in the interval $\pi/2$ to $\ln(4/\sqrt{1-k^2})\to\infty$. Within the limits of one period, i.e., under the variation of u from $-K(k)$ to $K(k)$, the function $\mathrm{sn}(u,k)$ changes from -1 to 1 through zero. For small k^2, the phase difference $\varphi(\theta)$ increases or decreases monotonically depending on whether $\eta = 1$ or $\eta = -1$, respectively (see Fig. 17).

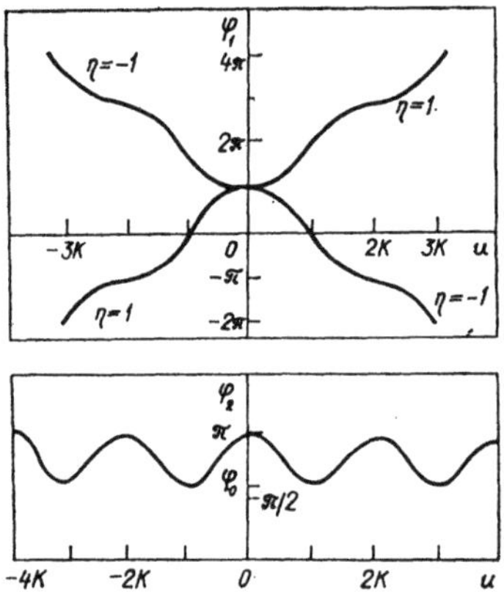

Figure 17. The function $\phi_{1\eta}(u) = 2\sin^{-1}[\eta sn(u,k)]+\pi$ for $k^2 \ll 1$, and the function $\phi_2(u) = 2\sin^{-1}[dn(u,k)]$ for $k_1^2 \ll 1$, and $\phi_0 = 2\sin^{-1}k_1$.

Interpreting the phase difference as an angle, its monotonic increase characterizes a rotation. For this reason, the solution of the type (40.17) is called a helical wave. Within Scott's mechanical model, this corresponds to a rotation of the pendulums. Approaching $k^2 = 0$, one has $\mathrm{sn}(u,k) = \sin u$. Then $\varphi(\theta)$ degenerates into a straight line. In the opposite limiting case, $k_1^2 = 1-k^2 \ll 1$, and the period of $\mathrm{sn}(u,k)$ approaches infinity. The function is transformed into a hyperbolic one, th$u$. For $\eta = 1$, the phase difference varies from 0 to 2π, and from 2π to 0, for $\eta = -1$. So, in this limiting case the excitation results in a kink or antikink, respectively.

Due to the fact that the phase difference changes by $2\pi\eta$ over the period l, the magnetic flow through a surface with length l and width of $1\,cm^2$ is equal to

$$\Phi = \frac{\Phi_0}{2\pi}\int_0^l \frac{\partial\varphi}{\partial\xi}d\xi = \eta\Phi_0, \;\; \eta = \pm 1. \tag{40.19}$$

The difference in electric potentials, and the magnetic field, at the junction are determined by the following expressions (in terms of dimensionless quantities)

$$v_z(\theta) = \frac{\partial\varphi}{\partial\tau} = -\frac{2W\gamma}{k}\mathrm{dn}(u,k), \;\; u = \gamma\theta/k,$$

$$h_y(\theta) = \frac{\partial\varphi}{\partial\xi} = \frac{2\gamma}{k}\mathrm{dn}(u,k),$$

where $\mathrm{dn}(u,k)$ is the Jacobian elliptic function of modulus k. The fields approach infinity as $W\to 1$, which is consistent with the tendency for the period (40.18) to vanish within that limit.

Since

$$\frac{\partial\varphi}{\partial\theta} = \frac{2\mathrm{dn}(u,k)}{k\sqrt{1-W^2}} \;\text{ and }\; 1-\cos\varphi = 2\mathrm{cn}^2(u,k),$$

the energy per period $4K(k)$ is given by the formula (see eqn. (29.26))

$$\gamma^{k-1}\mathfrak{S} = \int_0^{4K}\left\{\frac{2(1+W^2)\mathrm{dn}^2(u,k)}{k^2(1-W^2)}+\mathrm{cn}^2(u,k)\right\}du$$

$$= \frac{4}{k^2}\left\{\frac{3+2W^2}{1-\omega^2}E(k)-k_1^2K(k)\right\},$$

in agreement with the definition (39.6).

Case 3: $|E|<1$.
Assuming the integration constant E to satisfy the inequality $|E|<1$, one can rewrite eqn. (40.5) as follows:

$$\int_0^{\varphi(\theta)}\frac{d\varphi}{\sqrt{\sin^2(\varphi/2)-B}} = 2\eta\gamma\theta, \tag{40.20}$$

where

$$2B = 1-E>0, \;\; 0<B<1/2.$$

Designating

$$B = k_1^2 = 1-k^2, \tag{40.21}$$

and introducing the new variable $t = \sin(\varphi/2)$, one can transform eqn. (40.20) into

$$\int_1^{\sin(\varphi/2)} \frac{dt}{\sqrt{(1-t^2)(t^2-k_1^2)}} = \eta\gamma\theta. \tag{40.22}$$

Using the definition of the Jacobian elliptic function $\mathrm{dn}(u,k)$, one obtains from eqn. (40.22)

$$\sin(\varphi/2) = \mathrm{dn}(u,k), \quad u = \gamma\theta. \tag{40.23}$$

The function $\mathrm{dn}(u,k)$ is an even periodic function of the variable u with period $2K(k)$. It attains a maximum equal to unity at $u = 0$. Its minimum of k_1 occurs at $\pm K(k)$. Therefore, the phase difference varies from a maximum value of π to a minimum value $\varphi_0 = \arcsin k_1$. Such a solution can be called an oscillating solution. As far as Scott's mechanical model is concerned, this corresponds to the pendulum vibrations. For the oscillating excitation, the potential difference and the magnetic field are (in dimensionless units)

$$v_z = 2k\gamma W \mathrm{cn}(u,k), \quad h_g = -2k\gamma \mathrm{cn}(u,k).$$

40.2. Electric-Type Wave Solutions

Let us consider now the motion of excitations with velocities exceeding the characteristic velocity c_0. For velocities $W^2 > 1$, the motion of solitons should be accompanied by electromagnetic radiation - the so-called *Cherenkov radiation*. We present here the results obtained by neglecting this radiation.

For $W^2 > 1$, assuming $B = -E$, one obtains from eqn. (40.4)

$$\int_{\varphi_n}^{\varphi(\theta)} \frac{d\varphi}{\sqrt{\cos\varphi + B}} = \eta\sqrt{2}\gamma\theta, \quad \gamma \equiv (W^2-1)^{-1/2}. \tag{40.24}$$

Let us consider this equation for $B = 1$, $B > 1$, and $|B| < 1$.

Case 1: $B = 1$, or $E = -1$.

For $B = 1$, eqn. (40.24) takes the form

$$\int_0^{\varphi(\theta)} \frac{d\varphi}{\cos(\varphi/2)} = 2\eta\tilde{\gamma}\theta, \quad \eta = \pm 1. \tag{40.25}$$

Evaluating this integral, one finds

$$\mathrm{tg}[(\varphi+\pi)/4] = \exp(\eta\tilde{\gamma}\theta) \tag{40.26}$$

$$\varphi(\theta) = -\pi+4\mathrm{arctg}[\exp(\eta\tilde{\gamma}\theta)]. \tag{40.26a}$$

In this case, the derivative,

$$\varphi_\theta(\theta) = 2\eta\tilde{\gamma}\mathrm{sch}(\tilde{\gamma}\theta) \tag{40.27}$$

is depicted by a bell-like function. According to eqn. (40.26), as θ varies from $-\infty$ to ∞, the phase difference changes from $-\pi$ to π at $\eta = 1$, and in the opposite direction at $\eta = -1$, through zero at $\theta = 0$. The solution given at $\eta = 1$ can be interpreted as a kink, and, at $\eta = -1$, as an antikink.

The potential difference and the magnetic field are given by the following expressions in terms of dimensionless quantities

$$\mathrm{v}_z = 2\eta\tilde{\gamma}\mathrm{sch}(\tilde{\gamma}\theta),\ \ h_g = -2\eta W\tilde{\gamma}\mathrm{sch}(\tilde{\gamma}\theta).$$

Case 2: $B > 1$.
In this case, eqn. (40.24) may be rewritten as follows

$$\int_0^{\varphi(\theta)} \frac{d\varphi}{\sqrt{D-\sin^2(\varphi/2)}} = 2\eta\tilde{\gamma}\theta, \tag{40.28}$$

where

$$D \equiv (1+B)/2 > 1. \tag{40.29}$$

Introducing the parameter k^2 through the relation

$$k^2 = 1/D \tag{40.30}$$

and the new variable $t = \sin(\varphi/2)$, one can transform eqn. (40.28) into

$$\int_0^{\sin(\varphi/2)} \frac{dt}{\sqrt{(1-t^2)(1-k^2t^2)}} = \eta\tilde{\gamma}\theta/k. \tag{40.31}$$

According to the formula (40.16), the integral on the left-hand side of eqn. (40.31) defines the Jacobian elliptic function $\mathrm{sn}(u,k)$ with modulus k. From eqn. (40.31) it follows that

$$\sin(\varphi/2) = \eta\mathrm{sn}(u,k),\ \ u \equiv \tilde{\gamma}\theta/k \tag{40.32}$$

or

$$\varphi(\theta) = 2\eta \arcsin[\mathrm{sn}(u,k)]. \tag{40.32a}$$

Taking the properties of $\mathrm{sn}(u,k)$ into account, one sees that the solution (40.32) depicts a monotonically increasing (decreasing) function of the variable θ for $\eta = 1$ ($\eta = -1$) passing through zero at $\theta = 0$. The derivative of the phase difference becomes

$$\gamma_0(\theta) = k^{-1} 2\eta\tilde{\gamma}\mathrm{dn}(u,k). \tag{40.32b}$$

Substituting this and the relation

$$1 - \cos\varphi = 2\mathrm{sn}^2(u,k)$$

into eqn. (39.6) for the energy, one obtains

$$\tilde{\gamma}k^{-1}\mathcal{E} = 2\int \left[\frac{W^2+1}{k^2(W^2-1)} \mathrm{dn}^2(u,k) + \frac{1}{2}\mathrm{sn}^2(u,k) \right] du.$$

The periodic Jacobian elliptic function $\mathrm{sn}(u,k)$ possesses the period $4K(k)$ as a function of u. Hence, using eqns. (29.27) and (29.20), one finds the energy per period

$$\mathcal{E} = \frac{4}{\tilde{\gamma}k} \left[\frac{2W^2+1}{W^2-1} E(k) + K(k) \right],$$

where $E(k)$ and $K(k)$ are the complete elliptic integrals of the second and first kinds, respectively.

Case 3: $|B| < 1$.
If $|B| < 1$, the parameter D given by eqn. (40.29), is less than unity. For this reason, it is convenient to carry out the following transformation under the square root of the integral in eqn. (40.28),

$$D - \sin^2(\varphi/2) = D - 1 + \cos^2(\varphi/2).$$

Assuming

$$1 - D = k_1^2, \quad 0 < k_1^2 < 1, \tag{40.33}$$

and introducing the variable $t = \cos(\varphi/2)$, one can transform eqn. (40.28) as follows

$$\int_0^{\cos(\varphi/2)} \frac{dt}{\sqrt{(1-t^2)(t^2-k_1^2)}} = -\eta\tilde{\gamma}\theta. \tag{40.34}$$

Using the definition of the Jacobian elliptic function (see Section 29), one obtains

$$\cos(\varphi/2) = \mathrm{dn}(u,k), \tag{40.35}$$

where $\mathrm{dn}(u,k)$ is the periodic elliptic function of the modulus $k = \sqrt{1-k_1^2}$ which is even relative to the variable

$$u = \tilde{\gamma}\theta \tag{40.36}$$

and has the period $2K(k)$. From eqn. (40.35) one has

$$\varphi(\theta) = 2\arccos[\mathrm{dn}(u,k)]. \tag{40.37}$$

The function $\mathrm{dn}(u,k)$ has a maximum equal to unity at $u = 0, \pm 4K(k), \ldots$, and a minimum equal to k_1 at $u = \pm K(k)\ \pm 3K(k), \ldots$. The derivative of the phase difference is

$$\varphi_\theta(\theta) = -\tilde{\gamma}k\,\mathrm{cn}(u,k). \tag{40.38}$$

According to (40.35), $1-\cos\varphi = k^2\mathrm{sn}^2(u,k)$. Therefore, the energy per period becomes

$$\mathcal{E}_1 = \frac{1}{2}k^2\tilde{\gamma}^{-1}\int_0^{2K}\left[\frac{W^2+1}{W^2-1}\mathrm{cn}^2(u,k)+4\mathrm{sn}^2(u,k)\right]du.$$

Using eqns. (29.25) and (29.22), one finally obtains

$$\tilde{\gamma}k^{-1}\mathcal{E}_1 = \frac{W^2+1}{W^2-1}[E(k)-k_1^2K(k)]+2[K(k)-E(k)].$$

41. The Solution of the Sine-Gordon Equation by the Method of Separation of Variables

In 1971, Lamb [219] proposed to solve the sine-Gordon equation with the help of a transformation which produces a separation of variables,

$$\mathrm{tg}\left[\frac{1}{4}\varphi(\xi,\tau)\right] = \Phi(\xi)F(\tau). \tag{41.1}$$

Substituting formula (41.1) into eqn. (39.13) and carrying out the partial integration, one obtains a set of two first-order differential equations relative to the variables ξ and τ,

$$\Phi_\xi^2 = -p\Phi^4 + m\Phi^2 + q, \quad \Phi \equiv \Phi(\xi), \tag{41.2}$$

$$F_\tau^2 = -qF^4 + (m-1)F^2 + p, \quad F \equiv F(\tau),$$

where $p, q,$ and m are arbitrary integration constants. The transformation (41.1) was applied in [89,130] to obtain particular solutions of eqns. (41.2). Now we consider the solutions corresponding to some values of the integration constants in (41.2).

41.1. Single-Phase Solutions

Case 1: $p = q = 0$.

In this case eqns. (41.2) are reduced to the simple set of equations,

$$\Phi_\xi = \eta\sqrt{m}\,\Phi, \quad \eta = \pm 1, \tag{41.3a}$$

$$F_\tau = \eta\sqrt{m-1}\,F, \tag{41.3b}$$

depending only on the single parameter m. Therefore, the real solution (41.1) can be written as follows

$$\mathrm{tg}\left[\frac{\varphi(\xi,\tau)}{4}\right] = \mathrm{Re}\exp\left[\eta\sqrt{m}\left[\xi - \xi_0 - (\sqrt{(m-1)/m})\tau\right]\right]. \tag{41.4}$$

This expression gives three types of solutions corresponding to different values of the parameter m: (i) $m > 1$, (ii) $0 < m < 1$, and (iii) $m < 0$.

Let $m > 1$ be given. Using the relations

$$\sqrt{\frac{m-1}{m}} = W, \quad \sqrt{m} = \gamma \equiv (1 - W^2)^{-1/2} \tag{41.5}$$

one can express m in terms of the velocity W obeying the inequalities

$$0 \leqslant W^2 < 1. \tag{41.5a}$$

Then from eqn. (41.4) it follows that

$$\mathrm{tg}\frac{\varphi}{4} = \exp[\eta\gamma(\xi - \xi_0 - W\tau)], \tag{41.6}$$

or

$$\varphi(\xi,\tau) = 4\mathrm{arctg}\{\exp[\eta\gamma(\xi-\xi_0-W\tau)]\}. \tag{41.7}$$

The solution (41.6) is just the aforementioned solution given by eqn. (40.6). It characterizes a kink if $\eta = 1$ and an antikink if $\eta = -1$.

Assuming now that the parameter m satisfies the inequality

$$0 \leqslant m < 1,$$

one can write

$$\sqrt{\frac{m-1}{m}} = i, \quad \sqrt{m} = 1/\sqrt{1+\Omega^2} \equiv Q, \quad 0 \leqslant \Omega^2 < \infty. \tag{41.8}$$

From eqn. (41.4) one obtains

$$\varphi(\xi,\tau) = 4\mathrm{arctg}[e^{\eta\xi Q}\cos(Q\Omega\tau)]. \tag{41.9}$$

This solution describes a stationary wave oscillating in time with frequency $Q\Omega$ from the kink-type solution to the antikink-type one.

Let $m < 0$ be given. Introducing the parameters

$$\sqrt{\frac{m-1}{m}} = W, \quad \tilde{\gamma} \equiv 1/\sqrt{W^2-1}, \tag{41.10}$$

one obtains

$$\varphi(\xi,\tau) = 4\mathrm{arctg}\{\cos[\tilde{\gamma}(\xi-\tilde{W}\tau)]\}. \tag{41.11}$$

This is a periodic wave propagating with velocity $\tilde{W} > 1$. Its wavelength is

$$\lambda = \frac{2\pi}{\tilde{\gamma}} = 2\pi\sqrt{\tilde{W}^2-1}, \tag{41.12}$$

which vanishes at $\tilde{W} = 1$ and increases as $\tilde{W}^2 > 1$ does, too.

41.2. Double-Phase Solutions

Case 2: $p = 0, q = 1$.

If the relations

$$p = 0, \quad q = 1 \tag{41.13}$$

hold, the set of equations (43.2) takes the form

$$\Phi_\xi^2 = m\Phi^2+1,\quad F_\tau^2 = -F^4+(m-1)F^2. \tag{41.14}$$

Assuming $m>1$, there are two classes of solutions of the set of equations (41.14). The first one is

$$\Phi(\xi) = \frac{\mathrm{sh}\gamma\xi}{\sqrt{m}},\quad F(\tau) = \frac{\sqrt{m-1}}{\mathrm{ch}(\gamma W\tau)}. \tag{41.15}$$

According to (41.1) one therefore obtains

$$\varphi_{K-K}(\xi,\tau) = 4\mathrm{arctg}\left\{\frac{W\mathrm{sh}(\gamma\xi)}{\mathrm{ch}(\gamma W\tau)}\right\}, \tag{41.16}$$

where W and γ are given by (39.9). The subscripts $K-K$ indicate that the function (41.16) describes the interaction of two kinks moving with velocities W and $-W$. In fact, as $W|\tau|\to\infty$, the function (41.16) possesses the following asymptotes

$$\varphi_{K-K}(\xi,\tau)\to\begin{cases}4\mathrm{arctg}[We^{\gamma(\xi+W|\tau|)}-We^{-\gamma(\xi-W|\tau|)}], & \tau\to-\infty,\\ 4\mathrm{arctg}[We^{\gamma(\xi-W\tau)}-We^{-\gamma(\xi+W\tau)}], & \tau\to\infty.\end{cases} \tag{41.17}$$

Defining a shift ξ_0 of the coordinate ξ via the relation

$$W = \exp(-\gamma\xi_0), \tag{41.17a}$$

one can write the asymptotes (41.17) as

$$\varphi_{K-K}(\xi,\tau)\to\begin{cases}4\mathrm{arctg}[e^{\gamma(\xi-\xi_0+W|\tau|)}+e^{-\gamma(\xi-\xi_0-W|\tau|)}], & \tau\to-\infty,\\ 4\mathrm{arctg}[e^{\gamma(\xi-\xi_0-W\tau)}+e^{-\gamma(\xi-\xi+W\tau)}], & \tau\to\infty.\end{cases} \tag{41.18}$$

Each line of this expression describes kinks moving in opposite directions with velocities W. Their collision occurs at ξ_0 and $\tau=0$.

The other class of solutions of eqns. (41.14) are

$$\Phi(\xi) = \frac{l\mathrm{ch}(\gamma\xi)}{\sqrt{m}},\quad F(\tau) = \frac{-\iota\sqrt{m-1}}{\mathrm{sh}(\gamma W\tau)}. \tag{41.19}$$

Therefore,

$$\varphi_{K-A} = 4\mathrm{arctg}\left\{\frac{W\mathrm{ch}(\gamma\xi)}{\mathrm{sh}(\gamma W\tau)}\right\}. \tag{41.20}$$

One can write the asymptotes of this function in the following form

$$\varphi_{K-A} \underset{\tau\to-\infty}{=} 4\mathrm{arctg}[e^{\nu(\xi-\xi_0-W|\tau|)}+e^{-\nu(\xi+\xi_0+W|\tau|)}], \tag{41.21a}$$

$$\varphi_{K-A} \underset{\tau\to+\infty}{=} 4\operatorname{arctg}[e^{\nu(\xi-\xi_0-W|\tau|)}+e^{-\nu(\xi+\xi_0+W\tau)}]. \tag{41.21b}$$

So, the function (41.20) characterizes the collision of a kink and an antikink as they pass through each other. This process is schematically depicted in Fig. 18.

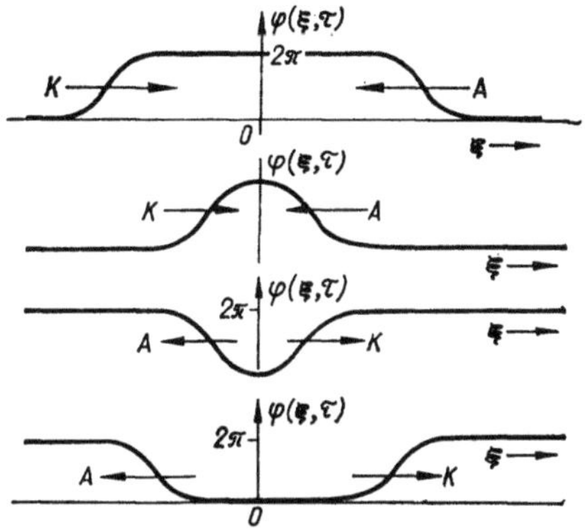

Figure 18. The collision of a kink (K) with an antikink (A).

Let the value of m which satisfies the inequality

$$0<m<1 \tag{41.22}$$

be given. The solutions of the set of eqns. (41.14) are then determined by the functions

$$\Phi(\xi) = \frac{i\operatorname{ch}(\sqrt{m}\,\xi)}{\sqrt{m}}, \tag{41.23}$$

$$F(\tau) = \frac{i\sqrt{1-m}}{\sin(\tau\sqrt{1-m})}.$$

Therefore,

$$\varphi(\xi,\tau) = 4\operatorname{arctg}\frac{\sin[\Omega\tau/\sqrt{1+\Omega^2}]}{\Omega\operatorname{ch}[\xi/\sqrt{1+\Omega^2}]} = 4\operatorname{arctg}\left\{\operatorname{tg}\mathrm{v}\frac{\sin(\tau\cos\mathrm{v})}{\operatorname{ch}(\xi\sin\mathrm{v})}\right\}, \tag{41.24}$$

where

$$\Omega = \sqrt{\frac{1-m}{m}} = (\operatorname{tg}\mathrm{v})^{-1}.$$

The solution (41.21) represents a localized pulsating object, or a bound state of a kink with an antikink which is known as a *breather*, or *breather-soliton*, or, finally, a *bion*. The breather-type solution satisfies the boundary conditions: $\varphi(\xi,\tau)\rightarrow\infty$ as $|\xi|\rightarrow\infty$. Hence, it is sometimes called a 0π-pulse. The internal frequency ω of pulsations of the localized breather is given by the formula

$$\omega = \cos\nu = \Omega/\sqrt{1+\Omega^2}. \tag{41.25}$$

The spatial spread of the pulsating region is inversely proportional to the quantity

$$Q = \sin\nu = 1/\sqrt{1+\Omega^2}. \tag{41.26}$$

Using eqns. (41.25) and (41.26), one obtains the equality

$$\omega^2+Q^2 = 1. \tag{41.27}$$

As $m\rightarrow 1$, Ω and ω approach zero, $\nu\rightarrow\pi/2$, and $Q\rightarrow 1$.

The transition from an immobile breather to that moving as a whole with velocity W is realized by means of the Lorentz transformation (39.8)

$$\tau\rightarrow\gamma(\tau-W\xi),\quad \gamma\equiv(1-W^2)^{-1/2},\quad \xi\rightarrow\gamma(\xi-W\tau).$$

Hence, the breather moving with velocity W is given by the function

$$\varphi(\xi,\tau) = 4\operatorname{arctg}\left\{\frac{\operatorname{tg}\nu\sin[(\tau-W\xi)\gamma\cos\nu]}{\operatorname{ch}[(\xi-W\tau)\gamma\sin\nu]}\right\}. \tag{41.28}$$

The frequency of pulsations in the mobile breather is

$$\omega W = \gamma\cos\nu = \frac{\gamma\Omega}{\sqrt{1+\Omega^2}}. \tag{41.29}$$

The total energy (39.6) and momentum (39.7) of a breather have been evaluated by Faddeev and Takhtadzian [117],

$$\mathcal{E}_{\mathrm{br}} = \frac{16\sin\nu}{\sqrt{1-W^2}},\quad \mathcal{P}_{\mathrm{br}} = \frac{16W\sin\nu}{\sqrt{1-W^2}}. \tag{41.30}$$

The breather's velocity is given by the ratio

$$W = \mathcal{P}_{\mathrm{br}}/\mathcal{E}_{\mathrm{br}}.$$

Hence the following relation

$$\mathcal{E}_{\mathrm{br}}^2 = \mathcal{P}_{\mathrm{br}}^2+M_{\mathrm{br}}^2 \tag{41.31}$$

holds. In eqn. (41.31), $M_{\rm br} = 16\sin\nu$ is the rest breather's mass.

According to (40.10), the total energy of a free kink and an antikink accounts for 16γ. Thus, accompanying the formation of a breather is an energy release of

$$E_{\rm binding} = 16|\gamma - \sin\nu|, \tag{41.32}$$

which can be interpreted as the binding energy. According to eqn. (41.25), as $m \to 1$ and $\nu \to \pi/2$, the binding energy (41.32) approaches zero, and the breather dissociates into a kink and an antikink.

41.3. Standing Waves

Case 3: $p \neq 0, q \neq 0$.

Applying, in this case, the following transformations

$$\Phi(\xi) = \left(\frac{q}{p}\right)^{1/4} \tilde{\Phi}(\xi) \text{ and } F(\tau) = \left(\frac{p}{q}\right)^{1/4} \tilde{F}(\tau), \tag{41.33}$$

one obtains the set of equations,

$$\tilde{\Phi}_\xi^2 = -\alpha\tilde{\Phi}^4 + m\tilde{\Phi}^2 + \alpha, \tag{41.34a}$$

$$\tilde{F}_\tau^2 = -\alpha\tilde{F}^4 + (m-1)\tilde{F}^2 + \alpha, \tag{41.34b}$$

$$\Phi(\xi)F(\tau) = \tilde{\Phi}(\xi)\tilde{F}(\tau),$$

the tilde superscript is henceforth removed.

If $\alpha > 0$, eqn. (41.34a) can be transformed as follows

$$\Phi_\xi^2 = \alpha(\Phi_0^2 - \Phi^2)(\Phi^2 + \Phi_1^2), \tag{41.35}$$

where

$$\Phi_0^2 = (2\alpha)^{-1}[m + \sqrt{m^2 + 4\alpha}], \quad \Phi_1^2 = (2\alpha)^{-1}[\sqrt{m^2+4\alpha} - m]. \tag{41.36}$$

Integrating eqn. (41.35), one obtains

$$\eta(\xi - \xi_0) = \frac{1}{\sqrt{\alpha}} \int_{\Phi_*}^{\Phi(\xi)} \frac{d\Phi}{\sqrt{(\Phi_0^2 - \Phi^2)(\Phi^2 + \Phi_1^2)}} = \tag{41.37}$$

$$= \frac{k}{\Phi_0\sqrt{\alpha}} \int_1^{\Phi(\xi)/\Phi_0} \frac{dt}{\sqrt{(1-t^2)(k^2t^2 + k_1^2)}},$$

where $\eta = \pm 1$,

$$t = \frac{\Phi(\xi)}{\Phi_0}, \quad \left[\frac{k_1}{k}\right]^2 = \left[\frac{\Phi_1}{\Phi_0}\right]^2, \quad k^2 = \sqrt{1-k_1^2}. \tag{41.38}$$

Inverting the integral on the right-hand side of eqn. (41.37), one has the Jacobian elliptic function $\mathrm{cn}(u,k)$. Hence, from eqn. (41.37) it follows

$$\Phi(\xi) = \Phi_0 \mathrm{cn}(u,k), \quad u \equiv \alpha\Phi_0(\xi-\xi_0)/k, \tag{41.39}$$

where Φ_0 is given by the formula (41.36). The modulus of the elliptic function is determined in terms of the parameters m and α,

$$k^2 = \frac{m+\sqrt{m^2+4\alpha}}{2\sqrt{m^2+4\alpha}}. \tag{41.40}$$

Replacing m by $m-1$ and $\xi-\xi_0$ by τ in eqn. (41.39), one can obtain the solution of eqn. (41.34b),

$$F(\tau) = F_0 \mathrm{cn}\left[\frac{F_0\sqrt{\alpha}}{q}\tau, q\right], \tag{41.41}$$

where

$$F_0 = \frac{m-1+\sqrt{(m-1)^2+4\alpha}}{2\alpha}, \quad q^2 = \frac{m-1+\sqrt{(m-1)^2+4\alpha}}{2\sqrt{(m-1)^2+4\alpha}}. \tag{41.42}$$

The general solution of eqn. (41.1) takes the form

$$\mathrm{tg}\left[\frac{\varphi(\xi,\tau)}{4}\right] = A\,\mathrm{cn}\left[\frac{\Phi_0\sqrt{\alpha}}{k}(\xi-\xi_0), k\right]\mathrm{cn}\left[\frac{F_0\sqrt{\alpha}}{q}\tau, q\right], \tag{41.43}$$

where

$$A = \Phi_0 F_0.$$

The function (41.43) depicts a standing periodic wave depending on the spatial ξ and time τ variables. This wave characterizes 'plasma oscillations' (see [97]). The function $\mathrm{cn}[k^{-1}\Phi_0\sqrt{\alpha}(\xi-\xi_0),k]$ is periodic relative to ξ with a period

$$l = 4kK(k)/\Phi_0\sqrt{\alpha}. \tag{41.44}$$

It varies between -1 and $+1$ over the period. The second factor in eqn. (41.43) has a temporal period of

$$T = 4qK(q)/F_0\sqrt{\alpha}. \tag{41.45}$$

In the limiting case, $\alpha \to 0$, both the moduli k and q approach unity. In this case, the spatial and temporal periods both tend to infinity, and the functions behave as follows

$$\mathrm{cn}(u,k) \to \mathrm{sch}u, \quad \mathrm{cn}(\mathrm{v},q) \to \mathrm{schv}$$

and take the form of bell-like curves with a maximum at $\xi = \xi_0$ and $\tau = 0$.

Real Josephson junctions are bounded, and this provides interest to the solutions satisfying specific boundary conditions. The simplest boundary conditions are the following

$$\left[\frac{\partial\varphi}{\partial\xi}\right]_{\xi=0} = \left[\frac{\partial\varphi}{\partial\xi}\right]_{\xi=1} = 0. \tag{41.46}$$

They correspond to the Josephson line with open ends. The boundary conditions enable the determination of ξ_0 and discrete values of the modulus k of the elliptic function $\mathrm{cn}(u,k)$.

Replacing q by $-r<0$ in the set of equations (41.2) and carrying out further the transformation

$$\Phi(\xi) = \left(\frac{r}{p}\right)^{1/4}\tilde{\Phi}(\xi), \quad F(\tau) = \left(\frac{p}{r}\right)^{1/2}\tilde{F}(\tau),$$

one obtains the set of equations (tilde omitted)

$$\Phi_\xi^2 = -\beta\Phi^4 + m\Phi^2 - \beta, \beta = \sqrt{rp} > 0, \tag{41.47}$$

$$F_\tau^2 = \beta F^4 + (m-1)F^2 + \beta,$$

which determines the second type of standing waves. Assuming

$$0 < m < 1 \quad \text{and} \quad 4\beta^2 < (m-1)^2$$

one transforms these equations as follows

$$\Phi_\xi^2 = \beta(\Phi_0^2 - \Phi^2)(\Phi^2 - \Phi_1^2), \tag{41.48a}$$

$$F_\tau^2 = \beta(F_0^2 - F^2)(F_1^2 - F^2), \tag{41.48b}$$

where

$$\Phi_0^2 = \frac{m+\sqrt{m^2-4\beta}}{2\beta}, \quad \Phi_1^2 = \frac{m-\sqrt{m^2-4\beta^2}}{2\beta}, \tag{41.49a}$$

$$F_0^2 = \frac{1-m+\sqrt{(1-m)^2-4\beta}}{2\beta}, \quad F_1^2 = \frac{1-m-\sqrt{(1-m)^2-4\beta}}{2\beta}. \tag{41.49b}$$

Solving eqn. (41.48a), one obtains the function of the spatial variable ξ,

$$\Phi(\xi) = \Phi_0 \mathrm{dn}[\Phi_0\sqrt{\beta}(\xi-\xi_0),k], \tag{41.50}$$

where $\mathrm{dn}(u,k)$ is the Jacobian elliptic function with the modulus k,

$$k = \sqrt{1-k_1^2}, \quad k_1^2 = \left[\frac{\Phi_1}{\Phi_0}\right]^2 = \frac{m-\sqrt{m^2-4\beta}}{m+\sqrt{m^2-4\beta}}. \tag{41.51}$$

Solving further eqn. (41.48b), one obtains the time-dependent function

$$F(\tau) = F_0 \mathrm{sn}\left[\frac{F_0\sqrt{\beta}\tau}{q}, q\right], \tag{41.52}$$

where $\mathrm{sn}(u,k)$ is the Jacobian elliptic function with the modulus q,

$$q^2 = \frac{1-m-\sqrt{(1-m)^2-4\beta}}{1-m+\sqrt{(1-m)^2-4\beta}}. \tag{41.35}$$

Therefore, the solution (41.1) of the sine-Gordon equation taking into account eqns. (41.47) takes the form

$$\mathrm{tg}\left[\frac{1}{4}\varphi(\xi,\tau)\right] = B\,\mathrm{dn}(u,k)\mathrm{sn}(\mathrm{v},q), \tag{41.54}$$

where

$$B = \Phi_0 F_0, \quad u \equiv \Phi_0\sqrt{\beta}(\xi-\xi_0), \quad \mathrm{v} \equiv \frac{F_0\sqrt{\beta}}{q}\tau.$$

The function $\mathrm{dn}(u,k)$ is a periodic function of ξ with period

$$l = 2K(k)/\Phi_0\sqrt{\beta}, \tag{41.55}$$

where $K(k)$ is the complete elliptic integral of the first kind. The solutions (41.54) are called 'breather oscillations' [97]. ξ_0 and discrete values of the modulus k of the elliptic function $\mathrm{dn}(u,k)$ are evaluated by means of the boundary conditions

$$\left[\frac{\partial\varphi}{\partial\xi}\right]_{\xi=0} = \left[\frac{\partial\varphi}{\partial\xi}\right]_{\xi=l}. \tag{41.56}$$

42. Josephson Tunnelling Effect in Superconductors

In superconductors at absolute zero temperature, all electrons near the Fermi surface are combined into pairs in the singlet spin state with vanishing total momentum. Because of this, the motion of electrons is correlated over distances of the order of the region of the pairing, $\sim 10^{-4} cm$. This distance is called the *coherence length*.

If two superconductors are separated by a thin dielectric layer of a thickness of the order of $10-20\,\text{Å}$ which is much less than the coherence length, due to the overlap of the wave functions of electrons belonging to neighboring superconductors, there occurs a tunnelling of electrons from one superconductor to the other. This effect was first predicted and revealed by the English physicist Josephson in 1962 and is known as the Josephson effect. Josephson demonstrated that a current satisfying the set of equations,

$$I = I_0 \sin\varphi, \tag{42.1}$$

$$\frac{d\varphi}{dt} = 2eV/\hbar, \tag{42.2}$$

flows through an insulating layer playing the role of a potential barrier.

In eqns. (42.1) and (42.2), e is the unit electric charge, I is the current density, I_0 is the maximum of I_0 depending on the material, temperature, and thickness of the isolating layer, φ is the real phase difference of the electron wave functions in the superconductors on both sides of the layer, V is the difference of electric potentials of the isolating layer which is sometimes called a junction.

From eqn. (42.2), it follows that the potential difference is conditioned by the penetration of an electron pair through the layer, or junction. Calculations demonstrate that the tunnelling process goes through the intermediate state in which pairs are dissociated. Hence, the probability of tunnelling is determined by transmitting of a single electron [83]. Equations (42.1) and (42.2) imply directly the possibility of a flow of current through a junction even at zero potential difference under constant φ. If the potential difference is nonvanishing, the current density through the junction becomes a time-dependent periodic function

$$I(t) = I_0 \sin(\omega t + \varphi_0) \tag{42.3}$$

with frequency ω proportional to the potential difference

$$\omega = 2eV/\hbar. \tag{42.4}$$

Let us now consider only extended Josephson junctions where the insulator layer has a large dimension in a single direction, which we assume to be the x-axis. We also assume that a constant magnetic field B is applied to the junction in a perpendicular direction (the y-axis), and that a potential difference V is applied across the layer (along the z-axis). In this case, the phase difference φ, current density I, and potential difference V are functions of x and independent of y. The phenomenological equation, describing the dependence of the phase difference on the variables x and t, has the form

$$c_0^2 \frac{\partial^2 \varphi}{\partial x^2} - \frac{\partial^2 \varphi}{\partial t^2} = \omega_0^2 \sin\varphi_0. \tag{42.5}$$

In eqn. (42.5)

$$c_0 = \omega_0 \Lambda_0 = (LC)^{-1/2} \tag{42.6}$$

is the velocity of an electromagnetic wave in the insulating layer in the absence of a current, and L and C are the inductance and capacitance, respectively, of a unit length of the junction along the x-axis. Such waves are called *Sweehart waves*. Their velocity, $c_0 \sim 10^{-2}$, is considerably less than that of light in vacuum, c. The quantities Λ_0 and ω_0 appearing in eqn. (42.6) are determined through a quanta of magnetic flux

$$\Phi_0 = \pi\hbar c/e \approx 2 \times 10^{-7} \text{ Gauss}/cm^2 \tag{42.7}$$

where

$$\Lambda_0 = \sqrt{\Phi_0 / 2\pi L I_0} \tag{42.8}$$

is the so-called 'Josephson penetration depth' which is equal to the characteristic distance, of the order of $10^{-2} cm$, over which the magnetic field penetrates inside the junction, and

$$\omega_0 = \sqrt{2\pi I_0 / C\Phi_0} \tag{42.9}$$

is the characteristic, or Josephson, frequency.

The phase difference $\varphi(x,t)$ defined by eqn. (42.5) characterizes completely the dynamics of the phenomena which takes place at the junction. In particular, the electric potential difference is expressed in terms of the phase difference as follows:

$$V_z = \frac{\Phi_0}{2\pi c}\frac{\partial\varphi}{\partial t}. \tag{42.10}$$

A similar equation is valid for the magnetic field:

$$B_y = \frac{\Phi_0}{2\pi c}\frac{\partial\varphi}{\partial x}. \tag{42.11}$$

Introducing the dimensionless length ξ and time τ via the relations

$$\xi = x/\Lambda_0 \quad \text{and} \quad \tau = t\omega_0, \tag{42.12}$$

one obtains the canonical form of eqn. (42.5) which determines the phase difference $\varphi(\xi,\tau)$ within the laboratory coordinate frame,

$$\varphi_{\xi\xi} - \varphi_{\tau\tau} = \sin\varphi. \tag{42.13}$$

In eqn. (42.13) and henceforth, the subscripts denote the derivatives relative to the corresponding variables. The function φ of ξ and τ characterizes a scalar field. By assumption, it satisfies the following boundary conditions: $\varphi(\xi,\tau)\rightarrow 0(\mathrm{mod}2\pi)$ as $|\xi|\rightarrow\infty$.

Representing the fields V_z and B_y in units of the following quantities,

$$V_0 = \Phi_0\Omega_0/2\pi c \quad \text{and} \quad B_0 = \Phi_0/2\pi\Delta_0, \tag{42.14}$$

the dimensionless fields $\nu_z = V_z/V_0$ and $h_y = B_y/B_0$ are given in terms of the phase difference $\varphi(\xi,\tau)$ by means of simple differentiation,

$$\mathrm{v}_z(\xi,\tau) = \varphi_\tau(\xi,\tau), \tag{42.15}$$

$$h_y(\xi,\tau) = \varphi_\xi(\xi,\tau). \tag{42.16}$$

The Josephson effect is a remarkable illustration of the application of the laws of quantum mechanics over a macroscopic scale in superconductors. It provides a study of the details of the long-range behavior in the superconducting state. Using this effect, one can measure with high accuracy the ratios of fundamental physical quantities, e.g. the electric charge e and Planck's constant $\hbar$. Generators of monochromatic coherent radiation of small power, $10^{-10}-10^{-13}\,W$, in the wavelength interval from a few millimeters to fractions of a millimeter, have been produced on the basis of the Josephson effect. A list of the original papers, reviews, and monographs devoted to the theoretical and experimental study of this effect would be rather lengthy. A detailed description of the Josephson effect is given in the books by Kulik and Yanson [83], Dmitrenko [59], and in the reviews [181,90,249,250].

In the case of the extended Josephson junction, displacement of the phase difference is related with that of the difference of electric potential and magnetic field. In agreement with eqns. (42.15) and (42.16), their dimensionless quantities are described by bell-like functions (see Fig. 14)

$$v_{z,\eta}(\xi,\tau) = -2\eta W\gamma \mathrm{sch}[\gamma(\xi-\xi_0 - W_\tau)],$$

$$h_{z,\eta}(\xi,\tau) = 2\eta\gamma \mathrm{sch}[\gamma(\xi-\xi_0 - W\tau)].$$

The magnetic flux through a junction of width $1cm$ is

$$\Phi = \int_{\infty}^{\infty} B_{y,\eta}(x,t)dx = B_0 \int_{-\infty}^{\infty} h_{y,\eta}(\xi,\tau)d\xi = \eta\Phi_0.$$

Therefore, the kink, $\eta = 1$, carries a single quanta of magnetic flux in the extended Josephson junction. For this reason, such an excitation is called a *fluxon*. If $\eta = -1$, the excitation is called an *antifluxon*. A fluxon is an extremely stable formation and can be fairly easy controlled with the help of external effects, which results in its use as a basic unit of information.

43. The Dynamics of Dislocations in Crystals

Solids having needle-like and layer-like structures are of current interest. A quasi-one-dimensional approach can be applied for describing some phenomena in such crystals. These quasi-one-dimensional structures arise in particular due to local breakings of the regular crystalline lattice. Dislocations in crystals refer to such local breakings.

Linear defects of crystalline structure, known as *dislocations*, are breakdowns of the regular crystal lattice along a certain direction over which the crystal regions, displaced relative to each other, are adjacent. In the particular case of rectilinear edge dislocation, the effect of the nondisplaced half-space of a crystal on the atoms arranged at the dislocation axis can be described via the Hamiltonian involving the interaction in the form of a periodic potential

$$V(u) = \omega_0 \sin^2 u.$$

The inclusion of such an interaction results in nonlinear equations involving the unknown function in sine form. Frenkel and Kontorova were the first who, in 1938, applied these equations to yield a description of the motion of dislocations in solids [124,182]. The theory of various physical phenomena based on the treatment of dislocations in crystals has been presented in numerous books, such as those by Frenkel [123] and Kosevich [79].

Dislocations were originally considered as immobile states of a crystal which were related to internal strain. A study of the dynamical properties of the motion of dislocations was first carried out in the aforementioned works by Frenkel and Kontorova, who developed the simplest microscopic theory of a one-dimensional atomic model of moving dislocations. Namely, for a model of dislocations they considered a chain of atoms of mass m displaced by distances u_n relative to the equilibrium positions of a crystal which is represented by a second chain of atoms rigidly fixed at an interatomic separation a.

The Hamiltonian operator of such the system takes the following form

$$H = \frac{1}{2}\sum_n [m u_n^2 + \kappa(u_n - u_{n-1})^2 + U(u_n)], \tag{43.1}$$

where

$$U(u_n) = U_0[1 - \cos(2\pi u_n / a)] \tag{43.2}$$

is the potential of the nth atom due to the fixed chain, and κ is the elastic constant of relative atomic displacements. The following set of equations of motion then results:

$$m\frac{d^2 u_n}{dt^2} + \kappa(2u_n - u_{n-1} - u_{n+1}) + \frac{\pi U_0}{a}\sin\left(\frac{2\pi u_n}{a}\right) = 0. \tag{43.3}$$

Applying the continuum approach by replacing $u_n(t) \to u(x,t)$, and introducing the function

$$\varphi(x,t) = 2\pi u(x,t)/a, \tag{43.4}$$

one obtains the nonlinear differential equation

$$\frac{\partial^2\varphi}{\partial x^2} - \frac{1}{c_0^2}\frac{\partial^2\varphi}{\partial t^2} = \frac{1}{\lambda_0^2}\sin\varphi, \tag{43.5}$$

where

$$c_0 = a\sqrt{\kappa/m} \tag{43.6}$$

is the longitudinal sound velocity, and

$$\lambda_0 \equiv \frac{\kappa^2}{2\pi}\sqrt{\kappa/U_0} \tag{43.7}$$

is the characteristic length.

Equation (43.5) is just the sine-Gordon equation. It has a stationary solution in the form of the domain wall (kink and antikink) travelling with a constant velocity W,

$$\varphi(x,t) = 4\mathrm{arctg}\left[\eta\exp\frac{x - x_0 - Wt}{\lambda_0\sqrt{1 - W^2/c_0^2}}\right], \tag{43.8}$$

where $\eta = 1$ corresonds to a kink, and $\eta = -1$ to an antikink. The quantity

$$2\lambda_0\sqrt{1 - W^2/c_0^2} \tag{43.9}$$

determines the 'width' of the domain wall. If the velocity W exceeds the longitudinal sound velocity c_0, the function $\varphi(x,t)$ becomes an oscillating function. In the opposite case, i.e., $W < c_0$, the function $\varphi(\zeta)$, with $\zeta = x - x_0 - Wt$, varies from -2π to 2π as ζ ranges over the interval $-\infty$, ∞ for $\eta = 1$. According to eqn. (43.4), this variation of $\varphi(\xi)$ corresponds to that of displacement from $-a$ to a.

Within the continuum approach, the Hamiltonian (43.1) takes the form

$$H = \frac{1}{2a}\int\left[m\left(\frac{\partial u}{\partial t}\right)^2 + \kappa a^2\left(\frac{\partial u}{\partial x}\right)^2 + 4U_0\sin^2(\varphi/2)\right]dk. \tag{43.10}$$

Substituting eqn. (43.8) into eqn. (43.10) and using eqn. (43.4), one obtains the explicit formula for the energy of a dislocation moving with velocity W,

$$\mathcal{E}(W) = \mathcal{E}_0/\sqrt{1 - W^2/c_0^2}, \tag{43.11}$$

where

$$\mathcal{E}_0 = 8a\sqrt{\kappa U_0} \tag{43.12}$$

is the energy of the dislocation at rest. The crystal becomes more plastic and less solid with decrease of $\mathcal{E}_0$. The dislocation carries the momentum

$$\mathcal{P}(W) = -\frac{m}{a}\int u_t u_x dx = \frac{M_0 W}{\sqrt{1-W^2/c_0^2}}, \tag{43.13}$$

where

$$M_0 = 8\sqrt{mU_0}/c_0 = \frac{8m}{a}\sqrt{U_0/\kappa}$$

is the mass of the dislocation. The energy of the momentum of the dislocation are related by the formula

$$\mathcal{E} = (W)\sqrt{\mathcal{E}_0^2 + c_0^2\mathcal{P}^2(W)}. \tag{43.14}$$

A similar relationship holds for the rest energy and mass,

$$\mathcal{E}_0 = Mc_0^2. \tag{43.15}$$

In deriving these formula we neglect the processes of energy dissipation. Taking into account the possibility of displacements of the 'crystal' atoms, one concludes that there occurs an exchange of energy between atoms under the motion of a dislocation. For an intensive heat motion, the average run length of a dislocation is comparable with the lattice constant, and the dislocation motion may be described by the diffusion equation. At present the theory of the energy losses experienced by a dislocation as it propagates has been well developed (for details see the review by Indelbom and Orlow [72]).

44. Bloch Wave Motion

The sine-Gordon equation has been widely applied to the study of the nonlinear dynamics of magnetization in ferromagnetics. A ferromagnetic having the so-called 'axis of easy magnetization' anisotropy possesses a preferred axis of magnetic anisotropy. The ground state of such a crystal corresponds to two directions of the magnetic moment: along ($\varphi=0$), and opposite to ($\varphi=\pi$), the axis (y) of easy magnetization. One can say that in this case there exists two vacuum states, i.e., the ground state is doubly degenerate.

Assuming the length of the magnetization vector to be constant, one can characterize the magnetization state of a given ferromagnetic at each spatial point by a single variable: the angle φ between the magnetization vector and the easy magnetization axis. We also assume that the angle φ refers to the yz plane, and varies along the x-axis. Hence, the problem is to seek the dependence of $\varphi(x)$ on the x-variable.

The double degeneracy of the ground state of the crystal suggests the states in which the angle φ takes different values at ∞ and $-\infty$, e.g. $\varphi(-\infty) = 0$ and $\varphi(\infty) = 2\pi$. The region separating these two states of magnetization is called a *domain wall*, inside of which the magnetization varies slowly from one value of φ to another. A travelling domain wall is called a *Bloch wave*.

Ferromagnetic magnetization is determined by spin exchange, and the orientation of magnetization momentum by the crystal anisotropy. In the case of large anisotropy, one can neglect the magnetic-dipole interaction between spins. Then the Hamiltonian function referred to the unit cross-section of the ferromagnetic an immobile domain wall can be represented by the sum of two terms:

$$H = \int \left[B\sin^3\varphi + A\left[\frac{\partial\varphi}{\partial x}\right]^2 \right] dx. \tag{44.1}$$

In eqn. (44.1), the first term describes the anisotropy energy, and the second term describes the exchange energy. A is strongly positive for ferromagnetics having an easy anisotropy axis. Varying the function (44.1), one obtains the following equation

$$\frac{d^2\varphi}{dx^2} = \frac{B}{2A}\sin(2\varphi), \tag{44.2}$$

which characterizes the crystal magnetization. Its particular solution is

$$\varphi = \pi n/2, \quad n = 0, \pm 1, \pm, \ldots \tag{44.3}$$

which corresponds to uniform magnetization.

We seek those solutions of eqn. (44.2) which satisfy the condition

$$\left[\frac{d\varphi}{\partial x}\right]_{\varphi=0} = 0. \tag{44.4}$$

Integrating this equation, one obtains

$$\frac{d\varphi}{dx} = \pm\sqrt{\frac{B}{A}}\sin(\varphi/2). \tag{44.5}$$

Integrating further, and imposing the condition $\varphi(x_0) = \pi$, we finally obtain

$$\mathrm{tg}[\varphi(x)/4] = \exp\left[\pm\sqrt{\frac{B}{A}}(x-x_0)\right], \tag{44.6}$$

where the 'plus' sign corresponds to the rest domain wall in the form of a kink, and the 'minus' sign to an antikink. In the case of the positive sign, φ varies from 0 to 2π inside the domain wall. The width of the domain wall is

$$L_0 = \approx \pi\sqrt{\frac{A}{B}} = \int_{-\infty}^{\infty}\sin\varphi dx. \tag{44.7}$$

Substituting eqn. (44.6) into (44.1), one determines the energy of the excitation at rest:

$$\mathcal{E}_0 = 4\sqrt{AB}. \tag{44.8}$$

This energy is basically concentrated within the domain wall.

If a constant external magnetic field is applied to the crystal and directed towards the easy magnetization axis, the region where the magnetization vector is parallel to the field is extended due to the region having an opposite magnetization. Therefore this results in the movement of the domain well due to the rotation of the magnetization vector. The equation of motion of the magnetization vector, i.e., the variation of its angle of rotation, is just the sine-Gordon equation [178]

$$\frac{\partial^2\varphi}{\partial x^2} - \frac{1}{c_0^2}\frac{\partial^2\varphi}{\partial t^2} = \frac{B}{2A}\sin(2\varphi), \tag{44.9}$$

where $c_0 = 2\gamma(2\pi A)^{1/2}$ is the limiting velocity of the Bloch wave, $\gamma = e/2m_ec$ is the gyromagnetic ratio of an electron with mass m_e and charge e, and c is the velocity of light in vacuum.

The solution of eqn. (44.9), which corresponds to the kink (antikink) motion with velocity W, has the form

$$\mathrm{tg}(\varphi/4) = \exp\left[\pm\sqrt{\frac{B}{A}}\xi\right], \tag{44.10}$$

where

$$\xi \equiv (x - x_0 - Wt) / \sqrt{1 - W^2 / c_0^2}. \tag{44.11}$$

Hence, the width L_0 of the domain wall decreases under the motion:

$$L(W) = L_0 \sqrt{1 - W^2 / c_0^2}.$$

The excitation energy becomes

$$\mathcal{E}(W) = \mathcal{E}_0 + \frac{1}{2} M_0 W^2, \tag{44.12}$$

where

$$M_0 = \frac{1}{2L_0\gamma^2} = \frac{\mathcal{E}_0}{8\pi\gamma^2 A} = \mathcal{E}_0 / c_0^2$$

is the excitation mass.

The dynamics of the magnetization of ferromagnetics has been studied by many authors. Weak excitations of a ferromagnetic are described by linear equations whose solutions are characterized by the superposition of magnons - quanta of plane spin waves. A complete description of spin waves is given in the book by Akhiezer et al. [10] in which the authors also demonstrate the fruitfulness of their approach for explaining a large number of physical phenomena in weakly excited ferromagnetics.

Magnons generate a quasi-ideal Bose gas. However, based on the exact quantum-mechanical calculation of the states of a one-dimensional spin chain without magnetic anisotropy, Bethe [140] demonstrated that, together with isolated spin waves, there exist magnon complexes travelling as a unique whole with a specific energy and momentum. The existence of the bound state indicates a breakdown of the linear approach.

If the degree of excitation of a ferromagnetic is not rather small, there arise substantially nonlinear effects. Kosevich, Akhiezer, Bar'yakhtar, Borovik, Ivanov, and other authors [12,13,70,71,80,178] have demonstrated that for studying nonlinear phenomena in ferromagnetics it is convenient to apply macroscopic equations of magnetization dynamics. In this case, nonlinear excitations are considered as bound states of many magnons. This approach enables an explanation of the basic types of nonlinear excitations in one- and three-dimensional ferromagnetics of different kinds. The main results of these investigations have been reviewed by Kosevich [80].

45. The Propagation of Ultrashort Optical Pulses

Advances in laser technology have enabled the production of coherent optical pulses of rather small duration, $10^{-9} - 10^{-12}\,s$. The passage of such pulses through two-level atomic systems with large relaxation times reveals some important and novel effects.

If the pulse duration is considerably shorter than the time of phase memory and other relaxation times of the atomic system, definite phase relationships between the incident pulse and the induced polarization of the medium take place during the time of passage of the pulse in a system. In this case, under the passage of a sufficiently intense pulse through a medium, its leading edge excites atoms while its back edge rehabilitates the energy of the transient pulse by means of the stimulated radiation. Under such conditions, the pulse propagates through a medium without attenuation. Its velocity is considerably less than the phase velocity of light in the medium. The electric field profile of such a steady-state pulse is described by a hyperbolic secant. The propagation of high-power ultrashort pulses through an atomic system without attenuation is known as self-induced transparency. This phenomenon was predicted and discovered by McCall and Hahn [230,231] in 1969, and discussed in detail in the reviews by Lamb [218,219] and Scott et al. [138].

The simplest way to study theoretically self-induced transparency is to consider an ensemble of noninteracting two-level atoms having an excitation frequency ω and transition dipole moment d. An electromagnetic wave having the same frequency propagates through this system,

$$E(x,t) = \mathcal{E}(x,t)\cos(\omega t - kx) \qquad (45.1)$$

where the amplitude $\mathcal{E}(x,t)$ depends weakly on x and t. Due to its influence, there arises a specific polarization in the medium

$$P(x,t) = \mathcal{P}(x,t)\sin(\omega t - kx). \qquad (45.2)$$

The state of the system in question is described by the following set of equations

$$\frac{\partial \mathcal{E}}{\partial x} + \frac{1}{c}\frac{\partial \mathcal{E}}{\partial t} = \alpha \mathcal{P}, \quad \alpha = \frac{2\pi\omega}{\hbar c} n_0 d^2, \tag{45.3}$$

$$\frac{\partial \mathcal{P}}{\partial t} + \frac{1}{T_1} = \mathcal{P} = \mathcal{E}N, \quad \frac{\partial N}{\partial t} + \frac{1}{T_2}N = -\mathcal{E}\mathcal{P}, \tag{45.4}$$

where n_0 is the density of atoms, $N = n/n_0$ is the normalized density of excited atoms, and T_1 and T_2 are the longitudinal and transverse relaxations of excited states, respectively. If the time interval of action of the light pulse is considerably less than T_1 and T_2, the set of eqns. (45.3) and (45.4) is replaced by the simpler set

$$\frac{\partial \mathcal{P}}{\partial t} = N\mathcal{E}, \quad \frac{\partial N}{\partial t} = -\mathcal{E}N, \quad \frac{\partial \mathcal{E}}{\partial x} + \frac{1}{c}\frac{\partial \mathcal{E}}{\partial t} = \alpha\mathcal{P}. \tag{45.5}$$

The following conservation law results

$$\mathcal{P}^2(x,t) + N^2(x,t) = 1. \tag{45.6}$$

The set of equations (45.5) may be solved using the following substitution

$$\mathcal{E}(x,t) = \frac{\partial \varphi(x,t)}{\partial t} \tag{45.7}$$

where the angular variable $\varphi(x,t)$ satisfies the sine-Gordon equation

$$\left[\frac{\partial^2}{\partial x \partial t} + \frac{1}{c^2}\frac{\partial^2}{\partial t^2}\right]\varphi(x,t) = \pm\sin\varphi(x,t). \tag{45.8}$$

In eqn. (45.8), the upper (plus) sign corresponds to the case when all atoms are initially excited (the amplifying medium), and the lower (minus) sign corresponds to the attenuating medium.

Measuring distance and time in units of $(c/\alpha)^{1/2}$ and $(\alpha c)^{1/2}$, respectively, one transforms eqn. (45.8) as follows

$$\frac{\partial^2 \varphi}{\partial x \partial T} + \frac{\partial^2 \varphi}{\partial T^2} = \pm\sin\varphi, \tag{45.9}$$

where

$$T = \sqrt{c\alpha t}, \quad X = x\sqrt{\alpha/c}. \tag{45.10}$$

The solution of eqn. (45.9), which determines the propagation of the envelope (45.7) of the electric field through the medium without damping, can be found in the works [218,219,230,231].

If the envelope velocity is given by $W = uc$, then, introducing the variable

$$\xi = \left(T - \frac{X}{u}\right) = \sqrt{c\alpha}\left(t - \frac{x}{uc}\right), \tag{45.11}$$

one can transform eqn. (45.9) into

$$\frac{d^2}{d\xi^2}\varphi(\xi) = \pm\frac{u}{u-1}\sin\varphi(\xi). \tag{45.12}$$

Equation (45.12) possesses two types of simple solution for an attenuating medium if the inequality

$$0 < u < 1 \tag{45.13}$$

holds. The first one is a kink, or 2π-pulse,

$$\varphi(\xi) = 4\mathrm{arctg}\left[\exp\left(\xi\sqrt{\frac{u}{1-u}}\right)\right]. \tag{45.14}$$

For this solution, the angle changes from 0 to 2π as ξ varies from $-\infty$ to ∞, and the electric field envelope $\mathcal{E}(x,t)$ propagates with a velocity W less than the velocity of light in the medium. The second solution,

$$\varphi(\xi) = 2\arcsin\left[\mathrm{sn}\left(\xi\sqrt{\frac{(B-1)u}{2(1-u)}}, k\right)\right], \tag{45.15}$$

is periodic. This is expressed in terms of the Jacobian elliptic function $\mathrm{sn}(u,k)$.

In the case of a positive sign, eqn. (45.12) describes an amplifying medium. It also has two types of solutions for $u > 1$, i.e., when the envelope velocity W exceeds the light velocity. These are the following:

(i) the kink-type solution,

$$\varphi(\xi) = 4\mathrm{arctg}\left[\exp\xi\sqrt{\frac{u}{u-1}}\right]; \tag{45.16}$$

(ii) and a solution in the form of the periodic Jacobian elliptic function $(B > 1)$

$$\varphi(\xi) = 2\arcsin\left[\mathrm{sn}\left(\xi\sqrt{\frac{(B-1)u}{2(u-1)}}\right)\right]. \tag{45.17}$$

Chapter 11

CONFORMATIONAL EXCITATIONS IN ONE-DIMENSIONAL MOLECULAR SYSTEMS

46. Quasi-One-Dimensional Systems Undergoing Structural Phase Transitions

Interest in the study of structural phase transitions in quasi-one-dimensional systems conditioned by displacements of the equilibrium positions of molecules has increased recently. The simplest theory [214] of such systems is based on the model of a quasi-one-dimensional chain of identical anharmonic oscillators interacting with each other as dipoles.

We assume that each oscillator corresponds to a particle of mass m moving in the anharmonic symmetric potential $V(u_n) = V(-u_n)$. Imposing cyclic boundary conditions, one can write the Hamiltonian of such a system of oscillators in the following form

$$H = \frac{1}{2}\sum_{n=1}^{N}\left\{m\left[\frac{du_n}{dt}\right]^2 + 2V(u_n) + \chi(u_{n+1} - u_n)^2\right\}. \tag{46.1}$$

This Hamiltonian describes a one-dimensional system of two sublattices A and B. Atoms having a mass larger than the A-th lattice are separated by equal distances l from each other. The positions of atoms of mass m in the sublattice B are characterized by their displacements u_n relative to the atoms of the lattice A. In eqn. (46.1), the parameter χ absorbs the dipole-dipole interaction of the displacements u_n of atoms of the sublattice B. This interaction leads to an instability of the system under study and also possible phase transitions.

Structural phase transitions occurring in ferro- and antiferroelectrics and in some other crystals have been described by the Hamiltonian (46.1) in Refs. [136,214,237,246]. The dynamics of such systems is often characterized by vibrations with sharply decreasing frequencies as a critical temperature is approached. Sometimes there arise ordered local regions, or clusters, and a so-called *central peak* is observed with a doublet of soft models in neutron scattering experiments [258].

The '*u*-four model' is usually applied in theoretical studies of these phenomena. According to this model, the potential field $V(u_n)$ is chosen to have the following form:

$$V(u) = \epsilon_0 \left[1-\frac{u^2}{a^2}\right]^2. \tag{46.2}$$

This has two minima at $u = \pm a$ separated by a potential barrier of height ϵ_0 at $u = 0$. The following set of equations corresponds to the Hamiltonian (46.1)

$$m\frac{d^2 u_n}{dt^2}+\chi[2u_n-u_{n+1}-u_{n-1}]+\frac{dV(u_n)}{du_n} = 0. \tag{46.3}$$

The properties of the set of equations described by the Hamiltonian (46.1) with the potential (46.2), depend strongly on the ratio of the potential barrier energy ϵ_0 to the coupling energy $-\chi a^2$ between displacements in neighboring nodes of a chain.

If the inequality

$$\chi a^2 \ll \epsilon_0 \tag{46.4}$$

holds, the quasi-independent motion of a particle of mass m within its potential well plays a basic role, while jumps into neighboring wells are less probable. In this case, one says that the system undergoes an 'order-disorder' transition. Crystals such as $NaNO_2$,$ClNH_4$,KCN, etc., belong in this class.

But if the following inequality

$$a^2\chi \gg \epsilon_0 \tag{46.5}$$

takes place, the collective self-consistent excitations of a large number of nodes of the lattice become dominant. We then refer to a system having a 'displacive-type' transition. $SrTiO_3$, $BaTiO_3$, $KNbO_3$ and others, are crystals of this type.

46.1. Order-Disorder Transitions

To begin, we consider that the inequality (46.4) is satisfied. At low temperatures, i.e., when the heat energy is rather small compared with the potential barrier energy ϵ_0, at each node of the lattice B the particles vibrate near one of the minima of the appropriate potential well. One can write

$$u_n = 2a\sigma_n + \xi_n, \quad \xi_n \ll a, \tag{46.6}$$

where σ_n is a pseudospin of a given nth particle which takes the values $+1/2$ or $-1/2$ depending on the sign of u_n. Within this approximation, the chain potential energy,

$$U = \sum_{n=1}^{N} \left\{ V(u_n) + \frac{1}{2}\chi(u_n - u_{n-1})^2 \right\} \tag{46.7}$$

is reduced to the sum of the independent components

$$U \approx Na^2\chi + \sum_{n=1}^{N} \frac{4\epsilon_0}{a^2}\xi_n^2 - 4\chi a^2 \sum_{n=1}^{N} \sigma_n\sigma_{n+1}. \tag{46.8}$$

Hence, this approach leads to the well-known one-dimensional Ising model with coupling constant $J = 4a^2\chi$ (the pseudospin formalism).

If the energy χa^2 of the coupling between the displacements at neighboring nodes of a lattice is rather small, the anharmonic vibrations of the atoms of a sublattice B at different nodes are really independent. Using eqn. (46.2), one can transform the equations of motion (46.3) into a set of independent ones,

$$m\frac{d^2u_n}{dt^2} + 4Au_n^3 - 2Bu_n = 0, \tag{46.9}$$

where

$$A \equiv \epsilon_0 / a^4, \quad B \equiv 2\epsilon_0 / a^2. \tag{46.10}$$

Integrating eqn. (46.9), and eliminating the index n, one obtains

$$\frac{1}{2}m\left(\frac{du}{dt}\right)^2 + Au^4 - Bu^2 = E, \tag{46.11}$$

where E is an integration constant having the dimension of energy. The formal solution of eqn. (46.9) becomes

$$\sqrt{\frac{2}{m}}\,t = \pm \int_{u_0}^{u(t)} \frac{dx}{\sqrt{E - Ax^4 + Bx^2}}. \tag{46.12}$$

Let us consider eqn. (46.12) for positive and negative values of the integration constant E.

Case 1: E is positive at high temperatures.
In this case, one can write

$$E - Ax^4 + Bx^2 = A(u_0^2 - x^2)(x^2 + u_1^2), \tag{46.13}$$

where

$$u_0^2 = \frac{\sqrt{B^2 + 4AE} + B}{2A} = a^2[\sqrt{1 + E/\epsilon_0} + 1], \tag{46.14}$$

$$u_1^2 = \frac{\sqrt{B^2 + 4AE} - B}{2A} = a^2[\sqrt{1 + E/\epsilon_0} - 1]. \tag{46.15}$$

Substituting eqn. (46.13) into (46.12) and inverting the integral, one can check that the function $u(t)$ is expressed in terms of the Jacobian elliptic function $cn(...)$,

$$u(t) = u_0 \mathrm{cn}(Dt, k) \tag{46.16}$$

of the modulus k,

$$k^2 = \frac{u_0^2}{u_0^2 + u_1^2} = \frac{\sqrt{1 + E/\epsilon_0} + 1}{2\sqrt{1 + E/\epsilon_0}}, \quad \frac{1}{2} \leqslant k^2 \leqslant 1. \tag{46.17}$$

Here

$$D \equiv \sqrt{\frac{2\epsilon_0}{ma^2}} \left[1 + \frac{E}{\epsilon_0}\right]^{1/4}. \tag{46.18}$$

The Jacobian elliptic function (46.16) is periodic in time. It varies from u_0 to $-u_0$ (a particle at its node passes from one potential well to another) with period T given by the relation

$$T \approx 4K(k)/D, \tag{46.19}$$

where $K(k)$ is a complete elliptic integral of the first kind. The appropriate vibrations are characterized by the cyclic frequencies,

$$\omega = \frac{2\pi}{T} = \frac{\pi D}{2K(k)}. \tag{46.20}$$

According to eqn. (46.17), the square of the modulus of the Jacobian function varies from 1 to $1/2$ in the interval $0 \leqslant E < \infty$. The complete elliptic integral $K(k)$

then varies from infinity to a value near 2. In the region where $E \approx 0$, the modulus $k^2 = 1$, and the function (46.16) is approximated by an hyperbolic function,

$$u(t) \approx u_0 \mathrm{sch}(Dt), \tag{46.21}$$

which has an infinite period ($\omega_1 \approx 0$ is a soft mode). As E increases, the complete elliptic integral $K(k)$ decreases while the vibration frequency grows.

Case 2: E is negative at low temperatures.
Assuming that the inequality

$$-\epsilon_0 \leqslant E \leqslant 0 \tag{46.22}$$

is satisfied, one can transform the polynomial of the integral (46.12) into the following one,

$$E - Ax^4 + Bx^2 = A(u_0^2 - x^2)(x^2 - u_2^2), \tag{46.23}$$

where u_0^2 is given by eqn. (46.14), and

$$u_2^2 = a^2(1 - \sqrt{1 + E/\epsilon_0}). \tag{46.24}$$

Substituting eqn. (46.23) into the integral (46.12), one can represent the function $u(t)$ in terms of the Jacobian elliptic function $dn(...)$,

$$u(t) = u_0 \mathrm{dn}(D_1 t, k) \tag{46.25}$$

where

$$k^2 = \frac{2\sqrt{1 + E/\epsilon_0}}{1 + \sqrt{1 + E/\epsilon_0}}, \quad 0 \leqslant k^2 \leqslant 1, \tag{46.26}$$

$$D_1 = \sqrt{\frac{2\epsilon_0}{ma^2}}[1 + \sqrt{E/\epsilon_0 + 1}]^{1/4}. \tag{46.27}$$

The Jacobian elliptic function $\mathrm{dn}(y,k)$ is an even periodic function relative to the variable y with period $2K(k)$, and varies over the interval

$$\sqrt{1 - k^2} \leqslant \mathrm{dn}(y,k) \leqslant 1. \tag{46.28}$$

Therefore, the function (46.25) describes the vibrations of a particle within a single potential well between the values $u_0\sqrt{1-k^2}$ and u_0. The period of vibration is

$$T = 2K(k)/D_1. \tag{46.29}$$

If $E \approx -\epsilon_0$, the square of the modulus of the elliptic function becomes negligible. The function (46.25) is thus approximated by the expression

$$u(t) = u_0\left[1-\frac{k^2}{2}\sin^2(D_1,t)\right], \quad k^2 \leqslant 1. \tag{46.30}$$

For $k^2 \ll 1$, $K(k) \approx \pi/2$, and the vibrational period is $T = \pi/D_1$.

46.2. Displacive Transitions

Consider the inequality (46.5). A vibrational state migrates over the chain and assures a substantially collective nature. In this case, one can apply the continuum approach by replacing the positions nl of nodes by a continuous variable x. The Hamiltonian (46.3) of the chain under study is then transformed as follows

$$H = \frac{1}{2l}\int_{-\infty}^{\infty} dx\left\{m\left[\frac{\partial u}{\partial t}\right]^2 + mc_0^2\left[\frac{\partial u}{\partial x}\right]^2 + 2\epsilon_0\left[1-\frac{u^2}{a^2}\right]^2\right\}, \tag{46.31}$$

where $c_0 = l(\kappa/M)^{1/2}$ is the sound velocity in the sublattice B.

The following equation of motion in partial derivatives corresponds to the Hamiltonian (46.31)

$$m\frac{\partial^2 u}{\partial t^2} - mc_0^2\frac{\partial^2 u}{\partial x^2} + 4Au^3 - 2Bu = 0, \tag{46.32}$$

where A and B are defined by eqn. (46.10).

We confine ourselves to treating further only those excitations which move along the chain with constant velocity $W < c_0$. Introducing the new variable,

$$\xi = (x - Wt)a \tag{46.33}$$

and the function $\varphi(\zeta) = u(x,t)/d$, one can transform eqn. (46.32)

$$\frac{d^2\varphi}{d\xi^2} - 2\gamma^2[\varphi^2 - \varphi] = 0, \tag{46.34}$$

where

$$\gamma^2 = \frac{2\epsilon_0}{m[c_0^2 - W^2]}. \tag{46.35}$$

In the particular case of vibrations of rather small amplitude, i.e., when the inequality

$$\varphi^3 \ll \varphi \ll 1$$

holds, eqn. (46.34) becomes

$$\frac{d^2\varphi}{d\xi^2} + 2\gamma^2\varphi = 0. \tag{46.36}$$

This has a plane-wave solution

$$\varphi_1(\xi) = \varphi_1 \sin(\sqrt{2}\gamma\xi + \alpha) = \varphi_1 \sin[q(x - Wt) + \alpha] \tag{46.37}$$

with wavenumber

$$q = \frac{2}{a}\sqrt{\epsilon_0 / m(c_0^2 - W^2)} \geqslant \frac{2}{ac_0}\sqrt{\epsilon_0 / m}$$

and frequency $\omega = Wq$ for $\varphi_1 \ll 1$. Such excitations describe ordinary phonons.

Taking anharmonicity into account, eqn. (46.34) possesses an exact particular solution,

$$\varphi(\xi) = \pm \operatorname{th}\gamma\xi, \tag{46.38}$$

or, in terms of the usual variables,

$$u(x,t) = \pm a \operatorname{th}[\gamma(x - Wt)/a]. \tag{46.39}$$

The upper sign corresponds to a kink, and the lower one to an antikink. The solution (46.39) coincides with the appropriate solutions of the sine-Gordon equation.

For a kink, the displacement in the region $x - Wt < 0$ is equal to $-a$ (the particle is located inside the left well), and in the region $x - WT > 0$ this accounts for a (the particle is inside the right well). A transition takes place in the region of the width

$$\Delta x \approx \frac{2\pi a}{\gamma} = \frac{\pi a}{\sqrt{\epsilon_0}}\sqrt{2m(c_0^2 - W^2)}. \tag{46.40}$$

The kink energy is concentrated in this spatial region which moves with velocity W. Substituting the formula (46.38) into eqn. (46.31), one has

$$\mathcal{E}(W) = \frac{2a}{3\gamma l}[m(c_0^2 + W^2)\gamma^2 + 2\epsilon_0]. \tag{46.41}$$

At small velocities, $W \ll c_0$, $\gamma = c_0^{-1}(2\epsilon/m)^{1/2}$, and hence, the kink (antikink) energy is represented by the sum of two terms,

$$\mathcal{E}(W) = \mathcal{E}(0) + m_{\text{sol}} W^2, \tag{46.42}$$

where

$$\mathcal{E}(0) = 4ac_0(m\epsilon_0)^{1/2}(3l)^{-1} \tag{46.43}$$

is the rest energy, and

$$m_{\text{sol}} = 2a(2m\epsilon_0)^{1/2}/3lc_0 \tag{46.44}$$

is the kink (antikink) mass.

A kink always moves with a velocity W less than the sound velocity c_0. The upper bound of the velocity which enables the continuum approach to be valid is determined by the following inequality

$$\pi a\sqrt{2m(c_0^2 - W^2)} > l\sqrt{\epsilon_0}.$$

The chain in question can possess several kinks and antikinks. If the average distance between them exceeds the kink 'width' given by (46.10), they are really noninteracting. Therefore, a system having a small density of kinks (antikinks) can be considered to be that of noninteracting particles, whose properties may be described on the basis of standard statistical physics. A detailed description of this procedure has been given by Krumhansl and Schriefer [214] who evaluated the free energy and the correlation function of a system of kinks, using the methods of statistical physics.

Kink-type, or antikink-type, excitations differ drastically from those of a phonon type. Phonons have definite values of wavenumbers and are not completely localized in coordinate space. Investigating their spatial motion, one should introduce wave packets which are 'blurred' in the course of time. On the other hand, kinks are rather well localized in space. In many respects, their properties are akin to particles.

The general solutions of eqn. (46.34) are represented by Jacobian elliptic functions. Integrating this equation, one obtains

$$\frac{1}{\gamma^2}\left[\frac{d\varphi}{d\xi}\right]^2 - \varphi^4 + 2\varphi^4 = C,$$

where C is an integration constant. The formal solution of the equation above can be written as

$$\int_0^{\varphi} \frac{dx}{\sqrt{C+x^4-2x^2}} = \pm\gamma\xi. \tag{46.45}$$

If the integration constant C obeys the inequality

$$0 \leqslant C \leqslant 1,$$

one has

$$C+x^4-2x^2 = (\varphi_0^2-x^2)(\varphi_1^2-x^2), \tag{46.46}$$

where

$$\varphi_0^2 = 1+\sqrt{1-C}, \ \varphi_1^2 = 1-\sqrt{1-C}. \tag{46.47}$$

Substituting eqn. (46.46) into the integral (46.45), one transforms the latter into the following form

$$\int_0^{\varphi/\varphi_1} \frac{dt}{\sqrt{(1-t^2)(t-k^2t^2)}} = \pm\gamma\xi\varphi_0, \tag{46.48}$$

where

$$k^2 = \frac{\varphi_1^2}{\varphi_0^2} = \frac{1-\sqrt{1-C}}{1+\sqrt{1-C}}. \tag{46.49}$$

Inverting the integral (46.48), one finds

$$\varphi(\xi) = \pm\varphi_1 \operatorname{sn}(\gamma\xi\varphi_0, k), \tag{46.50}$$

where $sn(...)$ is a Jacobian elliptic function depending on the modulus k whose value is determined by the integration constant C via eqn. (46.49). If $C \approx 0$, the inequalities $k^2 \ll 1$, $\varphi_1 \ll 1$, are satisfied, and $\varphi_0 \approx \sqrt{2}$. In this case, the function (46.50) is approximated by the expression

$$\varphi(\xi) = \varphi_1 \sin(\gamma\sqrt{2}\xi), \tag{46.51}$$

which is exactly the function given by eqn. (49.37).

As $C \to 1$, the modulus of the elliptic function (46.50) becomes close to unity, $\varphi_0 \approx \varphi_1 \approx 1$, and the function $\varphi(\zeta)$ is approximated as follows

$$\varphi(\xi) = \pm \mathrm{th}(\gamma\xi). \tag{46.52}$$

Equation (46.52) is simply eqn. (46.38) given earlier. Therefore, under approximate conditions, the function (46.50) describes ordinary phonons and kinks, as well as all possible intermediate excitations.

Chapter 12

SOLITONS AND PROTON MOTION IN HYDROGEN-BONDED MOLECULAR STRUCTURES

47. A Model of Proton Conductivity: Basic Equations

It has been determined experimentally that proton conductivity along one-dimensional chains of hydrogen bonds in solid alcohols, carbohydrates, and other compounds is higher by an order of three than that along the perpendicular direction. It is also known that proton mobility in ice is an order of magnitude than that of electrons in metals. For this reason an ice crystal is often considered as a 'protonic semiconductor'. According to modern concepts, proton mobility in ice is associated with its transfer along hydrogen bonds [23,94,96,113,114].

An ice crystal itself represents a continuous network of hydrogen bonds which link all the water molecules into a single giant molecule. Under normal conditions the latter has a structure in which each oxygen atom is surrounded at a distance of 2.76Å by four other oxygen atoms generating a regular tetrahedron (Fig. 19).

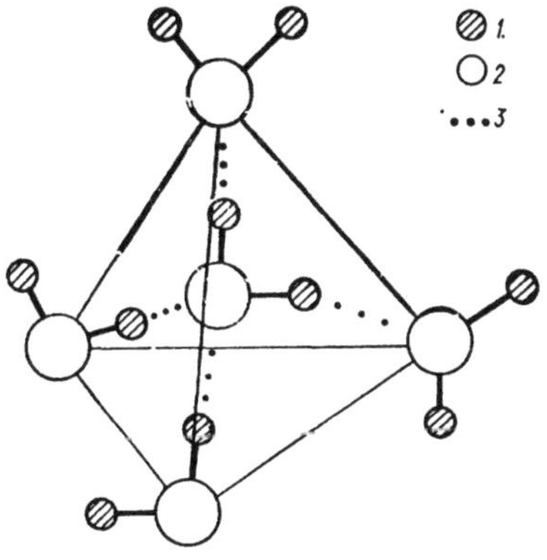

Figure 19. The structure of the elementary cell of an ice crystal: (1) hydrogen atoms; (2) oxygen atoms; ... hydrogen bonds.

Each water molecule is linked with neighboring molecules by four hydrogen bonds. A given proton lies on the line connecting neighboring oxygen atoms at a distance of 0.99Å from the oxygen atom with which a proton forms a covalent bond, and at a distance of 1.77Å from the other oxygen atom.

To study the process of proton transfer in ice crystals, one can consider one-dimensional chains of water molecules linked together by hydrogen bonds. These are called 'Bernal-Fowler filaments' [96]. These chains cannot be compared to rigid structures. Protons are translocated via hydrogen bonds between neighboring water molecules. Their branched pathway can be approximated by a quasi-one-dimensional curve known as the Bernal-Fowler filament. In liquid water, one cannot isolate the continuous Bernal-Fowler filaments which results in its comparatively low proton conductivity.

In 1979-1980 Antonchenko, Illyn, and Makovskii [5-7] carried out numerical simulations of liquid water in bounded volumes and thin films by using the Monte Carlo method. They demonstrated that the density of water is inhomogeneous in thin films (having a thickness of 10−20Å) bounded by solid walls. There arises in these layers a lamellar structure having a layer thickness of about 2Å. Planes of water molecules in these layers are oriented parallel to the walls which bound the given film. In relation to the ordered arrangement of water molecules and the hydrogen bonds between them, one should differentiate between the properties of water in narrow pores and in large volumes (bulk water). The ordered arrangement of hydrogen bonds along the walls of narrow pores prevents the formation of a loose hexagonal structure of ice. This is confirmed by numerous studies of glaciologists (G.Ya. Bliznin, N.A. Tjutjunov, A.A. Ananyan, and the others), who have revealed that in capillary cracks and microscopic pores of rocks, water can only freeze at a temperature lower than −45°C, and at considerably low temperatures in rather thin capillaries. These properties of the so-called film water are important for understanding physico-chemical processes in the frozen rocks of the Arctic. Due to the layer structure of near-surface water in narrow pores having a regular distribution of hydrogen bonds between water molecules, one can isolate there the continuous Bernal-Fowler filaments even at room temperature or higher. Therefore, near-surface water must be a good pathway for protons.

An important property of hydrogen bonding is that the proton potential energy curve in the hydrogen 'bridge' in molecules of water and alcohols has a double-

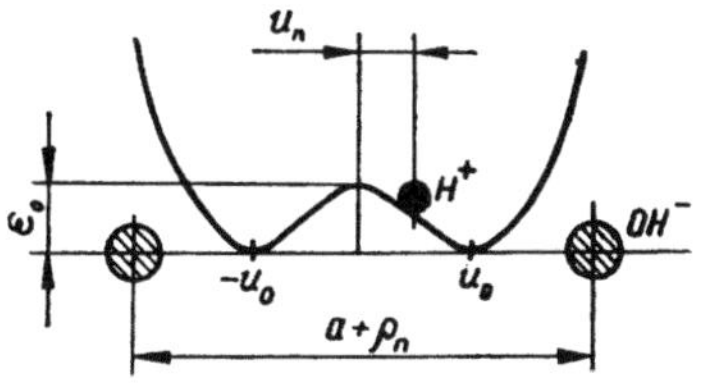

Figure 20. The potential well for protons in the hydrogen bridge.

minima form (Fig. 20) whose minima corresponds to two possible equilibrium positions of a proton [113,114].

Figure 21. The model for the transfering of the hydroxonium, H_3O^+, and hydroxyl, OH^-, ions along the water molecule chain: the motions of protons (1) in the potential wells between the hydroxyl ions (2) are denoted by arrows.

A study of proton conductivity in chains of water molecules, held by hydrogen bonds, is based on the concept that a proton can be translocated (Fig. 21) along the chain in the form of ionic defects, *hydroxonium* (H_3O^+) and *hydroxyl* (OH^-). These defects are formed under the dissociation of water molecules H_2O due to the transfer of a single proton to the neighboring molecule, in accordance with the chemical reaction:

$$2H_2O \rightleftarrows H_3O^+ + OH^-. \tag{47.1}$$

Together with the migration of ionic defects in the chain, there exists the possibility of motion of orientational Bjerrum defects, positive (D) and negative (L) (Fig. 22), generated by the rotation of water molecules [143]. For a positive Bjerrum defect, two protons lie between two hydroxyl ions, and for negative defect there are no protons. As shown in Fig. 22, the orientational Bjerrum defects can arise due to the rotation of a water molecule in the neighbor pair of water molecules around the OH-bond over 180° when one proton comes out of the plane. The second way of generating hte orientational defect is a rotation of one water molecule of the given pair over 120° in the plane of the figure. The processes of the generation of ionic and orientational defects illustrated in Figs. 21 and 22, are

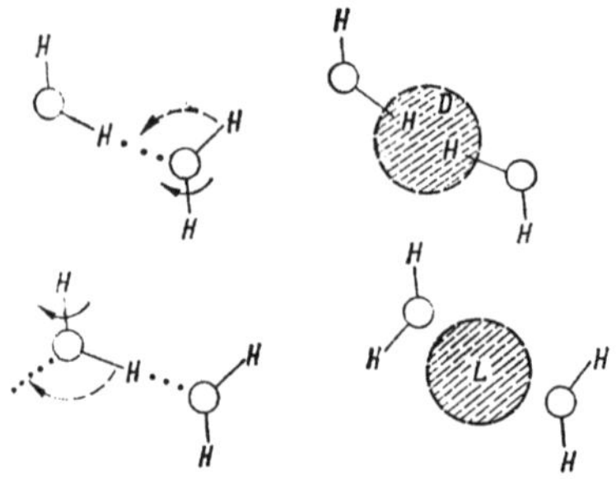

Figure 22. The formation of the positive (D) and negative (L) Bjerrum defects under a rotation of one water molecule around the $O-H$ bond.

rather simplified. We demonstrate further that the process of proton transfer via the hydrogen bonding barrier and the rotation of molecules, depicted in Fig. 22, actually occur over a certain extent of the chain. In other words, both these process are collective. The aforementioned figures correspond to the limiting case when the transition region is reduced to the dimension of a single hydrogen bridge. Nevertheless, we preserve further the notations 'Bjerrum defects' and 'ionic defects OH^- and H_3O^+', although this implies that, in fact, both these defects are extended. Inside the chain, the ionic (OH^- and H_3O^+) and the orientational (L and D) defects are generated only by pairs. At the ends of the chain, they can be formed one by one.

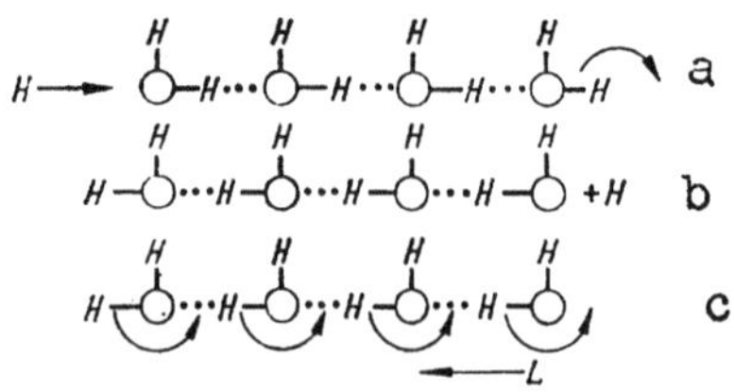

Figure 23. The motion of a proton which approaches at the left end of the chain - the state (a). The final state of the chain - the state (b). The transition of the chain to the initial state, a, is realized by the motion of the Bjerrum defect L from the right end of the chain to the left.

Figure 23 illustrates the motion of a proton (ionic defect H_3O^+) which approaches the left end of a given chain. The return of the chain into its initial state requires the motion of an orientational Bjerrum defect L from the right-hand end.

We present below the theory of protonic conductivity in a chain of water molecules, linked by hydrogen bonds, which has been developed by Antonchenko, Zolotariuk and the present author [43,134]. In this work, the Bernal-Fowler filament, represented by an infinite chain of water molecules, is chosen as the model system. In this chain, a single proton of each water molecule participates in the formation of hydrogen bonds, while the second proton does not contribute to the hydrogen bonding and is coupled with the oxygen atom by a covalent bond. The latter formation, an oxygen atom and this proton, is considered as a complete entity: the so-called hydroxyl ion OH^- (see Fig. 20). Therefore, we partition the given chain of water molecules into the following two sub-systems: the basic lattice formed by hydroxyl groups, and the protonic subsystem. If these subsystems interact with each other, the excited state of the whole system is two-component.

In the normal state of a given chain, each proton providing hydrogen bonding is coupled with an oxygen atom by a covalent bond, on the one hand, and by a hydrogen bond, on the other hand. These bonds interchange with each other due to proton transition through a potential barrier. So, we assume that the potential energy of the proton taking part in hydrogen bonding can be represented by a symmetric double-minima potential well (Figs. 20 and 24),

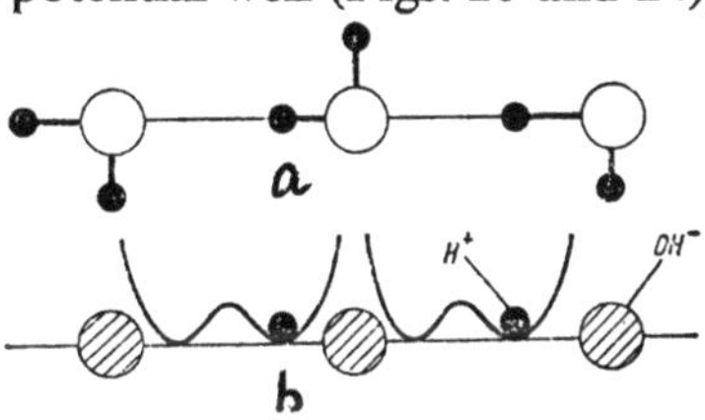

Figure 24. a) Water molecules linked via hydrogen bonds; b) The model of the one-dimensional system characterizing the position of proton H^+ in the chain of the hydroxyl ions OH^-.

$$U(u) = \epsilon_0(l - u^2/u_0^2)^2. \tag{47.2}$$

Here, ϵ_0 is the height of the potential barrier, u is the protonic displacement referred to the barrier, and $\pm u_0$ are the minima of potential. As mentioned above, the motion of oriented Bjerrum defects is conditioned by rotating water molecules. Such a rotation is related to overcoming the appropriate potential barrier, too. The theory of the motion of orientational Bjerrum defects is presented in Section 50.

As a whole, the chain is quasi-neutral in its normal state. If all the protons are in the left or right potential wells, i.e., if the protons form a regular sublattice, the positive charge is uniformly distributed over the chain. Under the local displacement of protons ($u \neq u_0$), the uniform distribution of that positive charge is broken. In those regions where the proton subsystem is contracted there arises an excess positive charge, while an excess negative charge arises in those regions where the proton sub-systems are extended.

We recall that a small fraction (10^{-7}) of water molecules are usually dissociated in liquid water into hydroxonium and hydroxyl ions, according to eqn. (47.1). The hydroxonium ion H_3O^+ corresponds to the state in which the protons are in the left potential wells on the right of a given OH^- group, and on the left of the OH^- group they are located in the right-hand wells (Fig. 21). The motion of hydroxonium is due to a jump of a proton from the left-hand to the right-hand well (a motion of a domain wall). The hydroxyl ion OH^- corresponds to the state of the chain in which, on the left of a given OH^- group, the protons are located in the left-hand potential wells, and in the right-hand wells, on the right of a given OH^- group (Fig. 21).

The Hamiltonian of a given system can be written as the sum of three terms

$$H = H_p + H_{OH} + H_{\text{int}}, \tag{47.3}$$

where

$$H_p = \sum_n \left\{ \frac{m}{2} \left[\left(\frac{du_n}{dt} \right)^2 + \omega_1^2 (u_{n+1} - u_n)^2 \right] + U(u_n) \right\} \tag{47.4}$$

is the Hamiltonian of the proton subsystem, and

$$H_{OH^-} = \sum_n \frac{M}{2} \left\{ \left(\frac{d\rho_n}{dt} \right)^2 + \Omega_0^2 \rho_n^2 + \Omega_1^2 (\rho_{n+1} - \rho_n)^2 \right\} \tag{47.5}$$

is the Hamiltonian of the basic lattice of OH^- ions of mass M. In eqns. (47.4) and (47.5), u_n is the displacement of a proton from the top of the potential barrier lying in the middle between the nth and $(n+1)$th ions OH^-, m is the proton mass, $m\omega_1^2$ is the elasticity coefficient of the proton subsystem, Ω_0 and Ω_1 are characteristic frequencies, and ρ_n is the relative displacement of OH^- ions. The third term on the right-hand side of eqn. (47.5) takes into account the dispersion of vibrations.

A change in distance between the hydroxyl ions deforms the potential well of the proton in the hydrogen bonding. In such a way, there originates an interaction between the proton and the hydroxyl subsystems. Taking into account such a coupling between the proton motion and the optical mode of the hydroxyl subsystem, one can represent the Hamiltonian of interacting of proton displacements and those of hydroxyl ions in the following form

$$H_{\text{int}} = \chi \sum_n \rho_n (u_n^2 - u_0). \tag{47.6}$$

Here, the parameter χ characterizes the intensity of this interaction. Hence, the complete Hamiltonian depends on two variables, u_n and ρ_n.

Designating the equilibrium distance between hydroxyls by a and introducing the continuum approach, $na \to x$, $u_n \to u(x)$, $\rho_n \to \rho(x)$, one can represent the Hamiltonian (47.3) as follows

$$H = a^{-1} \int dx \left\{ \frac{m}{2}(u_t^2 + c_0^2 u_x^2) + U(u) + \frac{M}{2}(\rho_t^2 + \Omega_0^2 \rho^2 + V_0^2 \rho^2) + \chi \rho (u^2 - u_0^2) \right\}, \tag{47.7}$$

where $c_0 = a\omega_1$ is the proton sound velocity in the light subsystem, and $V_0 = a\Omega_1$ is the characteristic velocity in the hydroxyl subsystem. The subscript of the functions u and ρ designates differentiation relative to the appropriate variable.

The equations of motion, corresponding to the Hamiltonian (47.7), take the form

$$u_{tt} - c_0^2 u_{xx} - \omega_0^2 (1 - u^2 / u_0^2) u + \frac{2\chi}{m} \rho u = 0, \tag{47.8}$$

$$\rho_{tt} + \Omega_0^2 \rho - V_0^2 \rho_{xx} + \chi M^{-1} (u^2 - u_0^2) = 0, \tag{47.8a}$$

where

$$\omega_0^2 = 4\epsilon_0 / m u_0^2 \tag{47.8b}$$

is a quantity describing the potential barrier. If the interaction between the displacements of protons and of hydroxyl ions from their equilibrium positions is absent ($\chi = 0$), the set (47.8) of coupled equations break down into independent equations. Equation (47.8) transforms into the nonlinear equation,

$$u_{tt} - c_0^2 u_{xx} - \omega_0^2 (1 - u^2 / u_0^2) u = 0, \tag{47.9}$$

which determines the dynamics of proton displacements. Its solution is studied in the next Section. If $\chi = 0$, eqn. (47.8a) transforms into a linear equation,

$$\rho_{tt} + \Omega_0^2 \rho - V_0^2 \rho_{xx} = 0. \tag{47.10}$$

This describes longitudinal optical waves with dispersion,

$$\Omega^2(k) = \Omega_0^2 + V_0^2 k^2, \tag{47.10a}$$

in the system of hydroxyl ions. A positive V_0^2 corresponds to a positive dispersion, and negative values of V_0^2 correspond to negative dispersion. The group velocity of motion of wave packets consisting of optical phonons is given by the formula

$$V_{gr} = kV_0^2 / (\Omega_0^2 + k^2 V_0^2)^{1/2}.$$

Hence, V_0 corresponds to the limiting magnitude of the group velocity of utmost large wavenumber k.

The set of equations (47.8) is valid for chains of water molecules as well as for other quasi-one-dimensional systems in which negatively charged proton complexes are linked by hydrogen bonds. In particular, for Bernal-Fowler chains consisting of molecules of water, one can use the following values of the parameters:

$$\epsilon_0 = 2eV, \qquad m = m_p, \qquad M = 17m_p,$$
$$\Omega = 600cm^{-1}, \qquad \omega_1 = 5\times 10^{14}Hz, \qquad c = 2.5\times 10^5 cm/s.$$

The values of the parameter χ and V_0 are less well known. One can suppose that their values lie in the following intervals:

$0{\cdot}05 < \chi < 0.3eV/\mathring{A}$ and $6.10^{-3} < V_0/_0 < 6{\cdot}10^{-1}$.

If the interaction between the sublattices is nonvanishing ($\chi \neq 0$), it is convenient to introduce a new dimensionless variable

$$\xi = (x - x_0 - Vt)/a \tag{47.11}$$

in eqns. (47.8) instead of the variables x and t. For the same reason, one can define the following dimensionless parameters

$$s = V/c_0, \ \sigma = V/a\Omega_0, \ \sigma_0 = V_0/a\Omega_0, \ \lambda^2 = \sigma_0^2 - \sigma^2, \tag{47.12}$$

$$\alpha = 2\epsilon_0/m\omega_1^2 u_0^2, \ \beta = \chi^2 u_0^2 / m\omega_1^2 M\Omega_0^2 \tag{47.13}$$

and the dimensionless functions

$$\varphi = u/u_0, \ \psi = \rho/\rho_0, \tag{47.14}$$

where

$$\rho_0 = \chi u_0^2 / M\Omega_0^2. \tag{47.14a}$$

The set of equations (47.8) is then transformed as follows

$$(1-s^2)\varphi_{\xi\xi}+2\alpha(1-\varphi^2)\varphi-2\beta\varphi\psi = 0, \tag{47.15}$$

$$\psi+\lambda^2\psi_{\xi\xi} = 1-\varphi^2. \tag{47.16}$$

The parameters α and β characterize the height of the potential barrier and the interaction of the basic sublattice with the proton one, respectively. The parameter ρ_0 determines the amplitude of the relative displacements of hydroxyls. The following inequalities

$$\omega_1>\Omega_1, \text{ i.e., } c_0>V_0 \text{ and } \alpha>\beta \tag{47.17}$$

are commonly satisfied.

The solutions of the set of equations (47.15) and (47.16) correspond to two-component solitons. The function $u(\xi) = u_0\varphi(\xi)$ describes the displacements of protons in hydrogen bonds, and the function $\rho(\xi) = \rho_0\psi(\xi)$ describes the relative contraction of distances between the neighboring hydroxyls. Using these solutions, one can evaluate the integrals of motion: viz., the total energy

$$E(V) = \int d\xi\left\{\frac{1}{2}m\omega_1^2(1+s^2)u_\xi^2+\epsilon_0\left[1-\frac{u^2}{u_0^2}\right]^2+\right. \tag{47.18}$$

$$\left.+\frac{M\Omega_0^2}{2}[\rho^2+(\sigma^2+\sigma_0^2)\rho_\xi^2]+\chi\rho(u^2-u_0^2)\right\}$$

and the total momentum

$$P(V) = \frac{V}{a^2}\int d\xi\{mu_\xi^2+M\rho_\xi^2\}. \tag{47.19}$$

In the dimensionless variables, these quantities are expressed as follows

$$E(V) = A\int d\xi\{(1+s^2)\varphi_\xi^2+\alpha(1-\varphi^2)^2+ \tag{47.20}$$

$$+\beta[\psi^2+(\sigma^2-\sigma_0^2)\psi_\xi^2-2(1-\varphi^2)\psi]\},$$

and

$$P(V) = mV\left[\frac{u_0}{a}\right]^2\int d\xi(\varphi_\xi^2+\gamma\psi_\xi^2), \tag{47.21}$$

where the abbreviated notations are used:

$$A \equiv \frac{1}{2} m\omega_1^2 u_0^2, \quad \gamma \equiv \frac{u_0^2 \chi^2}{mM\Omega_0^2}. \tag{47.22}$$

48. Proton Motion Neglecting the Interaction between Sublattices

Let us investigate the solutions of eqn. (47.10) which describe a molecular chain where we neglect the interaction between the displacements of protons and those of hydroxyls. In the dimensionless variables (47.12) and (47.13) this takes the form

$$\varphi_{\xi\xi} + 2\mu_0^2(1-\varphi^2)\varphi = 0, \tag{48.1}$$

where

$$\mu_0^2 = \alpha(1-s^2)^{-1}, \quad s^2 < 1. \tag{48.2}$$

A nonlinear equation of the type (48.1) has been studied in numerous papers dealing with condens matter physics (see, for example, Ref. [142]). Its solution has the simple form

$$\varphi_l(\xi) = l\,\mathrm{th}(\mu_0\xi), \quad l = \pm 1.$$

Hence, the function characterizing proton displacements in a given chain becomes

$$u_l(\xi) = lu_0\,\mathrm{th}(\mu_0\xi). \tag{48.3}$$

The function (48.3) with the positive sign ($l=1$) corresponds to a soliton (kink), and with the negative sign ($l=-1$) to an anti-sliton (antikink) (Fig. 25).

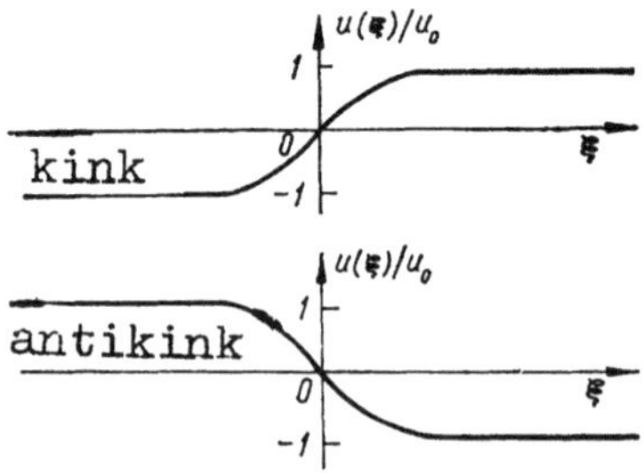

Figure 25. The arrangement of the seven α-helical segments of the bacteriorhodopsin molecule in a purple membrane.

The antisoliton describes an extended ionic defect (a contraction of the proton subsystem corresponding to an excess positive charge) which moves with a velocity V less than that of the 'proton sound' velocity c_0. Therefore, the transition of a proton from the left-hand potential well to the right-hand one is realized in a gradual way over the distance

$$\Delta\xi \sim \mu_0^{-1} = \omega_1 u_0 \sqrt{m(1-s^2)/2\epsilon_0}$$

rather than by a jump, as depicted in Fig. 20.

The positive sign in eqn. (48.3) corresponds to the function which characterizes the motion of the extended ionic defect with an excess negative charge, i.e., to a local extension of the proton subsystem. In a rigorous sense, along the chain there migrates neither hydroxonium nor hydroxyl ions, but a 'blurred' region of contraction and an extension of the average density of protons. The effect of 'blurring' is a result of collective and nonlinear interactions existing in the given system. It is this very fact which explains the exceptional stability of appropriate proton migrations.

The energy and momentum of a soliton and antisoliton, moving with velocity V, are just the same. Using eqns. (47.20) and (47.21) and assuming $\chi = 0$, one finds

$$\tilde{E}(V) = A \int d\xi\{(1+s^2)\varphi_\xi^2 + \alpha(1-\varphi^2)^2\} = \frac{\tilde{E}_0}{\sqrt{1-s^2}}, \tag{48.4}$$

where

$$\tilde{E}_0 = \frac{8A}{3}\sqrt{\alpha} = \frac{4u_0 c_0}{3a}\sqrt{2m\epsilon_0} \tag{48.4a}$$

is the energy of the rest soliton (antisoliton), and

$$\tilde{P}(V) = mV\left[\frac{u_0}{d}\right]^2 \int \varphi_\xi^2 d\xi = M_0 V(1-s^2)^{-1/2} \tag{48.5}$$

where

$$\tilde{M}_0 = \frac{4u_0}{3ac_0}\sqrt{2m\epsilon_0} \tag{48.5a}$$

is the mass of an immobile soliton (antisoliton). Hence,

$$\tilde{E}(V) = M_0 c_0^2 / \sqrt{1-s^2}. \tag{48.5b}$$

By virtue of eqn. (48.5a), one can rewrite the energy of a soliton as follows

$$E_0 = M_0 c_0^2, \tag{48.6}$$

$$E(V) = \sqrt{E_0^2 + c_0^2 p^2(V)}. \tag{48.7}$$

The formulas derived are valid if the velocities of the soliton and antisoliton are upper bound by the 'proton sound' velocity $c_0 = a\omega_1$. The validity of the continuum approach holds only at small velocities ($\mu \ll 1$).

49. Proton Motion Taking into Account the Interaction between Sublattice Displacements

If $\chi \neq 0 (\beta > 0)$, proton motion is described by the set of coupled equations (47.15) and (47.16). We first consider the motion of a proton having a fixed velocity $V = V_0$, where V_0 is the velocity entering into the dispersion law (47.9a). Due to eqn. (47.12), it follows that $\lambda = 0$.

Equation (47.16) then results in the following equation

$$\psi(\xi) = 1 - \varphi^2(\xi). \tag{49.1}$$

Substituting this into eqn. (47.15), one obtains the nonlinear equation

$$\varphi_{\xi\xi} + 2\mu_0^2(1-\varphi^2)\varphi = 0, \tag{49.2}$$

where

$$\mu_0^2 = (\alpha - \beta)/(1 - s_0^2), \quad \alpha > \beta, \tag{49.3}$$

$$s_0^2 \equiv (V_0/c_0)^2 < 1. \tag{49.4}$$

Comparing eqn. (49.1) with (48.1), one sees that an account of nonvanishing χ results in the replacement of α by $\alpha - \beta$. According to eqn. (47.13), the coefficient α is proportional to the height of the potential barrier. Therefore, coupling of the proton subsystem with the basic (OH^-) system leads to an effective lowering of the potential barrier by the factor

$$\eta^2 = 1 - \beta/\alpha < 1. \tag{49.5}$$

This lowering is independent of the sign of χ.

Therefore, for $\chi \neq 0$ and a given proton velocity $V = V_0$, the exact solution of the set of equations (49.1) and (49.2) describes the two-component soliton

$$u_l(\xi) = lu_0 \operatorname{th}(\mu_r\xi), \quad \mu = \sqrt{(\alpha-\beta)/(1-s_0^2)}, \quad l = \pm 1, \tag{49.6}$$

$$\rho = \rho_0\psi(\xi) = \frac{\chi u_0^2}{M\Omega_0^2}\operatorname{sch}^2(\mu_0\xi) \tag{49.7}$$

if $\alpha > \beta$, or equivalently,

$$\epsilon_0 > \chi^2 u_0^4 / 2M\Omega_0^2.$$

The function $u_l(\zeta)$ corresponds to a kink $(l=1)$ and an antikink $(l=-1)$ for the proton subsystem, while the function $\rho(\zeta)$ describes the distribution of local change of distances between hydroxyls. The kink (antikink) width, i.e., the region having a reduced (increased) density of protons, is given by the following relation

$$\Delta\xi \sim \frac{1}{\mu_0} = \sqrt{(1-s_0^2)/(\alpha-\beta)}. \tag{49.8}$$

Therefore, an account of the interaction of displacements of hydroxyl ions with protons $(\beta \neq 0)$ increases the separation in the chain over which a proton migrates from one partial well to another.

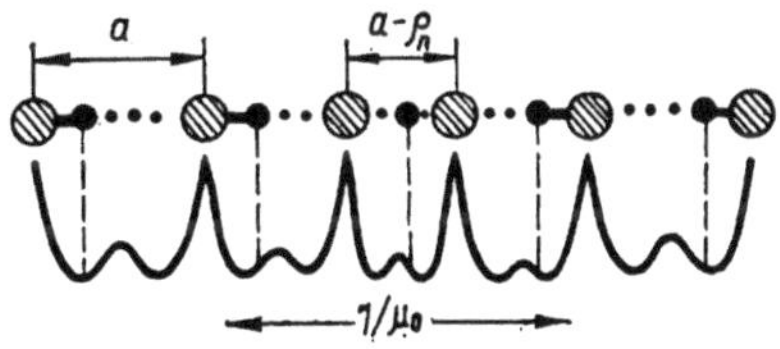

Figure 26. Schematic representation of the migration of protons (dark circles) along hydrogen-bonded chains with an account of the displacements of hydroxyl ions (shaded). μ_0^{-1} is the region of increased proton density.

Figure 26 presents a scheme of the proton migration in the hydrogen-bonded chain taking into account the displacements of the hydroxyl ions (shaded). In the region $1/\mu_0$ an increased density of protons is observed.

Let us derive expressions for the energy and momentum of a soliton (antisoliton) moving with velocity $V = V_0$. According to eqns. (47.20) and (47.21), one obtains

$$E(V_0) = \frac{8}{3}A\mu\left[1+\frac{4}{5}\gamma s_0^2\right], \tag{49.9}$$

$$P(V_0) = \frac{4}{3} m\mu \left[\frac{u_0}{a}\right]^2 \left[1+\frac{4}{5}\gamma\right] V_0, \tag{49.9a}$$

where μ_0 is given by (49.6). If $s_0^2 \ll 1$, these formulae can be simplified:

$$E(V_0) = E_0 + \frac{1}{2} M_s V_0^2, \quad P(V_0) = M_s V_0, \tag{49.10}$$

where

$$E_0 = \eta \frac{4u_0}{3a} \sqrt{2\epsilon_0 m c_0^2}, \quad M_s = 4\eta u_0 (3ac_0)^{-1} (2\epsilon_0/m)^{1/2} (1+8\gamma/5)]. \tag{49.11}$$

Hence, if $V = V_0$, the difference in the energies of the one-component soliton, given by eqn. (48.4a), and the two-component soliton becomes

$$\tilde{E}(V_0) - E(V_0) = \frac{8A\sqrt{\alpha}}{3\sqrt{1-s_0^2}} \{1-\eta(1+4\gamma s_0^2/3)\}, \tag{49.12}$$

$$\eta \equiv [(\alpha-\beta)/\alpha]^{1/2} \ll 1. \tag{49.12a}$$

This gain of energy is conditioned by a coupling of the proton subsystem with that of the hydroxyl subsystem. A lowering of energy facilitates the migration of protons along the chain of hydrogen bonds. However, the velocity V then becomes the maximum velocity of the two-component soliton. Formally, it follows that the parameter λ^2 appearing in eqn. (47.16) takes a negative value for $V > V_0$. For this reason, eqn. (47.16) has no spatially localized solutions. This conclusion has been confirmed numerically by Peyrard et al. [294], who demonstrated that solitons becomes unstable if their velocities are greater than the limiting velocity V_0. The displacements of hydroxyls in the heavy subsystem have no time to follow coincidently the proton motion, damp it and reduce its velocity up to a limiting velocity V_0.

If the velocity of a two-component soliton is less than the limiting one, we can write the solution of eqn. (47.16) in integral form

$$\psi(\xi) = 1 - \int \varphi^2(\xi_1) G(\xi-\xi_1) d\xi_1, \tag{49.13}$$

where Green's function $G(\zeta)$ is defined by the formula

$$G(\xi) = (2\lambda)^{-1} \exp(-|\xi|/\lambda). \tag{49.14}$$

Equation (49.13) determines the coupling between the displacements of the hydroxyls and the probability $\varphi^2(\zeta)$ of proton distribution in the given chain. If $\lambda^2 \neq 0$, this coupling becomes a nonlocal one instead of for the case $V = V_0$ when the Green's function transforms into a delta-function,

$$G_0(\xi) = \delta(\xi), \ \lambda^2 = 0. \tag{49.15}$$

Substituting the formula (49.13) into eqn. (47.15), one obtains the integro-differential equation

$$\varphi_{\xi\xi} + 2\mu_0^2(1-\varphi^2) - \frac{2\beta\varphi}{(1-s^2)}\left[1 - \int \varphi^2(\xi_1)G(\xi-\xi_1)d\xi_1\right], \tag{49.16}$$

which defines the function $\varphi(\zeta)$. Here

$$\mu_0^2 = \alpha(1-s^2)^{-1}. \tag{49.17}$$

The solution of eqn. (49.16) can be found by the variational method or the method developed by Laedke et al. [217].

At two-component soliton velocities which are less the limiting velocity V_0, the parameter λ^2 becomes small,

$$\lambda^2 = \sigma_0^2 - \sigma^2 = \frac{V_0^2 - V^2}{a^2\Omega_0^2} \ll 1. \tag{49.18}$$

This is a reason to seek the solution of eqn. (47.16) as an expansion series in powers of λ^2,

$$\psi(\xi) = \sum_{\nu=0}^{\infty} \lambda^{2\nu}\psi_\nu(\xi). \tag{49.19}$$

Within the zeroth order treatment ($\nu=0$), $\psi_0 = 1-\varphi^2$. Further, the next perturbation order gives

$$\psi(\xi) = 1-\varphi^2(\xi) - \lambda^2(\varphi^2)_{\xi\xi}. \tag{49.20}$$

Substituting this into eqn. (47.16), one obtains the differential equation

$$\varphi_{\xi\xi} + 2\mu^2[1-\varphi^2(\xi)]\varphi + \frac{1}{2}\kappa(\varphi^2)_{\xi\xi}\varphi = 0 \tag{49.21}$$

with a new parameter of smallness,

$$\kappa = 4\lambda^2\beta(1-s^2)^{-1} \ll 1, \tag{49.22}$$

$$\mu^2 = (\alpha-\beta)(1-s^2)^{-1}. \tag{49.23}$$

Multiplying eqn. (49.21) by φ_ξ and integrating further taking into account the boundary conditions

$$|\varphi(\pm\infty)| = 1, \quad \varphi_\xi(\pm\infty) = 0,$$

we obtain

$$\varphi_\xi^2 = \mu^2(1-\varphi^2)^2(1+\kappa\varphi^2)^{-1}. \tag{49.24}$$

Integrating this relation with the boundary condition $\varphi(0) = 0$, one finds

$$\xi = l\mu^{-1}\int_0^{\varphi(\xi)}(1-y^2)^{-1}\sqrt{1+xy^2}\,dy, \quad l = 1,2.$$

Inverting this integral, one obtains finally the expression

$$\varphi_l(\xi) = l\,\mathrm{th}(\mu_1\xi)\left\{1+\frac{1}{2}\kappa\,\mathrm{sch}^2(\mu_1\xi)\right\}, \tag{49.25}$$

where

$$\mu_1 = \mu\left[1+\frac{1}{2}\kappa\right]^{-1}, \quad l = \pm 1. \tag{49.26}$$

The positive sign of l gives a soliton (kink), and the negative sign yields an antisoliton (antikink).

Substituting eqn. (49.25) into (49.20) yields the function

$$\psi(\xi) = \mathrm{sech}^2(\mu_1\xi)\left\{1-\frac{\kappa}{2\beta}\left\{1-\frac{\kappa}{2\beta}[\alpha-\beta+(5\beta-3\alpha)]\mathrm{th}^2(\mu_1\xi)\right\}\right\}, \tag{49.27}$$

which determines the relative decrease in distance between hydroxyls,

$$\rho(\xi) = \frac{\chi u_0^2}{M\Omega_0^2}\psi(\xi). \tag{49.28}$$

Substituting further the solutions (49.25) and (49.28) with μ_1, given by relation (49.26), into the general formulae for the energy and momentum of the two-component soliton, eqns. (47.20) and (47.21), one obtains

$$E(V) = \left\{1+\frac{4}{5}\gamma s_0^2-\frac{1}{5}\kappa\left[\frac{1}{2}+\right.\right. \tag{49.29}$$

$$+\frac{2}{7}\left[\frac{\omega_1 s_0}{\Omega_0}\right]^2(20\alpha-17\beta)\Bigg]M_0 c_0^2(1-s^2)^{-1/2},$$

and

$$P(V) = \left\{1+\frac{4}{5}\gamma-\frac{1}{5}\kappa\left[\frac{1}{2}+\right.\right. \quad (49.30)$$

$$+\frac{2}{7}\left[\frac{\omega_1}{\Omega_0}\right]^2(20\alpha-17\beta)\Bigg]M_0 V(1-s^2)^{-1/2},$$

where

$$M_0 = 4u_0\eta(3ac_0)^{-1}(2m\epsilon_0)^{1/2}. \quad (49.31)$$

50. The Motion of Orientational Defects in Chains of Water Molecules

Figure 23 illustrates the migration of protons affected by an electric field or a concentration gradient along quasi-one-dimensional hydrogen-bonded chains of water molecules (see Section 47). Proton migration to the left-hand side of the chain causes successive jumps of protons (denoted by arrows in Fig. 23) from the left-hand potential wells of double-minima hydrogen bonds to the right-hand ones via the mechanism developed in Section 47. The result of these jumps is a free proton and the transition of the chain into the configuration shown in Fig. 23(b). This chains of water molecules enables the transfer of the new proton if it turns into the original configuration given in Figure 23(a). This is realized by the consecutive migration of a Bjerrum defect from right to left by rotation of water molecules around the vertical OH^- bonds (see Fig. 23(c)).

High proton conductivity of such chains is provided by the cooperative step-by-step migration of protons and orientational Bjerrum defects. At every step the Bjerrum defects go backwards along the chain into their original states. This mechanism of proton conductivity was first suggested by Sokolov [113] in 1955. According to the estimations given by Eigen and DeMayer [173], each process carries only a fraction of charge: $0.64e$ is transferred along the chain via the first process, and the remainder, $0.36e$, by the second process. The consecutive manifestation of the first and second mechanisms of migration results in a free proton arising from the right-hand of the chain, and the whole chain returns to the original state. We recall that the jump-like migrations of protons and orientational defects depicted in Fig. 21 can be considered to be a great idealization of the real processes. The migration occurs smoothly, covering some bridges of water-molecule chains rather than by jumps.

By Jaccard's estimations [204], mobility of ionic defects in ice accounts for about $10^{-3} cm^2 V^{-1} s^{-1}$, and that of orientational Bjerrum defects accounts for $10^{-4} cm^2 V^{-1} s^{-1}$. Therefore, the conductivity of protons is limited by the migration of Bjerrum defects. As the temperature increases, the velocity of migration of Bjerrum defects apparently does, too, and that of protons along the chains of hydrogen bonds decreases. This is the reason why the proton conductivity, conditioned by the joint effect of both these mechanisms, must increase with increasing temperatures up to about $30-40$°C and decrease further due to the breakdown of the ordered distribution of water molecules in the chain. A quantitative theory of temperature dependence of proton conductivity has not yet been developed.

A theory of migration of orientational defects (for brevity we use the term 'Bjerrum defect') along the hydrogen-bonded chain of water molecules has been developed by Sergienko [107-109] on the basis of the kink-soliton model. In his first works, he considered the simplest idealized chain model (a hypothetical configuration) plotted in Fig. 22, and also assumed that the elementary Bjerrum defect is realized by the rotation of a water molecule in the plane of Figure 22 by 120°.

A neutral water molecule possesses a permanent electric dipole moment $d = 1.87D$. Its effective length is about 0.2Å. The molecules in a given chain are separated from each other by a distance of $a = 2.7\text{Å}$. Their orientations in the plane, involving the molecular axis, are determined by the angles Φ_n between the direction of the dipole moment and the axis of the molecule.

The total energy H of orientational motions of molecules can be represented by the following sum

$$H = H_p + H_a + H_k, \tag{50.1}$$

where the form

$$H_p = \epsilon \sum_n \cos^2(2\Phi_n) \tag{50.2}$$

describes the interaction of each monomer dipole moment with its local surrounding. This function attains minima separated by the potential barriers of height ϵ at four angles Φ_n. The second term H_d in eqn. (50.1) characterizes the energy of the dipole-dipole interaction between neighboring molecules in the given chain and takes the following form

$$H_d = \frac{d^2}{a^3}\sum_n\{\cos(\Phi_{n+1}-\Phi_n)-3\cos\Phi_n\cos\Phi_{n+1}\}. \tag{50.3}$$

Finally, the third term,

$$H_k = \frac{1}{2}J\sum_n W_n^2 \tag{50.4}$$

describes the kinetic energy of rotations of molecules with the moment of inertia J. The total interaction H_p+H_d determines four minima corresponding to four equilibrium orientations of the dipole moments,

$$\Phi_{1,2} = \pm\Phi_0, \quad \Phi_{3,4} = \pi\pm\Phi_0 \tag{50.5}$$

where

$$\Phi_0 = \frac{1}{2}\arccos z_0, \quad z_0 \equiv 6V_0^2/a\omega_0^2 \tag{50.6}$$

and

$$V_0 = d/(2aJ)^{1/2} \tag{50.7}$$

gives the velocity of migration of free and consistent rotational vibrations of water molecules following the rotation of two bonds. The frequency ω_0 obeys the relation

$$\omega_0^2 = 4\epsilon/J. \tag{50.8}$$

We recall that a real change of dipole moments under the formation of the orientational Bjerrum defect in a chain occurs smoothly over some molecules, but not by jumps. In this case, the inequality

$$\Phi_{n+1}-\Phi_n \ll 2\pi$$

is valid. Hence, one can apply the continuum approach by introducing the function $\chi(x,t)$ depending on a continuous variable x and time t,

$$2\Phi_n(t) \rightarrow \chi(x,t). \tag{50.9}$$

By virtue of eqn. (50.1), one obtains the equation of motion

$$\chi_{tt} - V_0^2 \chi_{xx} - \omega_0^2 \sin(2\chi) + 12 V_0^2 a^{-2} \sin\chi = 0. \tag{50.10}$$

Derivatives of functions are henceforth designated by subscripts of the appropriate variables. Equation (50.10) coincides in form with the double sine-Gordon equation investigated by Bullough et al. [25].

Taking into account the translational symmetry of an infinite chain, or the cyclic boundary conditions for a finite chain, one can introduce the dimensionless coordinate

$$\zeta = (x - x_n - Vt)/a \tag{50.11}$$

for a system moving with constant velocity V. Relative to this system, the general solution of eqn. (50.10) is represented by the elliptic integral

$$\zeta = -\int dz \{12(1-z^2)/z_0(s^2-1)[C + 2z_0 z - z^2]^{1/2}\}, \tag{50.12}$$

where $s = V/V_0$, C is an integration constant, and

$$z \equiv \cos\chi, \quad z_0 = \epsilon V_0^2/a^2\omega_0^2 = `3d^2/4\epsilon a^3 \leqslant 1. \tag{50.13}$$

From the general solution (50.12) there follows two different classes of solutions: for $V > V_0$ and $V < V_0$. In the case of larger velocities ($V > V_0$), eqn. (50.12) describes the vibrational rotations of dipole moments around fixed positions (50.5) which are studied in Subsection 50.2. These vibrations are not related to reorientations of dipole moments from one equilibrium position to another and, thus, cannot describe the orientational defects.

The motion of orientational defects, i.e., the motion related to the consistent varying of equilibrium orientations, takes place only at velocities less than the characteristic velocity, given by the relation (50.7).

If $V < V_0$, the general solution (50.12) can be written as follows

$$\left[\frac{12}{z_0(1-s^2)}\right]^{1/2} \zeta = -\int \frac{dz}{\{(1-z^2)(z-z_1)(z-z_2)\}^{1/2}}, \tag{50.14}$$

where

$$z_{1,2} = z_0 \pm \sqrt{z_0^2 + C}. \tag{50.15}$$

Depending on the value of the integration constant C, eqn. (50.14) describes some types of motions of the dipole moments in the given chain (see [107,108]). We are interested only in those motions which are related to the reorientation of directions of dipole moments from one equilibrium position (50.5) to another. Such solutions corresponds to $C = -z_0^2$. In this case, from eqn. (50.14) it follows that

$$\mathrm{tg}\Phi(\zeta) = \pm \mathrm{ctg}\Phi_0 \mathrm{th}\mu\zeta \tag{50.16}$$

with

$$\mu = \left[\frac{3(1-z_0^2)}{z_0(1-s^2)} \right]^{1/2}. \tag{50.17}$$

For a positive sign, the function (50.16) describes the reorientation of dipole moments from the position $\pi - \Phi_0$ to Φ_0 over the region of chain enveloping $N_0 = 2\pi/\mu$ water molecules. This migration in the system corresponds to that of the positive Bjerrum defect distributed over $aN_0 = 2\pi a/\mu$ (a thickening of positive dipole ends). For the opposite sign, the function (50.16) describes the reorientation of a dipole moment from Φ_0 to $\Phi_0 - \pi$. Such a reorientation is just a negative Bjerrum defect.

50.1. The Motion of Orientational Defects in an Electric Field

As mentioned in Section 47, the formation of a primitive orientational Bjerrum defect is realized in two ways:

(a) by the rotation of a water molecule in the plane of Fig. 22 by 120°;

(b) by the rotation of a water molecule around the OH-bond perpendicular to the axis of the chain by 180°.

In the second case, the other OH-bond comes out of the figure plane. We now consider this case of orientational defect formation.

Within the model, the positions of dipole moments are characterized by two angles: (i) by the angle Φ_0 between a dipole moment and the fixed OH-bond perpendicular to the chain axis, and (ii) by the angle ψ of rotation of the dipole moment projection in the plane perpendicular to the fixed OH-bond.

In the preceding Section, it was assumed that $\psi = 0$ and the angle Φ is varied only. Now we fix the angle $\Phi = \Phi_0$ and vary ψ.

The dipole-dipole interaction energy is then represented as follows

$$H_d = -\sum_n \frac{d^2 \sin^2\Phi_0}{2a^3}\{\cos(\psi_{n+1}-\psi_n)+3\cos(\psi_{n+1}+\psi_n)\}. \tag{50.18}$$

The local field acting on a dipole moment takes the form

$$H_p = \epsilon\sum_n \sin^2\psi_n. \tag{50.19}$$

Both these interactions define two equilibrium positions

$$\psi_{n1}^0 = 0 \text{ and } \psi_{n2}^0 = \pi, \tag{50.20}$$

separated by a potential barrier of height ϵ. The kinetic energy of rotations of molecules in the chain is given by

$$H_k = \frac{1}{2}J_0\sum_n \dot{\psi}_n^2, \tag{50.21}$$

where J_0 is an inertia moment of a molecule relative to a rotation around the fixed OH-bond. Let us assume that the effect of the permanent electric field of strength E upon the chain of molecular dipoles is determined by the formula

$$H_E = -pE\sum_n \cos\psi_n, \quad p = d\sin\Phi_0. \tag{50.22}$$

The sum of the expressions (50.18), (50.19), (50.21), and (50.22) defines the equation of motion. To study the steady-state motion in the field, one can introduce a dissipation term proportional to the angular velocity ψ_t in the equation. Within the continuum approach, one obtains the following equation

$$\psi_{tt} - V_0^2\psi_{xx} + \Omega_0^2\sin(2\psi) + \mathcal{E}_0\sin\psi - \eta J_0^{-1}\psi_t = 0, \tag{50.23}$$

where the abbreviated notations are used

$$V_0 = p/(2a^3J_0)^{1/2}, \quad \Omega_0^2 = 2(\epsilon+3p^2/a^3)/J_0, \quad \mathcal{E}_0 = pE/J_0. \tag{50.24}$$

Let us assume that dissipation completely compensates the effect of the electric field, and in the given chain there is established a stationary regime - the migration of an orientational defect with a constant velocity V. Introducing the dimensionless variable

$$\zeta = (x-x_0-Vt)/a, \tag{50.25}$$

one can represent the solution of eqn. (50.23) by the function

$$\psi(\zeta) = \arccos[\pm \mathrm{th}(\mu\zeta)] \tag{50.26}$$

with

$$\mu = \alpha\Omega_0\sqrt{2}/V_0[1-V^2/V_0^2]^{1/2} \tag{50.27}$$

and

$$\alpha = \eta J_0\Omega_0\sqrt{2}/p. \tag{50.28}$$

The velocity of the steady-state is determined by an electric field strength E and a dissipation parameter η via the relation

$$V = V_0/(1+\alpha^2/E^2)^{1/2}. \tag{50.29}$$

For a strong dissipation ($\eta\to\infty$), the velocity of the steady migration is rather small. If dissipation is absent, the velocity of the steady migration of an orientational defect may increase up to the limiting velocity V_0. Such a limiting velocity cannot, however, be realized since the structure of the chain is broken; in particular, the double-minima potential vanishes with increasing the electric field E. To determine this magnitude of the electric field, one takes into account that the maximum of the potential barrier between the positions of minima, given at $\psi = 0$ and π, is expressed in terms of ψ_{max} which obeys the relation

$$\cos\psi_{max} = -\epsilon_0/2\Omega_0^2 = -pE/2(\epsilon+3p^2/a^3). \tag{50.30}$$

The limiting electric field magnitude at which the hydrogen-bonding potential becomes a single-minimum one, is determined by the condition

$$\cos\psi_{max} = 1.$$

Hence, the limiting electric field magnitude is

$$E_{max} = 2p^{-1}(\epsilon+3p^2/a^3). \tag{50.31}$$

50.2. The Interaction of Orientational Bjerrum Defects

In order to study the interaction of extended orientational Bjerrum defects in the absence of an external electric field and dissipation, one should find double-phase solutions of eqn. (50.23) given at vanishing $\mathcal{E}_0$ and η. In this case, the equation takes the form

$$\psi_{tt} - V_0^2\psi_{xx} + \frac{1}{2}\Omega_0^2\sin(2\psi) = 0 \tag{50.32}$$

where V_0 is the characteristic velocity in the given chain. Introducing the dimensionless variables

$$\xi = x/\Lambda_0 \ \text{ and } \ \tau = \Omega_0 t \tag{50.33}$$

with

$$\Lambda_0 = V_0/\Omega \tag{50.34}$$

one can transform eqn. (50.32) to the canonical form

$$\psi_{\xi\xi} - \psi_{\tau\tau} = \frac{1}{2}\sin(2\psi), \quad \psi = \psi(\xi,\tau). \tag{50.35}$$

This equation differs from the sine-Gordon equation treated in Section 39 by the double ψ-function under the sine. The theory of this equation was developed by Sergienko [107-109].

To investigate the single- and double-phase solutions of eqn. (50.35), one can use the method of separation of variables presented in Section 41. Let us now introduce the transformation of the ψ-function by means of the relation

$$\mathrm{tg}\left[\frac{1}{2}\psi(\xi,\tau)\right] = \Phi(\xi)F(\tau). \tag{50.36}$$

Substituting (50.36) into eqn. (50.35) results in the set of two first-order differential equations for the variables

$$\Phi_\xi^2(\xi) = -p\Phi^4(\xi) + m\Phi^2(\xi) + q, \tag{50.37}$$

$$F_\tau^2(\tau) = -qF^4(\tau) + (m-1)F^2(\tau) + p,$$

where $p, q,$ and m are arbitrary integration constants. Single-phase solutions of this set of equations correspond to vanishing values of parameters p and q. In this case, the set of equations (50.37) is simplified

$$\Phi_\xi = \delta\sqrt{m}\,\Phi, \quad \delta = \pm 1, \quad F_\tau = \delta\sqrt{m-1}\,F. \tag{50.38}$$

This set of equations depends only on the single parameter m and possesses the following real solution

$$\mathrm{tg}\left[\frac{1}{2}\psi(\xi,\tau)\right] = \mathrm{Re}\exp\left\{\delta\sqrt{m}\left[\xi-\xi_0-\sqrt{\frac{m-1}{m}}\tau\right]\right\}. \tag{50.39}$$

Let us consider the particular case when $m>1$. The unique parameter m is expressed in terms of the dimensionless velocity W

$$[(m-1)/m]^{1/2} = W, \quad \sqrt{m} = \gamma = (1-W^2)^{-1/2}. \tag{50.40}$$

The latter obeys the inequality $0 \leqslant W < 1$ and is measured in units of the characteristic velocity V_0, i.e., $W = V/V_0$. By virtue of eqn. (50.40), one can rewrite the solution (50.39) as follows

$$\psi(\xi,\tau) = 2\mathrm{arctg}\{\exp[\eta\gamma(\xi-\xi_0-W\tau)]\}. \tag{50.41}$$

This describes a change of ψ by π instead of that by 2π determined by the single-phase solution of the traditional sine-Gordon equation (39.11).

Negative values of m in the function (50.39) correspond to a velocity V exceeding the characteristic velocity V_0. The appropriate dimensionless velocity $U = V/V_0 > 1$ is expressed in terms of the parameter $m<0$ via the relation

$$[(m-1)/m]^{1/2} = U.$$

In this case, the solution (50.39) takes the form

$$\psi_n(\xi,t) = 2\mathrm{arctg}\left\{\cos\left[\frac{\xi-U\tau}{\sqrt{U^2-1}}\right]\right\}.$$

This describes a periodic wave propagating with velocity $U>1$ along ζ. A wavelength of such excitation,

$$\lambda = 2\pi\sqrt{U^2-1}$$

increases from zero at $U=1$ with increasing velocity U. The function $\psi_n(\zeta,\tau)$ describes the periodic vibrations near a zeroth value, but not the transitions between the equilibrium positions.

Double-phase solutions of the set of eqns. (50.37) correspond to the parameters $p=0$ and $q=1$. In this case, there remains a unique free parameter m in the set of equations (50.37) which transforms into the following

$$\Phi_\xi^2 = m\Phi^2 + 1, \quad F_\tau^2 = -F^4 + (m-1)F^2. \tag{50.42}$$

If $m > 1$, this describes the kink-kink interaction, since, in this case, it possesses in fact the solutions

$$\Phi(\xi) = m^{-1/2}\mathrm{sh}\xi, \quad F(\tau) = \frac{\sqrt{m-1}}{\mathrm{ch}(\tau W\sqrt{m})}. \tag{50.43}$$

Here m is expressed in terms of γ and the velocity $W = V/V_0$ due to eqn. (50.40). Substituting eqn. (50.43) into (50.36), one obtains the function of the double-phase solution

$$\psi_{K,K}(\xi,\tau) = 2\mathrm{arctg}\frac{\mathrm{sh}(W\xi)}{\mathrm{ch}(\gamma W\xi)}. \tag{50.44}$$

Introducing the quantity ξ_0 by means of the relation

$$W = \exp(-\gamma\xi_0), \tag{50.45}$$

one obtains the asymptotes of the function $\psi_{K,K}$:

$$\frac{1}{2}\psi_{K,K}(\xi,\tau) \Rightarrow \begin{cases} \mathrm{arctg}\{e^{\gamma(\xi-\xi_0+W|\tau|)} + e^{-\gamma(\xi-\xi_0-W|\tau|)}\}, & \tau \to -\infty \\ \mathrm{arctg}\{e^{\gamma(\xi-\xi_0-W\tau)} + e^{-\gamma(\xi-\xi_0+W\tau)}\}, & \tau \to \infty. \end{cases} \tag{50.46}$$

Each line in (50.46) describes a pair of kinks moving in opposite directions with velocity W. Their collision occurs at $\zeta_0 = 0$ and $\tau = 0$, as is illustrated in Fig. 27.

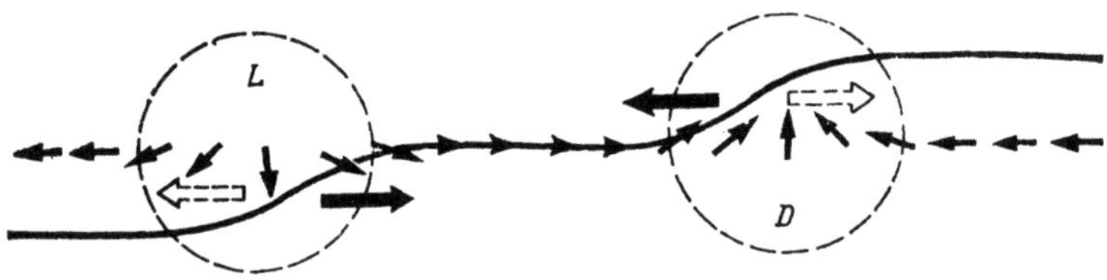

Figure 27. Collision of kinks.

Small arrows denote projections of directions of the dipole moments of molecules onto the plane involving the ξ-th axis and being perpendicular to the fixed OH-bond of each molecule of this model. The right-hand kink carries a positive defect D (a thickening of the ends of the electric dipole projections), and the left-hand one carries a negative defect (rarefaction of dipole ends). At $\xi = 0$, the defects collide elastically and travel in opposite directions. The direction of movement of kinks before the elastic collision is depicted in Fig. 27 by the solid and heavy arrows and, afterwards by the shaded arrows.

For $m > 1$, the set of equations (50.42) admits the following solutions, too,

$$\Phi(\xi) = \frac{i\,\mathrm{ch}(\gamma\xi)}{\sqrt{m}}, \quad F(\tau) = \frac{-i\sqrt{m-1}}{\mathrm{sh}(\gamma W\tau)}.$$

These define the real function

$$\psi_{KA}(\xi,\tau) = 2\mathrm{arctg}\left\{\frac{W\mathrm{ch}(\gamma\xi)}{\mathrm{sh}(\gamma W\tau)}\right\}. \tag{50.47}$$

An analysis of its asymptotic behavior under conditions of $|\xi|$ and $|\tau|$ approaching infinity demonstrates that this describes the kink-antikink collision illustrated in Fig. 28.

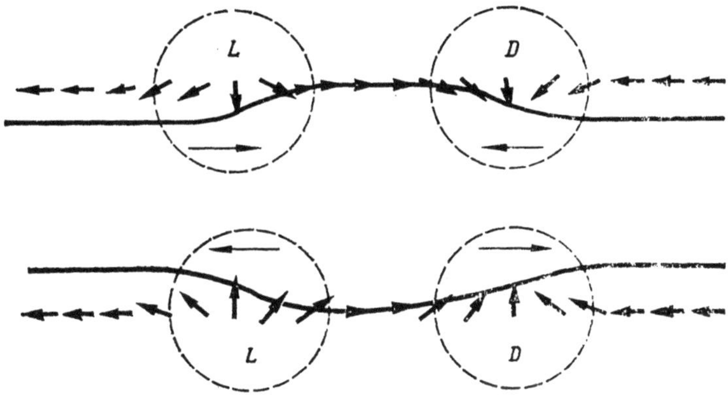

Figure 28. Collision of anti-kinks.

By the arrows, the top of Fig. 28 shows an arrangement of dipole moments of the antikink-positive orientational defect migrating from the right to the left, and of the kink (negative L-defect) migrating in the opposite direction. During collision, they pass through each other, and deviations of the dipole moment projections are displaced from the region $(-\pi,0)$ to the region $(0,\pi)$. Hence, after collision the kink carries a positive D-defect, while the antikink carries a negative (L) one, which leads to the repulsion of orientational defects from each other; however, their carriers, the kink and antikink, pass through each other.

So, according to eqn. (50.32) involving a double function under the sine, in the process of collision kink and kink are elastically repulsed, while kink and antikink pass through each other. As demonstrated in Subsection 41.2, the kink-kink and kink-antikink collisions are not differential qualitatively if they are described by the sine-Gordon equation involving the function $\psi(\xi,\tau)$ under the sine.

The set of equations (50.38) admits also the third type of solutions if the parameter m obeys the inequality

$$0 < m < 1. \tag{50.48}$$

In this case, the solutions are

$$\Phi(\xi) = \frac{i\,\mathrm{ch}(\xi\sqrt{m})}{\sqrt{m}}, \quad F(\tau) = \frac{i\sqrt{1-m}}{\sin(\tau\sqrt{1-m})}. \tag{50.49}$$

The complete function $\psi_0(\xi,\tau)$ is given by the expression

$$\psi_0(\xi,\tau) = 2\mathrm{arctg}\left\{\frac{\mathrm{tg}\nu\sin(\tau\cos\nu)}{\mathrm{ch}(\xi\sin\nu)}\right\}, \tag{50.50}$$

where

$$\Omega = \sqrt{\frac{1-m}{m}} = \frac{1}{\mathrm{tg}\nu}. \tag{50.51}$$

Such a solution satisfies the boundary conditions $\psi_0(\xi,\tau) = 0$ as $|\xi| \to \infty$. Hence, it describes the localized, immobile, and pulsating formation representing a bound state of a positive (D) orientational defect with a negative (L) one, i.e., of kink with antikink. This formation is called a bion, or breather (see Section 41). The internal frequency of the localized bion is $\omega = \cos\nu = \Omega(1+\Omega^2)^{-1/2}$. The spatial dimension of the pulsating region is inversely proportional to the quantity

$$Q = \sin\nu = (1+\Omega^2)^{-1/2}.$$

It is obvious that the simple relation

$$\omega^2 + Q^2 = 1 \tag{50.52}$$

holds.

Transformation from immobile bion to the bion, moving with a constant velocity W as a whole, precisely of the Lorentz type (see Section 39):

$$\tau \to \gamma(\tau + W\xi), \quad \xi = \gamma(\xi - W\tau), \quad \gamma \leqslant (1-W^2)^{-1/2}. \tag{50.53}$$

Therefore, the latter is described by the function

$$\psi_0(\xi,\tau) = 2\mathrm{arctg}\left\{\frac{\mathrm{tg}\nu\sin[(\tau - W\xi)\gamma\cos\nu]}{\mathrm{ch}[(\xi - W\tau)\gamma\sin\nu]}\right\}. \tag{50.54}$$

The frequency of its pulses is determined by

$$\omega_0 = \gamma\cos\nu = \gamma\Omega(1+\Omega^2)^{-1/2}. \tag{50.55}$$

Finally, its total energy

$$\mathcal{E}_n = \frac{1}{2}\int d\xi\left\{\psi_\tau^2+\psi_\xi^2+\frac{1}{2}[1-\cos2\psi]\right\}$$

and total momentum

$$\mathcal{P}_0 = -\int d\xi\psi_\tau\psi_\xi$$

take the form

$$\mathcal{E}_0 = \frac{16\sin\nu}{\sqrt{1-W^2}}, \quad \mathcal{P}_0 = \frac{16W\sin\nu}{\sqrt{1-W^2}}. \tag{50.56}$$

By virtue of eqn. (50.56), one obtains the formula for the bion velocity

$$W = \mathcal{P}_0/\mathcal{E}_0.$$

The soliton models for structural defects in the hydrogen-bonded network of liquid water and diverse polymorphic forms of ice have recently been developed by Kashimori et al. [296], Pnevmatikos [297], Weberpals and Spatschek [298], Halding and Lomdahl [299], Kryachko [300,301,303,304], and Yanovitskii and Kryachko [302].

51. The Migration of Protons in Near-Surface Layers of Water

The lamellar structure of water in near-surface layers and narrow pores, and the high proton conductivity along these layers, may be essentially important in some phenomena. For example, the lamellar structure of water in narrow pores of some materials and their large proton conductivity may serve as the basis of a new technology of detection of pores in some industrial products.

The phenomenon of high near-surface conductivity may also serve to justify theoretically Skulachev's hypothesis concerning the migration of protons on the surface of biological membranes through the surface layers of water surrounding the membranes (Moscow University, 1976-1980).

Proton pumps (see the next Section) produce a higher concentration of protons at some sites of membranes of mitochondria and chloroplasts. Normally, there should be an equilization of the concentration of protons and electric potential over the entire membrane surface.

In the seventies, during the investigation of the three-dimensional structure of mitochondria in unicellular organisms by the observation of parallel microscopic sections by means of an electronic microscope, giant net mitochondria were revealed whose dimensions are comparable with those of cells. Within a single microscopic section this was manifested in the form of tens or hundreds of small rounded mitochondria. Bakeeva observed a unique network of extended mitochondria penetrating an entire cell of higher animals; in particular, the muscle cells of rat. Net-like mitochondria in the muscles of animals have been revealed by Bubenuer (West Germany), and Gote and Padicula (France) in 1966, but they used single sections and could not determine the spatial structure of the whole mitochondria network.

The existence of an extended network of mitochondria in cells requires elucidation of the molecular mechanism of the equilization of the potential and concentrations of protons over their surfaces. Special studies have been carried out in Skulachev's Laboratory to determine the nature of this phenomenon. Some coworkers of this Laboratory (A. Glagolev, T. Glagoleva, T. Potapova, and A. Chailakhanjan) chose the thread-like cyanobacteria -trihomes- as the object of invstigation. These multicellular organisms, consisting of hundreds of cells, are linked via an end-to-end manner in the form of filaments. The cytoplasma of each cell is surrounded by a membrane which involves proton pumps. Under illumination, these pumps produce a difference of potentials and concentrations of protons on the cytoplasmatic membrane. Proton potential plays the role of an energy source for the motion of bacteria, or in other words, this demonstrates a proton motor. These investigators also demonstrated that under the illumination of only 4-5% of the total length of the bacteria, the proton potential arises over the entire length. They measured the distribution of the proton potential over the length of a molecule by means of specific extracellular electrodes following the excitation of a single end by light. It appeared that the protonic potential propagates rapidly along the bacteria to a distance of a few millimeters. The investigators of this astonishing experiment attempted to explain this rapid propagation of the proton potential along the thread-like cyanobacteria with the help of the so-called 'cable hypothesis' proposed by V. Skulachev in 1968. According to this, the propagation of the potential along the membrane surface is realized in a manner similar to that in an electric cable. However, it seems to be difficult to accept this interpretation of these very important experiments. The electrical resistance of a cytoplasmatic membrane is rather large. This membrane endures electric fields of $10^5\ V/cm$. Protons cannot migrate along the interval layers of the membrane. Migration transverse to this membrane is realized only through the protein proton pathways (channels), or by means of proton ionophones (disconnectors). It appears that an equalization of proton potentials and concentrations over the extended membrane surface takes place along the near-surface layers of the structuralized water surrounding the given membrane. Such water is characterized by a large proton conductivity.

All living organisms consist of two-thirds of water. Therefore, water is one of the basic components of life. A large fraction of cellular water is in close contact with the surfaces of membranes of numerous organells, cytoplasmatic membrane, microfilaments and microtubes of cells. The properties of near-surface water of thickness $30-40\text{Å}$ differ considerably from those of bulk water, even in the case when these surfaces do not carry an electric charge. Unfortunately, these peculiarities are beyond the interest of the traditional description of the role of the cellular water. The specific properties of near-surface water must certainly be taken into account in studies of the functions of enzymes in contact with water.

51.1. Proton Pumps in Cellular Membranes

The transport of protons plays an important role in bioenergetics, and the processes of transferring vibrational energy and electrons. According to the generally accepted hypothesis of the English biochemist P. Mitchell, the intermediate stage of realizing the energy of oxidation of food-stuff substrates in the mitochondria of cells is the energetization of conjugated membranes. Such a state is achieved by an increase in the concentration of protons on their external surface by means of specific proton pumps. The difference in electrochemical potentials, arising in this way, is utilized by these membrane for the synthesis of ATP molecules, initiating the rotational motion of flagells of bacteria, and some other phenomena.

Mitchell's hypothesis has been multiply confirmed; however, a properly-based understanding of the molecular mechanism of the activity of proton pumps is still lacking. This is related to the fact that the organization and the structure of proteins and lipids surrounding them and forming a proton pump have been studied insufficiently. To date, the proton pumps of the purple membranes of Halobacteria have been well studied. These pumps are activated by light.

In March 1976, Stoeckenius (University of California) reported his study (as long ago as 1965) of micro-organisms, *Halobacteria*, which utilize light energy without the participation of chlorophyll. Up to that time, the photosynthesis process without the involvement of chlorophyll was unknown.

Halobacteria live only in very salty lagoons, lakes, and ponds (with high concentrations of $NaCl$ and Mg^{2+} salts) in which salt is extracted by evaporation. They possess a cylindrical rod-shaped form with a diameter of about 0.5 microns and a length of 5 microns and are enclosed by a glycoprotein wall. Tights providing a mobility of bacteria in water are attached to its both ends.

A violet (bright purple) fraction, consisting of single-type proteins and lipids in the ratio 3:1 and which gives the main colour of the bacteria, is a part of the membrane of Halobacteria. In undamaged bacteria, this fraction forms islets (violet and purple patches) on the membrane surface. These icelets have a two-dimensional hexagonal crystalline structure. Each unit cell contains three protein molecules of the same type with a molecular weight of 26,000. Triples of molecules, triads, are combined into the regular hexagons. The greater part of the surface of these molecules is hydrophobic and immersed in the lipid layer of the membrane.

A coloured grouping, the residue of a retinal molecule (aldehyde of vitamin A), is involved in the protein chain of each molecule. By their structure, the molecules of the violet patches are close to those of rhodopsin. The latter is a vision pigment of the retina in animal eyes which are formed by a complex of the retinal molecule with the protein opsin. For this reason, the protein, revealed in violet patches of Halobacteria, is called *bacteriorhodopsin*.

In the bacteriorhodopsin molecule, the aldehyde end of the retinal molecule is bounded to the amino group of the lysine amino acid residue of the protein molecule with the release of a water molecule. Than, between the nitrogen of the lysine residue and the carbon of the aldehyde end of the retinal molecule there forms a covalent bond called a Schiff base.

The absorption of light by the bacteriorhodopsin molecule in the 570 nm region is due to the presence of a bond of the Schiff-base type. When light is absorbed, a proton is detached from the nitrogen atom of the Schiff base, i.e., there occurs a deprotonation. Since the final state of this proton refers to the continuum spectrum, the absorption band in the red region is rather broad. After removal of the proton, the absorption spectrum of the bacteriorhodopsin molecule shifts to the blue region of $412\,nm$. After the proton has been reattached (protonation), the absorption spectrum shifts back to the red region of $570\,nm$.

Skulachev and his collaborators investigated the time evolution of the absorption process of a light impulse (of duration $3 \times 10^{-8}\,s$) by the bacteriorhodopsin molecule. They succeeded in isolating three evolution phases. Phase I having a during of less than $10^{-7}\,s$ is characterized by the absorption of red light and corresponds to the photo-deprotonation, i.e., to the removal of a proton from the nitrogen atom of the Schiff base due to light and its escape into the exterior of the membrane. As the results of deprotonation, the absorption spectrum of the bacteriorhodopsin molecule shifts to the blue region. The lifetime of this state of the bacteriorhodopsin molecule accounts for the tens of microseconds. This is Phase II. Phase III having a duration of tens of milliseconds is related to the population of the vacation site in the Schiff base by a proton (protonation). In this phase, the molecule transforms into to initial state.

In 1973, Oesterhelt and Stoeckenius demonstrated that, in the cellular membrane of Halobacteria, the violet (purple) islets act as protonic pumps pumping out protons from cytoplasma to the exterior of the cell under the action of light. Associated with this, a proton concentration gradient and an electric potential difference are produced across the membrane. The free energy of the electrochemical proton gradient is used further for the synthesis of ATP molecules in the other regions of the membrane containing the enzymes of ATP-ase.

Danon and Stoeckenius (1973) showed that Halobacteria can synthesize ATP molecules by two routes using either light energy or the chemical energy of oxidation. Light, as well as respiration, generates a transverse electrochemical potential on the membrane.

Intensive studies of Halobacteria and the unique properties of its membranes have been carried out in many laboratories all over the world. In the Soviet Union, great success was achieved by Skulachev's Laboratory where, in particular, the electric potential was measured arising on the membrane under the work of the protonic pumps inducing proton migration.

Of great interest are the experiments by Stoeckenius and Racker at Cornell University with artificial lipid vesicles (liposomes) into which were incorporated fragments of purple membranes. When these vesicles are illuminated with light, the *pH* of the external medium increases. Therefore, under the action of light, protons come out of the lipid vesicles. Because bacteriorhodopsin was the only protein in the lipid vesicles and because it had the same activity in vesicles formed from different lipids too, Stoeckenius and Racker concluded that it is the bacteriorhodopsin which utilizes the light energy for generating the proton gradient. Under the action of light upon the bacteria cells, the medium was oxidated, i.e., there occurred a decrease in *pH* caused by the migration of protons outside the cells. The difference in the action of light upon bacteria cells and lipid vesicles indicates the reverse arrangement of the bacteriorhodopsin molecules in cells and artificial lipid vesicles.

The independent arrangement of the enzymes of ATP-ase and the proton pumps in the lipid membrane was confirmed by the Racker's experiments. He attached the ATP-ase enzyme, taken from calf heart mitochondria, to the lipid vesicles containing bacteriorhodopsin. Under illumination, these vesicles synthesized ATP molecules. This experimentally established that bacteriorhodopsin is the basic element of the photosynthesis without chlorophyll which takes place in Halobacteria. Under illumination of bacteria by a light, protons escape from the Schiff base of the molecules into the medium.

The primary structure of the bacteriorhodopsin molecule, which contains 248 amino acid residues, was firstly interpreted by Yu. Ovchinnikov and his collaborators in 1978. In particular, they showed that bacteriorhodopsin involves the hydrophilic radicals of tyrosine, serine and threonine containing polar groups. As a whole, the bacteriorhodopsin molecule is hydrophobic and embedded entirely in the lipid layer of the membrane. Bacteriorhodopsin is the simplest example of the known proton pumps. It differs from all other respirator and light-dependent proton pumps in that the inherent proton transfer is not coupled with the transport of electrons.

Based on X-ray and electron microscopic data, R. Henderson and his co-workers (1975) demonstrated that the bacteriorhodopsin molecule crosses the membrane seven times by α-helical and approximately parallel segments, or 'columns' having a length of about $4\,nm$ separated from each other by a distance of

$\sim 1\,nm$. These helical segments are embedded into the lipid layer of the membrane and are joined to each other by hydrophilic proton filaments arranged outside the membrane (Fig. 29).

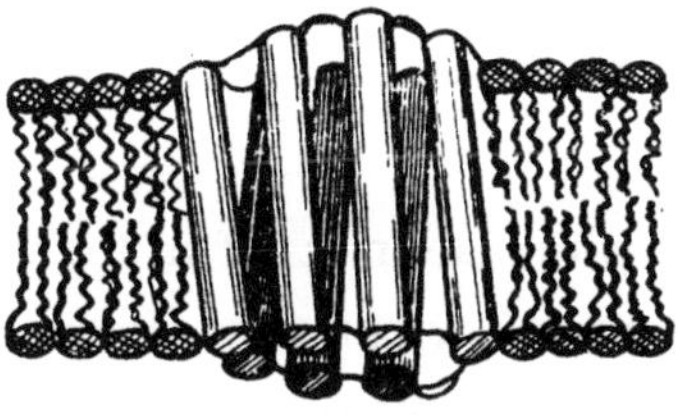

Figure 29. The arrangement of the seven α-helical segments of the bacteriorhodopsin molecule in a purple membrane.

In the cytoplasmatic membrane of bacteria, every three bacteriorhodopsin molecules are clusterized. In the centre of a cluster there is a channel whose diameter is about $1-2\,nm$. The hydrophilic radicals of the bacteriorhodopsin molecule are directed inside this channel filled by water molecules. The exterior of the whole complex is hydrophobic.

Taking into account the high proton conductivity of water in narrow channels, one can propose the following hypothesis for the mechanism of a proton pump in the membrane of Halobacteria (Davydov [49,158a,169]). Under the absorption of light, a proton is ejected from the Schiff base and leaves the purple membrane from outside (deprotonation of Schiff base). This is the first phase corresponding to the broad red band of light absorption conditioned by the proton transition from a bound state to a free state, beyond the membrane. Deprotonized (without a proton) molecules of bacteriorhodopsin absorb blue light. Vacancies forming in the Schiff base are occupied by protons. The second and third aforementioned phases correspond to this protonation process. Protons come from cytoplasma via migration (apparently, in the form of hydroxonium ions, H_3O^+) along the chains of hydrogen bonds of molecules of the structuralized water in the central channel of the bacteriorhodopsin molecule. The state of the hydrogen-bonded chain changes after each passage of a proton, and returns to the initial one after the migration of Bjerrum defects.

One can also suggest a second way of protonation when vacancies in the Schiff base of the bacteriorhodopsin molecule are occupied by protons belonging to the neighboring dissociated molecules of water. In this case, hydroxyl ions migrate along the chains of hydrogen bonds of the structuralized water in the internal channel to the interior of the membrane due to the concentration gradient of hydroxyl ions. Attaching protons from the cytoplasma, the hydroxyl ions form water molecules. After each passage of an hydroxyl ion, the reconstructed chains of water molecules return to the original state due to the migration of Bjerrum defects.

Both these processes of protonation result in the occupation of a vacant place in the Schiff base by protons of cytoplasma. However, they occur with different velocities. This is apparently the reason for the existence of two protonation phases.

Important results were obtained by Skulachev and his co-workers by the isotopic substitution of water by its deuterium analogue (D_2O). It appeared that the time of protonation (Phases II and III) increased considerably under this substitution. One can assume that these results confirm the aforementioned hypothesis, according to which under protonation of vacancies in the Schiff bases, the protons (deuteria) migrate along the chains of hydrogen bonds of molecules of ordinary (heavy) water filling the narrow central channel of the bacteriorhodopsin molecule in which the stable hydrogen-bonded chains are formed.

The important role of hydrogen bonds in the process of the migration of protons in purple membranes was suggested in 1978 by Nagle and Morowitz [233], [234], who assumed, however, that the hydrogen bonds were formed between the hydroxyl groups of hydrophilic radicals of serine, threonine, tyrosine, and the carboxylic groups of asparagine and glutamine acids present in the bacteriorhodopsin molecule. To justify this assumption, one should suggest that the hydroxyl radicals are arranged so closely so as to form a continuous network of hydrogen bonds. The next failure is that it is difficult to explain the way by which a system of hydrogen bonds between rigidly fixed radicals returns to its original state.

The model of the present author [49,158a] is based on the fact that the migration of proton is realized via the hydrogen bonds between the structuralized layers of water molecules in the narrow hydrophilic channel of the molecule. The initial state is rehabilitated due to the rotation of water molecules, or the migration of the orientational Bjerrum defect. Hydroxylic radicals in the central channel of the molecule plays an important role in filling this channel by water molecules only. The small diameter of the channel provides a spatial and orientational ordering of water there, and also the realization of continuous chains of hydrogen bonds along its axis.

51.2. The Role of Structuralized Water in Cartilage Tissues

A so-called cartilage tissue, or *cartilage* occupies a specific place among different tissues of animals. It performs many important functions in an organism, and differs considerably from the other tissues. A cartilage does not contain nerves, blood-vessels, and a lymphatic system.

A cartilage plays a role in creating the shape of a body and serves further as a covering and bearing for bones. It withstands high pressures and provides a free or unhindered sliding of bones relative to each other. The knitting together of bones after a fracture begins by generating cartilage. Physical and chemical peculiarities of the structure of cartilage provides its strength and high elasticity, allowing it to return to its original form and structure after deformation.

A cartilage consists of cells of chondrocyte where huge macromolecules are synthesized. These macromolecules fill the intercellular volume and form a substance which determines the cartilage properties and basically consists of collagen and *proteoglycane* molecules.

The strength of cartilage is due to the fibrillar organization of rigid protein molecules of a collagen. This rod-like macromolecule having a diameter of $1.4\,nm$ is formed by folding the three left protein helices into the right helix. Such molecules are joined into the interval $10-200\,nm$. A network of collagen fibrils plays the same role in the cartilage as that of a steel covering in architectural constructions.

Between the collagen fibrils there are arranged giant macromolecules having a molecular weight of $250{,}000-300{,}000$. These are called *proteoglycanes*. A hyaluronic acid serves as the frame of such a molecule. Lengthy polypeptide chains, called central beams, pull of its principal filament. More than 100 chains of central proteins are joined with a single molecule of hyaluronic acid, which is also coupled with a large number of polysaccharide molecules, about 50 molecules of keratinsulfate and 100 of chondrotinesulfate. The total mass of the whole aggregate accounts for nearly 10,000000.

The regions of intracellular substance without molecules are filled by water. The latter is the third and, apparently, the basic component of cartilage tissue. Sugar molecules, present in the aggregates, and proteoglycanes, carry negative electric charge. Water molecules, possessing permanent electric dipole moments, form the solvation shells within the electric field of the proteoglycanes. According to the modern concepts, the elasticity of cartilage results from the interaction of water molecules with each other and also with the segments of polysaccharide chains which carry electric charges, too. When a cartilage is under pressure, water is ejected from the negatively charged regions, which further approach each other. But the charge-charge repulsive forces prevent their closer contact. When the pressure disappears, the water returns to its original place.

One needs to assume that the explanation of the elastic properties of a cartilage on the basis of electrostatic interactions only is incomplete. Because of the high dielectric permittivity of water, these interactions are strongly screened and non-negligible at small separations only.

Investigations carried out by Antonchenko, Illyn, and Makovskii [134a] at the Institute of Theoretical Physics, Kiev have demonstrated that the electrostatic interaction play an auxiliary role in the elastic properties of cartilage. The forces arising under the structural reconstruction of water caused by a change in the distances between macromolecules of cartilage tissue are of greater importance. It was found that near any surface (including the unchanged surfaces) water molecules are arranged via the layers parallel to boundary. The molecular planes and hydrogen bonds are also oriented mainly parallel to the boundary. This, near the huge macromolecules forming cartilage tissue, water appears partially to the spatially and orientationally ordered system. Under load, the macromolecules in a cartilage approach each other, and the degree of structural ordering of water molecules increases. As was demonstrated numerically, due to the reconstruction of the water structure there arise repulsive forces which increase exponentially between the boundaries of a water system decrease. These are mostly manifested under conditions of the maximum pressure of a cartilage tissue and apparently determine its elastic properties. Hence, the structural reconstruction of water, occurring under the variation of the distances between the cartilage fibrils, seems to be the basic reason which determines its unique elastic properties.

In the absence of blood-vessels, under loading and further weakening, the diffusion of water provides the feeding of the living cells of cartilage tissue synthesizing molecules filling the intercellular volume. It is known that the cartilages of articulations become thin and fragile under lingering immobility. This is one of the reasons for the necessity of active movement to ensure the vital activity of vertebral animals.

Chapter 13

THREE-DIMENSIONAL SOLITONS (POLARONS) IN IONIC CRYSTALS

52. The Landau and Pekar Concepts of Autolocalization of the Electron in Ionic Crystals

Of the various phenomena involving autolocalization of particles (excitons or electrons) in crystals, that of electrons in ionic crystals is of specific interest. In this case, autolocalization results from the long-range Coulomb interaction of an electron with the vector field of polarization arising in a crystal due to the influence of the same electron.

The concept of electron autolocalization in the ideal ionic crystal was first proposed by Landau [86,220] and intensively developed by Pekar [98,99], Tyablikov [118-120], Fröhlich [184,185], and others. As shown by Landau, electron autolocalization arises in an ionic crystal with inertial polarization if the electron creates a local polarization due to its own field. This local polarization is equivalent to the potential well of an electron. Inside a rather deep potential well, the electron occupies a state with a discrete energy, and supports the lattice polarization by its mean field. Such an autolocalized state is called a *polaron*.

The Landau-Pekar polarons are three-dimensional non-topological solitons in ionic crystals. For this reason, the comprehensive treatment of such formations is of great interest from the point of view of elucidating the general conditions under which non-topological solitons are formed in three-dimensional systems with long-range Coulomb-type interaction.

Enormous efforts have been applied recently to the study of the possibility of formation of solitons in two- and three-dimensional media. Considerable success has been achieved in investigating the so-called *topological solitons*. Such solitons may arise in media possessing several states corresponding to the zeroth energy (a degenerate vacuum). For instance, in an isotropic ferromagnetic the vacuum state, i.e., the state having the lowest energy, corresponds to the total magnetization for which the magnetization vector can take arbitrary directions. *Kinks* and *antikinks*, studied in the foregoing Chapters, refer to topological solitons.

It was shown by Dzyaloshinskii and Ivanov [58] that topological solitons localized in three dimensions can exist in a ferromagnetic. Rebbi [242] has discussed the general conditions for the formation of topological solitons, described by nonlinear equations with self-interaction, in media having a different number of dimensions. In particular, the field characterizing the states of two- and three-dimensional solitons has to possess a vector nature with the appropriate two or three components.

Wave functions of non-topological solitons must vanish onto an infinite sphere (a nondegenerate vacuum) since the excitation is localized in a small spatial region. However, analytical investigations of the possibility of formation of non-topological solitons in two- and three-dimensional media lead to serious difficulties. Recently, Bogoliubskii [20] has demonstrated that a transition from a one- to a three-dimensional system results, as a rule, in lowering the stability of scalar, as well as spinor, solitons. In particular, he showed that for a complex scalar field with a self-interaction proportional to the fourth power of the field, charged three-dimensional spherically symmetric solitons are unstable, in contrast to the one-dimensional ones. The instability of spherically symmetric solutions, or non-topological solitons, which approach zero at infinity, in three-dimensional space has been pointed out by Zakharov [64] and Petviashvili [101].

Using the phase plane method, Anderson [133] studied the problem of the existence of three-dimensional non-topological solitons in some nonlinear fields. This approach was applied by Ivanov and Kosevich [70], [71] who have considered the bound states of a large number of magnons in ferromagnetics, and showed that in a three-dimensional crystal, the bound state arise iff the amplitude of their mutual attraction exceeds a certain critical value. The conditions for the formation of topological and non-topological (dynamical) solitons are discussed in the review by Kosevich [80].

The substantial difference in the processes of exciton autolocalization in one- and three-dimensional crystals was first pointed out by Rashba [102,103] in 1957. This problem has been discussed by the present author [35], too. He showed that within the continuum approach (long-wave excitations) autolocalization becomes possible in one-dimensional systems with a local, or contact interaction of excitons with acoustic phonons. The detailed analysis of the conditions under which excitons are autolocalized in different types of crystals has been recently presented in the review by Rashba [241]. In the next Section we outline the adiabatic theory of the interaction of an electron with optical phonons of the polarization of an isotropic ionic crystal as developed by Pekar.

52.1. Adiabatic Theory of Pekar's Polaron

Applying a variational principle, Pekar [98,99] developed the polaron theory by considering the strong interaction of an excess electron with displacements of ions in an isotropic ionic crystals. An electron state is described within the effective mass approach which takes the crystal fields into account by a change of the electron mass, $m \to m^*$. A crystal is treated as a continuum medium described by the polarization vector $\vec{P}(\vec{r})$. As a first step, the kinetic energy of motion of ions is neglected (the adiabatic approximation). The electron energy in the crystal referred to the energy of the bottom of the conduction band, and the polarization energy, can be represented by the sum of three terms,

$$E\{\psi,\mathbf{P}(r)\} = \int \psi^*(\mathbf{r}) \left[-\frac{\hbar^2}{2m} \nabla^2 \right] \psi(\mathbf{r}) d^3r + 2\pi\tilde{\epsilon} \int \mathbf{P}^2(\mathbf{r}) d^3r - \qquad (52.1)$$

$$- \int \mathbf{P}(\mathbf{r}) \mathbf{D}(\mathbf{r}) d^3r,$$

where

$$\mathbf{D}(r) = e \int |\psi(\mathbf{r}_1)|^2 \frac{\mathbf{r}_1 - r}{|\mathbf{r}_1 - r|^3} d^3r_1 \qquad (52.2)$$

is the induction vector of the electric field at the point $\vec{r}$ created by the electron located at $\vec{r}_1$. The first two terms in eqn. (52.1) determine the energy of the electron and the polarization field, and the last term describes their interaction. In eqn. (52.1)

$$\tilde{\epsilon} = \frac{\epsilon_0 \epsilon_\infty}{\epsilon_0 - \epsilon_\infty}$$

is the effective dielectric permittivity expressed in terms of the macroscopic dielectric constants, $\epsilon_0 = \lim_{q,\omega\to 0}\epsilon(q,\omega)$ and $\epsilon_\infty = \lim_{q\to 0}\epsilon(q,\omega)$.

Equation (52.1) can be viewed as the functional of the unknown functions $\psi(\vec{r})$ and $\vec{P}(\vec{r})$. In the ground state, it should attain the absolute minimum relative to independent variations of $\psi(\vec{r})$ and $\vec{P}(\vec{r})$ constrained by the normalization condition

$$\int |\psi(\mathbf{r})|^2 d^3r = 1. \tag{52.3}$$

In eqns. (52.1)-(52.3) the integration is carried out over a large volume V of the crystal. Assuming the variation of the functional (52.1) conditioned by the charge of $\vec{P}(\vec{r})$ under fixed $\psi(\vec{r})$ to be vanishing, one obtains the relation between the polarization and induction vector,

$$\mathbf{P}(\mathbf{r}) = \mathbf{D}(\mathbf{r})/4\pi\tilde{\epsilon}, \tag{52.4}$$

where $\vec{D}(\vec{r})$ is given in terms of $\psi(\vec{r})$ via eqn. (52.2).

Substituting eqn. (52.4) into the functional (52.1), one finds the functional which depends on $\psi(\vec{r})$ only,

$$J\{\psi(\mathbf{r})\} = \frac{\hbar^2}{2m^*}\int |\nabla\psi|^2 d^3r - \frac{1}{8\pi\tilde{\epsilon}}\infty\mathbf{D}^2(\mathbf{r})d^3r. \tag{52.5}$$

The functional (52.5) determines the energy of the crystal in which the polarization is consistent with the state of electron motion described by the function $\psi(\vec{r})$. In particular, within the effective mass approach, the electron is the conduction band is described by the function

$$\psi_k(\mathbf{r}) = \frac{1}{\sqrt{V}}\exp(i\mathbf{kr}), \tag{52.6}$$

which depends on the wavenumber $\vec{k}$. In this state, there exists only a uniform polarization of crystal, and the functional becomes positive

$$\mathbf{J}\{\psi_k\} = \hbar^2\mathbf{k}^2/2m^* > 0. \tag{52.7}$$

However, one can fairly easily verify that the state $\psi(\vec{r})$, which corresponds to the electron motion inside a bounded region of the crystal, is more energetically preferable to be occupied by the electron, compared with that in the conduction band described by the plane wave (52.6). The most preferable state can be found by minimizing the functional (52.5) subject to the normalization condition (52.3).

In order to minimize the functional (52.5), Pekar applied the direct variational principle by approximating $\psi(\vec{r})$ via the trial function involving some variational parameters. For an isotropic crystal, in the ground state of a polaron, the function $\psi(\vec{r})$ must be spherically symmetric, i.e., it depends on a single variable $r = |\vec{r}|$. Choosing the trial function, which satisfies the normalization condition (52.3), to have the following form

$$\psi_0(r) = \frac{1+r/r_0}{\sqrt{7\pi r_0^3}} \exp(-r/r_0), \tag{52.8}$$

and minimizing the functional (52.5), one finds the effective radius of the polaron,

$$r_0 = 2\hbar^2 \epsilon / m^* e^2. \tag{52.9}$$

The polaron energy is

$$E = J\{\psi_0(r)\} = -0.054 E_a m^* / m\epsilon, \tag{52.10}$$

where

$$E_a = -\frac{me^4}{\hbar^2} \approx 27.2\, eV$$

is the atomic unit of energy (1 hartree). Using eqns. (52.2) and (52.4), one obtains the polarization of the crystal corresponding to the state (52.8),

$$\mathbf{P}_0(\mathbf{r}) = \frac{e}{4\pi\tilde{\epsilon}} \int |\psi_0(\mathbf{r}_1)|^2 \frac{(\mathbf{r}_1 - \mathbf{r})}{|\mathbf{r}_1 - \mathbf{r}|^2} d^3 r_1, \quad e > 0. \tag{52.11}$$

The inertial polarization (52.11) creates a scalar spherical electric field in the crystal in which the electron moves. This field is given by the formula

$$\varphi(\mathbf{r}) = -e \int \frac{\mathrm{div}\mathbf{P}(\mathbf{r}_1)}{|\mathbf{r} - \mathbf{r}_1|} d^3 r_1 = \begin{cases} -\dfrac{2e}{\tilde{\epsilon} r_0}(0.28 - r^2/2r_0^2) & \text{if } r \ll r_0 \\ -\dfrac{e}{\tilde{\epsilon} r} & \text{if } r \gg r_0. \end{cases} \tag{52.12}$$

Due to translational symmetry, the polaron, or electron with a polarization field, is able to migrate over the crystal as a unique whole with a certain effective mass: the so-called *polaron mass*. The effective mass of the polaron for its translational motion was first evaluated in 1948 by Landau and Pekar [8] within the perturbation theory treatment. In order to evaluate the effective mass of a polaron, one needs to determine the dependence of the crystal energy on the polaron velocity v to within terms of order v^2. Thus, besides the potential energy of polarization, one should take its kinetic energy into account. As a first step, one assumes the polaron velocity to be small, i.e., it obeys the inequality

$$v \ll \Omega_0 r_0, \tag{52.13}$$

where r_0 is the effective radius of the polaron given by eqn. (52.9) and Ω_0 is the frequency of the optical vibrations of ions. If the polaron moves along the x-axis with a constant velocity v, its wave function, ψ, polarization $\vec{P}$, and induction $\vec{D}$ are functions of the variable

$$\xi = \{x - vt, y, z\}.$$

For small velocities v, eqn. (52.4) is transformed as follows

$$\mathbf{P}(\xi) = (4\pi\tilde{\epsilon})^{-1}\left[\mathbf{D}(\xi) - \beta\frac{\partial^2 \mathbf{D}}{\partial t^2}\right], \tag{52.14}$$

where β is a parameter. The application of the operator $\partial/\partial t$ to the functions depending on ξ is equivalent to that of the operator $v\partial/\partial x$. Hence, one can rewrite eqn. (52.14) in the following form

$$\mathbf{P}(\xi) = (4\pi\tilde{\epsilon})^{-1}\left[\mathbf{D}(\xi) - \beta v^2\frac{\partial^2 \mathbf{D}}{\partial x^2}\right]. \tag{52.15}$$

Replacing the relation (52.4) by (52.14) in the functional (52.5), one obtains

$$J_v\{\psi\} = \frac{\hbar^2}{2m^*}\int|\nabla\psi|^2 d^3r - \frac{1}{8\pi\tilde{\epsilon}}\int \mathbf{D}^2 d^3r - \frac{\beta v^2}{8\pi\tilde{\epsilon}}\int\left\{\frac{\partial \mathbf{D}}{\partial x}\right\}^2 d^3r. \tag{52.16}$$

To find the total energy of the crystal for fixed ψ, one has to add the kinetic energy of polarization,

$$\frac{3\beta v^2}{8\pi\tilde{\epsilon}}\int\left[\frac{\partial \mathbf{D}}{\partial x}\right]^2 d^3r,$$

to that given by eqn. (52.16). Therefore, the explicit formula for the total energy in terms of the function ψ becomes

$$E_v = J\{\psi\} + \frac{\beta v^2}{8\pi\tilde{\epsilon}}\int\left[\frac{\partial \mathbf{D}}{\partial x}\right]^2 d^3r. \tag{52.17}$$

Pekar's choice of parameter β is Ω_0^{-2}. If the inequality (52.13) is satisfied, the third term in eqn. (52.16) can be viewed as a small perturbation correction. Hence, evaluating the energy (52.27), one can choose $\psi(r)$ to be $\psi_0(r)$, given by eqn. (52.8) and minimizing the functional (52.16) with $v = 0$. Within this approximation, the polaron energy is represented by the following two terms,

$$E_v = J\{\psi_0\} + \frac{1}{2} M v^2, \tag{52.18}$$

where $J\{\psi_0\}$ is the total energy of the crystal for the rest polaron, and

$$M = (4\pi\tilde{\epsilon}\Omega_0^2)^{-1} \int \left[\frac{\partial D(\psi_0)}{\partial x} \right]^2 dx \tag{52.19}$$

is its effective mass. Substituting eqn. (52.8) into (52.19), one obtains

$$M = 5.8 \cdot 10^{-3} \left[\frac{m^* e^2}{\hbar \epsilon} \right]^3 \frac{e}{\tilde{\epsilon}\Omega_0^2} \approx 9 \cdot 10^3 \left[\frac{m^*}{m} \right]^3 / \tilde{\epsilon}^4 \Omega_0^2 . \tag{52.20}$$

The application of perturbation theory to the continuum spectrum and the introduction by Landau and Pekar of eqn. (52.14) with $\beta = \Omega_0^{-2}$ require stronger justification, to which the following works were devoted. The most consistent translational-invariant adiabatic perturbation theory for polarons was developed by Bogoliubov [18] in 1950 and Tyablikov [118-120] in 1951. The Bogoliubov approach has been advanced by Tavkhelidze, Khrustalev et al. [115,116]. In all these papers, the kinetic energy of ions is chosen to be a small perturbation correction. In order to characterize the motion of the centre of the polarization well, Bogoliubov and Tyablikov introduced three additional degrees of freedom. This complication requires three additional conditions to be imposed to constrain these degrees of freedom of the system. The great complexity of this theory prevented its wide practical application.

In 1976, Gross [192] developed the translational-invariant perturbation theory relative to the inverse powers of a dimesionless parameter of coupling of an electron with the polarization field without introducing extra variables. Within the Gross approach, the state of an electron and the polarization field is described relative to the centre of the polarization well. In all the aforementioned theoretical approaches dealing with polarons, an account of the kinetic energy of the polarization of a crystal is carried out via the perturbation treatment. Hence, the possible change of spherical symmetry of the polarization cloud due to electron motion is neglected. Furthermore, only optical phonons without dispersion are considered. The latter constraint seems to be rather crucial. Rigorously speaking, polarons cannot be mobile in crystals without the spatial dispersion of optical phonons. An account of the motion of polarons requires undoubtedly an introduction of the spatial dispersion of longitudinal optical phonons into the theory. In the next Section, we develop with some modifications the approach proposed by Enol'skii and the author [55] which takes spatial dispersion and the motion of polarons into account (see also [295]).

53. Generalized Theory of Polarons

Polarons are two-component autolocalized excitations (solitons) arising under the coupling of two fields, one of which is the field of an excess electron in an ionic crystal, and the other is the field of the inertial longitudinal polarization of the crystal. The latter is conditioned by displacements of ions due to the Coulomb field of an electron. For an isotropic crystal, the energy band of quanta of the longitudinal polarization (optical phonons) is characterized by the following range of frequencies

$$\Omega(k) = \sqrt{\Omega_0^2 + k^2 V_0^2}.$$

For typical ionic crystals, Ω_0 accounts for $5 \times 10^{13} s^{-1}$. The second term under the root accounts for the spatial dispersion. We consider further crystals having positive dispersion, $V_0^2 > 0$. Optical phonons with negative spatial dispersion, or a negative effective mass, cannot form polarons. The spatial dispersion of quasiparticles determines their collective nature and possible motion.

The group velocity of propagation of wave packets of optical phonons is given by the formula

$$v_{gr} = k V_0^2 [\Omega_0^2 + k^2 V_0^2]^{-1/2}.$$

Thus, the parameter V_0 characterizing spatial dispersion is just the group velocity of ultimately short waves ($k \to \infty$). Using the notation of the foregoing Section, one can represent the Hamiltonian function of an electron in the conduction band of a crystal in the following form within the effective mass approach,

$$H_{el} = \int \psi^*(r,t)\left[-\frac{\hbar^2\nabla^2}{2m^*}\right]\psi(\mathbf{r},t)d^3r. \tag{53.1}$$

The wave function of the electron is normalized by the condition

$$\int |\psi(r,t)|^2 d^3r = 1. \tag{53.2}$$

It is noteworthy that since the motion of the polaron is taken into account, the wave function $\psi(\vec{r},t)$ becomes time dependent. The interaction of the electron with the polarization field $\vec{P}(\vec{r},t)$ created by an electron can be written in two equivalent forms,

$$H_{int} = -e\int \varphi(\mathbf{r},t)|\psi(\mathbf{r},t)|^2 d^3r = -\int \mathbf{P}(\mathbf{r},t)\mathbf{D}(\mathbf{r},t)d^3r. \tag{53.3}$$

Here e is the effective electron charge, and $\varphi(\vec{r},t)$ is the potential resulting form polarization and determined by the equation

$$\nabla_r\varphi(\mathbf{r},t) = 4\pi\mathbf{P}(\mathbf{r},t). \tag{53.4}$$

The vector

$$\mathbf{D}(\mathbf{r},t) = e\nabla_r\int |\psi(\mathbf{r}_1,t)|^2\frac{d^3r_1}{|\mathbf{r}-\mathbf{r}_1|} \tag{53.4a}$$

defines the induction of the electric field of the electron.

An excess electron, moving in a homogeneous polarizable medium, causes the displacement of ions from their equilibrium positions. The Hamiltonian function of the vector field of the resulting polarization $\vec{P}(\vec{r},t)$ can be represented as follows

$$H_p = \frac{2\pi\tilde{\epsilon}}{\Omega_0^2}\int d^3r\left[\left(\frac{\partial \mathbf{P}}{\partial t}\right)^2 + \Omega_0^2\mathbf{P}^2 - V_0^2\mathbf{P}\nabla_r^2\mathbf{P}\right]. \tag{53.5}$$

This formula differs from that used by Pekar (see Subsection 52.1), i.e., $2\pi\tilde{\epsilon}\int\vec{P}^2d^3\vec{r}$, by the terms which take into account the kinetic energy of polarization (the first term) and the dispersion of optical phonons (the third term). From the Hamiltonian (53.5), one obtains directly the equation of motion

$$\frac{\partial^2}{\partial t^2}\mathbf{P} + \Omega_0^2\mathbf{P} - V_0^2\nabla_r^2\mathbf{P} = 0. \tag{53.6}$$

In particular, it has solutions of plane-wave type which correspond to longitudinal optical phonons with a dispersion law given by eqn. (53.1). Summarizing eqns.

(53.1), (53.4), and (53.5), one obtains the total Hamiltonian of the system

$$\mathcal{H} = H_p + H_{\rm el} + H_{\rm int}, \tag{53.7}$$

from which one finds the appropriate equations of motion

$$\left[i\hbar\frac{\partial}{\partial t} + \frac{\hbar^2}{2m^*}\nabla_r^2 + e\varphi(\mathbf{r},t) \right]\psi(\mathbf{r},t) = 0, \tag{53.8}$$

$$\left[\frac{\partial^2}{\partial t^2} + \Omega_0^2 - V_0^2\nabla_r^2 \right]\varphi(r,t) = \frac{e\Omega_0^2}{\tilde{\epsilon}}\int |\psi(\mathbf{r}_1,t)|^2 \frac{d^3r_1}{|\mathbf{r}-\mathbf{r}_1|}. \tag{53.9}$$

Due to translational symmetry, we seek the solutions of this set of equations in the form of excitations moving with constant velocity v. Let the z-axis be directed towards the velocity v, and let the real functions $\varphi(\rho)$ and $G(\rho)$ be defined by the relations

$$\varphi(\mathbf{r},t) = \varphi(\rho), \quad \psi(\mathbf{r},t) = a^{-3/2}G(\rho)\exp[i(\mathbf{kr}-\omega t)] \tag{53.10}$$

where ρ is a dimensionless vector having the components

$$\xi = x/a, \quad \eta = y/a, \quad \xi = (z - z_0 - {\rm v}t)/a, \tag{53.11}$$

where a is the lattice constant,

$$k = m^*{\rm v}/\hbar$$

is the wavenumber, and

$$\hbar\omega = E_{\rm el} + E_{\rm int}$$

is the energy of the electron interacting with the polarization field.

The transformation (53.10) suggests that the envelope $G(\rho)$ of the electron wave function and the polarization field $\varphi(\rho)$ move with velocity v as a unique whole, that is possible iff the velocity v is less than the maximum allowed velocity V_g of electron within the conduction band. For this reason, we henceforth assume the condition

$${\rm v} < V_g \equiv \hbar/am^* \tag{53.12}$$

to be valid. The second necessary condition requires that the polaron velocity should be less than the limiting group velocity V_0 of optical phonons. So, as a unique whole, the polaron can be mobile if its velocity is less than both the velocities V_g and V_0. $V_0 < V_g$ is usually the case; hence, a polaron must be at rest in the absence of spatial dispersion ($V_0=0$). Any movement should be accompanied by the detachment of the polarization field from an electron due to a time retardation effect. This process requires the expense of energy.

Using eqn. (53.10), one transforms eqn. (53.9) into the following

$$\mathbf{R}\varphi(\rho) = -\frac{e}{a\tilde{\epsilon}}\int\frac{G^2(\rho_1)d^3\rho_1}{|\rho-\rho_1|}, \tag{53.13}$$

where the operator $\mathbf{R}$ is defined by the formula

$$\mathbf{R} = 1+\sigma^2\frac{\partial^2}{\partial\xi^2}-\sigma_0^2\nabla_0^2, \tag{53.14}$$

and

$$\sigma_0^2 = V_0^2/a^2\Omega_0^2,\ \ \sigma^2 = v^2/a^2\Omega_0^2. \tag{53.15}$$

Now we are only interested in the autolocalized state of the electron. In this case, the right-hand side of eqn. (53.13) decreases more rapidly than $|\rho|^{-1}$. For this reason, we seek those particular solutions of eqn. (53.13) which satisfy the boundary condition

$$\varphi(|\rho|\to\infty) = 0. \tag{53.16}$$

One can represent the solution of eqn. (53.13) through the Green's function,

$$\varphi(\vec{\rho}) = -\frac{e}{a\tilde{\epsilon}}\int\frac{W(\vec{\rho}-\vec{\rho}_1)G(\vec{\rho}_1)d^3\rho_1d^3\rho_2}{|\vec{\rho}_2-\vec{\rho}_1|}; \tag{53.17}$$

the latter being defined by the integral

$$W(\vec{\rho}) = (2\pi)^{-3}\int\frac{\exp(i\vec{q}\vec{\rho})d^3q}{1-\sigma^2q_z^2+\sigma_0\vec{q}^2}. \tag{53.18}$$

Its evaluation requires choosing the index of poles in accordance with the causality principle. If

$$0\leqslant s^2\equiv\sigma^2/\sigma_0^2\leqslant 1, \tag{53.19}$$

Green's function takes the form

$$W_1(\vec{\rho}) = \frac{\exp[-\sqrt{\zeta^2+(1-s^2)(\xi^2+\eta^2)}/\sigma_0\sqrt{1-s^2}]}{4\pi\sigma_0\sqrt{\zeta^2+(1-s^2)(\xi^2+\eta^2)}}. \tag{53.20}$$

In the opposite case, $s^2 > 1$, it becomes

$$W_2(\vec{\rho}) = \frac{\cos[\sqrt{\zeta^2-(s^2-1)(\xi^2+\eta^2)}\,/\,\sigma_0\sqrt{1-s^2}\,]}{2\pi\sigma_0\sqrt{\zeta^2-(s^2-1)(\xi^2+\eta^2)}} \tag{53.21}$$

if $\zeta < 0$, $\zeta^2 < (s^2-1)(\xi^2+\eta^2)$, and vanishes if these inequalities are not satisfied.

The function $\varphi(\vec{\rho})$ characterizes the polarization potential and can be derived by means of the Green's function (53.21). This oscillates with a constant amplitude as $\rho \to \infty$. Hence, it does not satisfy the boundary condition (53.16). Furthermore, the correct $\varphi(\rho)$ which obeys this boundary condition, can be expressed only in terms of the Green's function (53.20). In other words, spatially localized solutions exist if their velocities are less or equal to the maximum group velocity V_0 of the polarization waves. This excitation becomes immobile in the absence of the dispersion of optical phonons, i.e., $V_0 = 0$.

Substituting the formula (53.17) with $W(\vec{\rho}) = W_1(\vec{\rho})$ into eqn. (53.8), one obtains the integro-differential equation for the real envelope $G(\vec{\rho})$ of the function (53.10),

$$\left[\frac{\hbar^2}{2m^*a^2}\nabla_\rho^2 + \Lambda + \frac{e^2}{a\tilde{\epsilon}}\int\frac{W(\vec{\rho}-\vec{\rho}_2)G^2(\vec{\rho}_1)d^3\rho_1 d^3\rho_2}{|\vec{\rho}_1-\vec{\rho}_2|}\right]G(\vec{\rho}) = 0, \tag{53.22}$$

where $\nabla_\rho^2 = \partial^2/\partial\xi^2 + \partial^2/\partial\eta^2 + \partial^2/\partial\zeta^2$, and

$$\Lambda = \frac{\hbar^2}{2m^*a^2}\int[\nabla_\rho G(\vec{\rho})]^2 d^3\rho - \frac{e^2}{a\tilde{\epsilon}}\int\frac{W(\vec{\rho}-\vec{\rho}_2)G^2(\vec{\rho}_1)d^2\rho_1 d^3\rho_2}{|\vec{\rho}_1-\vec{\rho}_2|}$$

is the energy of the electron in the potential well $\varphi(\vec{\rho})$ moving with velocity V, referred to the bottom of the conduction band. In solving eqn. (53.22), one can obtain the unknown function via the variational principle, by minimizing the functional

$$J\{G\} = -\frac{\hbar^2}{2m^*a^2}\int d^3\rho\, G(\vec{\rho})\Big[\nabla_\sigma^2 G(\vec{\rho}) + \tag{53.23}$$

$$+\gamma G(\vec{\rho})\int\frac{W(\vec{\rho}-\vec{\rho}_1)G^2(\vec{\rho}_1)}{|\vec{\rho}_1-\vec{\rho}_2|}d^3\rho_1 d^3\rho_2\Big].$$

In eqn. (53.23),

$$\gamma = \frac{e^2}{a\tilde{\epsilon}}\frac{m^* a^2}{\hbar^2} \tag{53.24}$$

is the dimensionless parameter proportional to the ratio of the Coulomb energy to the doubled energy of the electron with momentum $\hbar/a$ in the conduction band. $\gamma = 7.2$ for typical ionic crystals, such as $NaCl$ and NaI, with the following parameters: $\Omega = 5.10^{13} Hz$, $a = 5\text{Å}$, and $m^* = 3m$.

Assuming the dispersion to be absent and the polaron to be at rest, so; that is $\sigma_0 = \sigma = 0$, the function (53.20) becomes a δ-function. Then the functional (53.23) transforms into that proposed by Pekar [98] having the form of the zeroth-order functional in the context of the Bogoliubov-Tyablikov theory [18,118-120].

Using the relation

$$\frac{\partial \vec{P}}{\partial t} = -\frac{v}{a}\frac{\partial \vec{P}}{\partial \zeta},$$

one can express the energy of polarization of the crystal in terms of the variable $\vec{\rho}$ as follows

$$E_p(v) = \frac{2\pi\tilde{\epsilon}a^3}{\Omega_0^2}\int\left[v^2\left(\frac{\partial \vec{P}}{\partial \zeta}\right)^2 + \Omega_0^2\vec{P}^2 + \frac{V_0}{a^3}(\nabla_0\vec{P})^2\right]a^3\rho. \tag{53.25}$$

If the function $G_0(\vec{\rho})$ minimizes the functional (53.23) under the auxiliary condition

$$\int G_0^2(\vec{\rho})d^3\rho = 1,$$

then, using eqns. (53.4) and (53.17), one obtains the polarization vector field

$$\vec{P}(\vec{\rho}) = \frac{e}{4\pi a^2\tilde{\epsilon}}\nabla_\rho\int\frac{W(\vec{\rho}-\vec{\rho}_2)G_0^2(\vec{\rho}_1)}{|\vec{\rho}_1-\vec{\rho}_2|}d^3\rho_1 d^3\rho_2. \tag{53.26}$$

Using the identity

$$\int W(\vec{\rho}-\vec{\rho}_1)W(\vec{\rho}-\vec{\rho}_2)d^3\rho = \mathfrak{L}W(\vec{\rho}_1-\vec{\rho}_2),$$

where

$$\mathfrak{L} \equiv 1+\sigma^2\frac{\partial^2}{\partial\sigma^2}+\sigma_0^2\frac{\partial^2}{\partial\sigma_0^2},$$

one obtains the total energy, $E_{el}+E_p+E_{int}$, of the polaron moving with a constant velocity:

$$E(\mathrm{v}) = \frac{1}{2}m^*\mathrm{v}^2 + \frac{\hbar^2}{2m^*a}\int d^3\rho[\nabla_\rho G(\vec{\rho})]^2 + \tag{53.27}$$

$$+\frac{e^2}{8\pi\tilde{\epsilon}}\int\left\{\vec{D}(\vec{\rho}_1)\vec{D}(\vec{\rho}_2)[\mathfrak{L}-2]+\sigma^2\frac{\partial\vec{D}(\rho_1)}{\partial\zeta_1}\frac{\partial\vec{D}(\vec{\rho}_2)}{\partial\zeta_2}\mathfrak{L}+\right.$$

$$\left.+\sigma_0^2\nabla_{\rho 1}\vec{D}(\vec{\rho}_1)\nabla_{\rho 2}\vec{D}(\vec{\rho}_2)\mathfrak{L}\right\}W(\vec{\rho}_1-\vec{\rho}_2)d^3\rho_1 d^3\rho_2.$$

The formulas obtained are valid for all velocities of the polaron for which the application of the continuum approach is still justified and the inequalities (53.12) and $\mathrm{v} \leqslant V_0$ are satisfied too, the latter being provided by the spatial dispersion of optical phonons.

53.1. The Direct Variational Principle

The direct variational principle can be applied to minimize the functional (53.22). Using the fact that when an electron moves, the polarization accompanying it has cylindrical symmetry, one can choose the following trial normalized function

$$G(\vec{\rho}) = \left[\frac{2}{\pi}\right]^{3/4}\alpha^{1/2}\beta\exp\{-\alpha^2\zeta^2-\beta^2R^2\},\ \ R^2 \equiv \xi^2+\eta^2, \tag{53.28}$$

which depends on two variational parameters α and β. Substituting this into the functional (53.23), one obtains the function

$$J(\alpha,\beta) = \frac{\hbar^2}{2m^*a^2}[\alpha^2+2\beta^2-\gamma\alpha^2\beta^4 j(\alpha,\beta)\pi^3], \tag{53.29}$$

where

$$j(\alpha,\beta) = \int\frac{W_1(\vec{\rho}_3)}{|\vec{\rho}_1|}\exp\left\{-\frac{\alpha^2}{2}[(\zeta_1+\zeta_2+\zeta_3)^2+(\zeta_2-\zeta_1-\zeta_3)]^2]-\right. \tag{53.30}$$

$$\left.-\frac{1}{2}\beta[(R_1+R_2+R_3)^2+(R_2-R_1-R_3)^2]\right\}\times d^3\rho_1 d^3\rho_2 d^3\rho_3.$$

If $0 \leqslant s^2 \leqslant 1$, the function $W_1(\vec{\rho})$ entering into eqn. (53.30), is given by the formula (53.20). If $s^2 > 1$, Green's function has the form (53.21). Hence, as mentioned above, the function $\varphi(\vec{\rho})$, obtained with the help of the integral (53.17), does not describe the spatially localized solutions. In this case, applying the variational principle, Enol'skii and the author [54] obtained solutions which oscillate at infinity. Such solutions can describe the nonstationary ones which are localized during the rather small interval for which the polarization field cannot detach a mobile electron. The exact expressions (53.29) and (53.30), describing localized states, are too complicated for analytical calculations. This is the reason for using approximate formulas which are valid with some values of the parameters of the system under study.

In view of the continuum approach applied, let us consider only the excitations whose localization region exceeds the lattice constant. The dispersion of optical phonons is hence small, so the inequality

$$\sigma_0^2 \ll 1 \tag{53.31}$$

is valid. If the inequality

$$\sigma^2 \leqslant \sigma_0^2 \tag{53.32}$$

is fulfilled simultaneously, one can use a simpler expression for the Green's function $W_1(\vec{\rho})$

$$W_1(\vec{\sigma}) = \left[1-\sigma^2\frac{\partial^2}{\partial\zeta^2}+\sigma_0^2\nabla_\rho^2+\ldots\right]\delta(\xi)\delta(\eta)\delta(\zeta). \tag{53.33}$$

In the following Subsection, we obtain the function $G(\vec{\rho})$, the energy $E(\mathrm{v})$, and the other quantities characterizing the mobile polaron in the case when eqn. (53.33) is justified.

53.2. The Energy and Wave Function of a Slow Polaron

Let us suggest that the polaron motion is slow if the inequality (53.32) takes place. For typical ionic crystals $a\Omega_0 = 2.6\times 10^6 cm.S^{-1}$. Hence, the inequality (53.32) is satisfied even for electron velocities which are not so small.

Using the approximate formula (53.33) in eqn. (53.30), one obtains that the wave function (53.28) minimizes (53.29) if

$$\beta = \alpha_0(1-6\alpha_0^2\sigma_0^2)+\frac{12}{7}\alpha_0^3\sigma^2, \tag{53.34}$$

$$\alpha = \beta\left[1+\frac{12}{7}\alpha_0\sigma^2\right]^{1/2}, \quad \alpha_0 \equiv \gamma/\sqrt{\pi}. \tag{53.35}$$

Therefore, the square of the wave function is

$$G_0^2(\vec{\rho}) = \left[\frac{2}{\pi}\right]^{3/2} \alpha\beta^2 \exp\left[-2\beta^2\left[\vec{\rho}^2 + \frac{12}{7}\alpha_0^2\sigma^2\zeta^2\right]\right]. \tag{53.36}$$

So, in the case of a immobile polaron the probability of the spatial distribution of the polaron is spherically symmetric. A positive dispersion decreases the electron localization. For a mobile polaron, constant values of $G_0(\vec{\rho})$ are on a surfaces having the form of an oblate ellipsoid of revolution with a symmetry axis directed towards the polaron velocity. An increase of electron velocity is accompanied by a strengthening of its localization.

The total polaron energy involving the polarization energy can be represented as follows

$$E(\mathrm{v}) = E(0) + \frac{1}{2}M_{\mathrm{eff}}\mathrm{v}^2, \tag{53.37}$$

where the energy $E(0)$ of the polaron at rest is

$$E(0) = \mathscr{E}_0 - \frac{a^2}{8\pi\tilde{\epsilon}}\int |\vec{D}^2 + \sigma_0^2(\nabla_\rho\vec{D})]^2 d^3\rho \approx \tag{53.38}$$

$$\approx \mathscr{E}_0 - 0.053E_a\frac{m^*}{m\tilde{\epsilon}}(1 - 0,142\gamma^2\sigma_0^2),$$

where m is the mass of a free electron, $E_a = ma^4/\hbar^2$ is the atomic unit of energy. In eqn. (53.37), the effective polaron mass M_{eff} is

$$M_{\mathrm{eff}} = m^* + \frac{a}{4\pi\tilde{\epsilon}\Omega_0^2}\int\left[\left[\frac{\partial\vec{D}^2}{\partial\zeta}\right] - 2\sigma_0^2\left[\frac{\partial}{\partial\zeta}\nabla_\rho\vec{D}\right]^2 + 3\sigma^2\left[\frac{\partial^2\vec{D}}{\partial\zeta^2}\right]^2\right]d^3\rho = \tag{53.39}$$

$$= m^* + 4e^2\gamma^2(81\pi^2a^3\Omega_0^2)^{-1}\left[1 + \frac{\gamma^2}{5\pi}\left[\frac{187}{21}\sigma^2 - 18\sigma_0^2\right]\right].$$

It follows from eqn. (53.39) that as the polaron velocity increases, its effective mass does too, in accordance with the strengthening of its localization. One should notice that in the absence of dispersion, $\sigma_0 = 0$, and ultimately a small electron velocity, $\mathrm{v} \to 0$, the formula (53.39) is just that obtained via a rather heuristic way by Landau and Pekar [88].

Let us recall that a spatial electron localization takes place if the velocity does not exceed the limiting group velocity V_0 of optical phonons. At larger velocities only nonlocalized stationary states described by plane waves are possible. In these cases the amplitude functions $G(\vec{\rho})$ and the polarization field are independent on the spatial variable $\vec{\rho}$. Such excitations are uniformly distributed over the whole crystal with negligible density. In these states, the electron motion is described by wave packets. These are accompanied by a small 'coat' of virtual phonons replacing each other continuously.

Unlike the well-known approaches, the method of solving the problem of the mobile polaron (a three-dimensional soliton) proposed in this Chapter is not based on perturbation theory with respect to the kinetic energy of ions. Taking into account the dispersion of optical phonons enables the investigation of the nonlocality of the interaction conditioned by that dispersion. This method can be also applied to study those systems having a strong Coulomb interaction consisting of vector and scalar inertial fields with dispersion. We also demonstrate that such two-component autolocalized excitations (polarons) in ionic crystals can be mobile if their velocities are less then or equal to the limiting group velocity of optical phonons provided by their spatial positive dispersion. In order to evaluate in a consistent way the effective polaron mass, even at infinitely small velocities, one should take the spatial dispersion into account. Otherwise, the polarons are immobile.

Chapter 14

THE GENERATION OF SOLITONS AND BISOLITONS

54. The Generation of a Soliton at the Boundary of a Soft Molecular Chain

The explanation of the mechanism of the generation of solitons is one of the central problems of the theory of solitons in molecular systems. Due to the topological stability considered in Section 5, solitons can be generated only at boundaries of a molecular chain. In this Section, we study the molecular mechanism of the formation of solitons and bisolitons within the simple model proposed by the author and Ermakov [290].

Let us consider an infinite quasi-one-dimensional molecular chain consisting of two parts: 'soft' and 'rigid' subchains joined at $x = 0$. We suggest that quasiparticles (excitons or electrons) with an effective mass m^* propagate over a rigid subchain with its local deformation, this subchain being in the negative region of the x-axis. The motion of quasiparticles are described by the following Schrödinger equation,

$$\left[i\hbar\frac{\partial}{\partial t}+\frac{\hbar^2}{2m^*}\frac{\partial^2}{\partial x^2}\right]\psi_1(x,t) = 0, \quad x \leqslant 0. \tag{54.1}$$

We also suggest that free excess particles which, travelling along the rigid subchain, encounter the soft subchain, generate soliton states, i.e., collective two-component excitations. It is assumed that these particles have an effective mass m and interact with the local deformation of intermolecular separations a of the molecules with mass M. The motion of such excess particles, interacting with the field $\rho(x)$ of the local variations of intermolecular separations, is described by Schrödinger equation

$$\left[i\hbar\frac{\partial}{\partial t}-H\right]\psi_2(x,t) = 0, \quad x > 0. \tag{54.2}$$

Here

$$H = \mathscr{E}_0 + U(x) - \frac{\hbar^2}{2m}\frac{\partial^2}{\partial x^2} \tag{54.3}$$

is the appropriate Hamiltonian, and $\mathscr{E}_0$ determines the energy of the bottom of the energy band of the free quasiparticles. In eqn. (54.3), the term

$$U(x) = -\sigma\rho(x)/2a \tag{54.3a}$$

plays the role of the interaction energy of a free quasiparticle with a local deformation of the equilibrium positions of molecules in a chain caused by this quasiparticle. The elasticity of the chain is characterized by the longitudinal sound velocity, $V_0 = a(\kappa/M)^{1/2}$, depending on the longitudinal elasticity coefficient κ.

We begin the study of stationary states by assuming that the flow of quasiparticles with the energy $E = \hbar^2k^2/2m$ is generated at $x = -\infty$ (the rigid subchain). This flow is partially reflected at the boundary $x = 0$. The stationary motion of a quasiparticle is described by the following wave function

$$\psi_1(x,t) = a^{-1/2}\Phi_1(x)\exp(-iEt/\hbar) \tag{54.4}$$

in the region of negative x,

$$\Phi_1(x) = De^{ikx} + \mathrm{R}e^{-ikx}. \tag{54.5}$$

We further suggest that the density of flow of incident particles obeys the inequality $D^2 \leqslant 1$. The subflow of particles which passes into the soft chain is described in its stationary state by the function

$$\psi_2(x,t) = a^{-1/2}\Phi_2(x)\exp(-iEt/\hbar), \tag{54.6}$$

which satisfies eqn. (54.2). The relation among the functions (54.4) and (54.6) is determined by the conditions of their matching at the boundary $x = 0$.

The probability of finding a quasiparticle in the soft subchain which depends on the density and energy of the incident flow of quasiparticles is given by the integral

$$I = \int_0^\infty |\psi_2(x)|^2 dx = \frac{1}{a}\int_0^\infty |\Phi_2(x)|^2 dx. \tag{54.7}$$

If the velocity of a mobile soliton is considerably less than that of the longitudinal sound velocity, one can represent the elastic deformation of the soft chain as follows

$$W = \frac{\kappa}{2a}\int_0^\infty \rho^2(x)dx. \tag{54.8}$$

Then the total energy of the local deformation and the quasiparticles interacting with it becomes

$$E = W + \int_0^\infty \psi_2^*(x,t)H\psi_2(x,t)dx. \tag{54.9}$$

Requiring that the total energy (54.9) does not involve linear terms relative to $\rho(x)$, one obtains the consistent deformation field,

$$\rho(x) = \sigma|\Phi|(x)|^2/2\kappa a. \tag{54.10}$$

Hence, the deformation energy (54.8) takes the form

$$W = \frac{\sigma^2}{8\kappa a^3}\int_0^\infty |\Phi(x)|^4 dx. \tag{54.11}$$

Using formula (54.10), one transforms eqn. (54.2), describing the quasiparticle interacting with the deformation field, into the nonlinear Schrödinger equation for the envelope of the wave function (54.6)

$$\left[-\frac{\hbar^2}{2m}\frac{\partial^2}{\partial x^2}+(\mathcal{E}_0-E)-G|\Phi(x)|^2\right]\Phi(x) = 0, \tag{54.12}$$

where

$$\tilde{\epsilon} = \epsilon_0 + W, \text{ and } G = G^2/4MV_0^2 \tag{54.13}$$

is the nonlinear parameter (at small velocities).

Equation (54.12) is just that for the envelope of the wave function of slow solitons examined in Section 5 where the coupling parameter $\chi = \sigma/2a$ is used. Representing the effective quasi-particle mass through the energy of the resonance, or the exchange interaction of neighboring molecules, $m = \hbar^2/2a^2J$, and introducing the following dimensionless quantities,

$$\gamma = [(\mathcal{E}_0-E)/J]^{1/2}, \; g = \frac{G}{2J} = \frac{ma^2G}{\hbar^2}, \; \xi = \frac{x}{a}, \tag{54.14}$$

one obtains the solutions of eqn. (54.12) localized in the chain,

$$\Phi_2(x) = \gamma \exp(i\varphi)/\sqrt{g}\,\mathrm{ch}[\gamma(\xi-\xi_0)]. \tag{54.15}$$

Substituting this function into eqn. (54.7), one can express ξ_0, characterizing the position of the maximum of the function $\Phi_2(x)$ in terms of the energy of incident quasiparticles, the dimensionless coupling parameter and probability I, as

$$\xi_0 = (2\gamma)^{-1}\ln[gI/(2\gamma-gI)]. \tag{54.16}$$

The probability I and the phase φ are determined by matching the functions ψ_1 and ψ_2, and their derivatives at $x = 0$,

$$\varphi = \mathrm{arctg}\left\{\frac{\gamma-gI}{ka}\right\}, \quad k = \hbar/\sqrt{2m^* E} \tag{54.17}$$

and

$$I = 4D^2/(2\gamma-gI)\left[1+\left\{\frac{gI-\gamma}{ka}\right\}^2\right], \quad 2\gamma > gI, \tag{54.18}$$

$$W = JI\gamma g(1-gI/3\gamma). \tag{54.18a}$$

If the inequality

$$\gamma \ll ka \tag{54.19}$$

holds, eqn. (54.18), which gives the probability I through the density of flow of incident particles, has the form

$$(2\gamma-gI)I = 4D^2. \tag{54.20}$$

We introduce a restriction by only considering the positive solutions of eqn. (54.2),

$$I_{1,2} = \gamma g^{-1}[1\mp(1-4gD^2/\gamma^2)^{1/2}]. \tag{54.21}$$

Assuming the relation

$$4gD^2 = \gamma^2 \tag{54.22}$$

to be valid, both solutions coincide, and

$$I_1 = I_2 = \gamma/2g. \tag{54.23}$$

Let us designate the appropriate value of D^2 by the index m, that is

$$D_m^2 \equiv \gamma^2/4g = (\epsilon_0 - E_{cr})/4gJ. \tag{54.24}$$

Equation (54.24) gives the relation between each energy E_{cr} and the critical density D_m^2 of the flow of incident particles. Assuming that this is this case, one demonstrates that the maximum of the function (54.15) is at the boundary of the soft subchain if $\xi = 0$. At small flow densities of particles, possessing the energy E_{cr} corresponding to the critical value D_m, which satisfy the inequality

$$D^2 < D_m^2 = (\epsilon_0 - E_{cr})/4gJ, \tag{54.25}$$

the positive solutions (54.21) take the form

$$I_1 = 2D^2/\gamma, \quad I_2 = \frac{2\gamma}{g} - 2D^2/\gamma. \tag{54.26}$$

One should notice that the function (54.15) with maximum located in the region of negative ξ near the boundary of the chain correspond to the unique solution I_1. In this case, the soft subchain is only covered by the exponentially decaying part of the curve $|\Phi_2(x)|$. As the density of the particle flow increases to a critical value D_m^2 under the fixed energy, the maximum of the function (54.15) is displaced towards the boundary $\xi = 0$. If $D^2 \leqslant D_m^2$, there exist in the system stationary states in which a complete reflection of the incident flow from the boundary $\xi = 0$ occurs.

Each critical value D_m^2, for which the stationary states still exist, corresponds to the critical energy of the flow of incident particles,

$$E_{1,cr} = \tilde{\epsilon}_0 \ - 4gJD_m^2, \tag{54.27}$$

in accordance with eqn. (54.24). Hence, a lower critical energy corresponds to a larger density of flow of incident particles. The probability $I = 1/2$ corresponds to the critical energy E_{cr} of the start of formation of a single soliton in the soft chain. From eqn. (54.23) it follows that

$$E_{cr} = \tilde{\epsilon}_0 - g^2 J/4.$$

Bearing in mind eqn. (54.27), one concludes that this critical energy corresponds to the density $D_m^2 = g/16$ of the flow of incident particles. Using eqn. (54.18a), one obtains $W = g^2 J/16$, and therefore,

$$E_{1,\mathrm{cr}} = \epsilon_0 - g^2 J / 12. \tag{54.27a}$$

If the density of flow of incident particles is equal to D_m^2, and the energy exceeds the critical energy, $E_{1,\mathrm{cr}}$, there arise nonstationary states in which the solitons are generated inside the soft subchain. These solitons break off from its boundary and pass to the region of positive ξ. Such states are studied in the next Subsection.

54.1. The Dependence of Wave Functions. Soliton Generation

In order to investigate the conditions under which solitons are generated in the soft chain, one needs to consider the time dependence of the appropriate wave functions. For this purpose, let us suggest that the function of the flow of quasi-particles in the rigid chain has the form

$$\Psi_1(x,t) = a^{-1/2}[D(t)e^{ikx} + R(t)e^{-ikx}]\exp\left\{-\frac{iEt}{\hbar}\right\}, \tag{54.28}$$

where the amplitude $D(t)$ of the incident wave varies slowly in the course of time in the neighborhood of D_m,

$$\frac{\partial D}{\partial t},\ \frac{\partial R}{\partial t} \ll \frac{E}{\hbar}. \tag{54.29}$$

For the soft chain, eqn. (54.12) transforms as follows

$$\left[i\hbar\frac{\partial}{\partial t} + \frac{\hbar^2}{2m}\frac{\partial^2}{\partial x^2} - \mathcal{E}_0 + G|\Psi_2(x,t)|^2\right]\Psi_2(x,t) = 0. \tag{54.30}$$

One can look for its solution in the following form

$$\Psi_2(x,t) = a^{-1/2}\Phi_2(x,t)\exp\{i[\varphi(x,t) - Et/\hbar]\}, \tag{54.31}$$

where the envelope $\Phi_2(x,t)$ depends weakly on time. Substituting this into eqn. (54.31), one obtains the set of equations

$$a^2\frac{\partial^2\Phi_2}{\partial x^2} + 2g\Phi_2^3 - \Gamma^2(t)\Phi_2 = 0, \tag{54.32}$$

$$2\frac{\partial\varphi}{\partial x}\frac{\partial\Phi^2}{\partial x} + \frac{\partial^2\varphi}{\partial x^2}\Phi_2 + \frac{2m}{\hbar}\frac{\partial\Phi}{\partial t} = 0. \tag{54.33}$$

The function

$$\Gamma^2(t) = \left[\mathcal{E}_0 - E + \hbar\frac{\partial\varphi}{\partial t}\right]/J \tag{54.34}$$

appearing into eqn. (54.32) depends weakly on the coordinates and time.

Equation (54.32) possesses the solution

$$\Phi_2(x,t) = \Gamma(t)/\sqrt{g}\,\mathrm{ch}\{\Gamma(t)[\xi-\xi_0(t)]\}, \tag{54.35}$$

which decreases at infinity. In eqn. (54.35) the time-dependent quantity $\xi_0(t)$ determines the maximum of the function. Substituting the function (54.35) into the integral

$$I(t) = \frac{1}{a}\int_0^\infty |\Psi(x,t)|^2 dx, \tag{54.36}$$

which give the probability of finding a soliton in the soft chain, one obtains the explicit form of the function

$$\xi_0(t) = [2\Gamma(t)]^{-1}\ln\left[\frac{gI(t)}{2\Gamma(t)-gI(t)}\right]. \tag{54.37}$$

Using eqn. (54.10) and the weak dependence of the phase $\varphi(x,t)$ on time, from eqn. (54.33) one has

$$\frac{\partial\varphi}{\partial\xi} = m\mathrm{v}/\hbar, \quad \mathrm{v} \equiv d\xi_0/dt,$$

where $V = a\mathrm{v}$ is the velocity of motion of the maximum of the function (54.35). Differentiating eqn. (54.37) on time and taking the weak time dependence of $I(t)$ into account, one finds the following equality

$$V = a\frac{d\xi_0}{dt} = a[2\Gamma(t)-gI(t)]^{-1}I^{-1}(t)\frac{d}{dt}I(t), \quad 2\Gamma > gI. \tag{54.38}$$

The explicit form of the equation for the phase $\varphi(x,t)$ at $x = 0$ and the function $I(t)$ is determined by matching the wave functions Ψ_1 and Ψ_2 and their derivatives at the point $x = 0$. Introducing the new dimensionless function $z(t)$ via the formula

$$z^2(t) = -1+2\Gamma(t)/gI(t) \tag{54.39}$$

and using eqns. (54.31), (54.36), and (54.38), one obtains the set of equations

$$z^2-z-4kaD\sin\varphi/I^2g^{3/2} = 0, \tag{54.40}$$

$$\frac{\partial\varphi}{\partial t} = \hbar^{-1}J\left[\frac{1}{4}g^2I^2(1+z^2)-\gamma^2\right], \tag{54.40a}$$

$$\frac{dI}{dt} = \hbar k g I^2 z^2 (am)^{-1} \left\{ \frac{2D\,|\cos\varphi|}{|z|\,/\sqrt{g}} - 1 \right\}. \tag{54.40b}$$

Analyzing these equations in the case of the constant value

$$D < D_m \equiv \gamma / 2g$$

one concludes that they transform into the aforementioned eqns. (54.17) and (54.18). If

$$D > D_m$$

the solution of eqn. (54.30) has the form (54.31) for which the phase φ and position ξ_0 depend on time.

The function $z(t)$ varies in time periodically. At $t = t_0$, where the phase in eqn. (54.40a) vanishes, and the function $z^2(t_0)$ is equal to unity, the maximum of the function (54.35) corresponds to the point $\xi = \xi_0 = 0$. As time further increases, the function $z^2(t)$ decreases, and the maximum position of the function (54.35) is displaced into the interior of the soft chain, e.g. when the function $z^2(t)$ attains the value $1/3$, the maximum position of the function (54.35) is displaced to the point $\xi_0 = 3\ln(3/4gI)$. At this point in time, the soliton is completely formed in the interior of the soft chain. If further moves with velocity

$$V = \frac{\hbar k}{m} \left(\frac{E - E_{1,\mathrm{cr}}}{D_m} \right) \tag{54.41}$$

and takes off from the boundary of the soft chain. Simultaneously, the function $z^2(t)$ makes a jump, increasing to a value greater than unity, and then there begins further the process of forming a new soliton, which provides a decreasing z^2 up to unity, and so on.

Given the density of the flow of electron in the rigid chain and energy which determines the quantity D_m, the solitons are generated through a time interval

$$T \approx \frac{3am \ln 3 \sqrt{J(\epsilon_0 - E_{1,\mathrm{cr}})}}{4\hbar k (E - E_{1,\mathrm{cr}})}. \tag{54.42}$$

According to eqn. (54.24), a decrease of energy of the incident electrons results in an increase in the critical value of D_m. Solitons are generated only at those densities of flow of the incident particles which exceed $D_m^2 (D^2 > D_m^2)$. From eqn. (54.27) it follows that this inequality is equivalent to the energy inequality, $E > E_{1,m}$. For densities less than D^2_m, the soliton is elastically reflected from the soft chain, and soliton generation drops.

54.2. Bisolitonic Generation

We have studied before the generation of isolated solitons corresponding to vibrational and chargeless excitations, or electrosolitons with a unit electric charge. However, Brizhik and the present author [21,22,287] have demonstrated that, in molecular chains which can attach excess electrons, the electron pairs are energetically more profitable. They are in a singlet spin state and coupled with the total chain deformation. Such collective formations with a zeroth spin are called bisolitons. These carry a double electric charge and compare with Bose particles. The binding energy released under bisolitonic formation is four times greater than the energy of generation of two separated solitons. For this reason, slow bisolitons are much too stable compared with isolated ones.

It has been shown in various aforementioned papers that the Schrödinger equation, describing the motion of slow bisolitons, has the form of eqn. (54.12) in which the dimensionless nonlinearity parameter is two times greater than the value of g given by the formula (54.14). For this reason, when a flow of electron pairs of density $2D^2$ arises at the boundary of the soft chain, bisolitons are generated starting with the critical energy

$$E_{2,\mathrm{cr}} = \epsilon_0 - \frac{2}{3} g^2 J. \qquad (54.43)$$

$E_{2,\mathrm{cr}}$ is considerably lower than the similar value given by eqn. (54.27), for the generation of isolated electrosolitons.

For a fixed density of incident electrons having the energy in the interval $E_{1,\mathrm{cr}}$, $E_{2,\mathrm{cr}}$, bisoliton formation appears to be a two-stage process. At the very start, a single electron enters the soft chain. Its state is described by the tail of the wave function (54.15) whose maximum decreases in the region of negative ξ. This electron remains to be attached to the boundary of the soft chain. If a second electron with the same energy and opposite spin enters the soft chain, there occurs a reconstruction of the local chain deformation. Due to the nonlinearity of the interaction of electrons with the deformation, the nonlinearity parameter, incorporated into the Schrödinger equation for paired electrons, is doubled. The bisoliton, generated under the energy of the flow of electrons exceeding $E_{2,m}$, detaches from the boundary and moves along the soft chain. One can say figuratively that entering the soft chain, the first electron expects the second one having an opposite spin to disturb locally the chain and thus form a bisoliton. The latter further takes off from the boundary of the soft chain and moves along it with a double electric charge and without friction.

Summarizing, one concludes that the energy conditions for the generation of bisolitons at the boundary of the soft chain are considerably profitable compared with those of the formation of isolated electrosolitons. One seems that this fact is of great importance for elucidating the molecular mechanism of high-T_c superconductivity which has recently been revealed in compounds having a small number of charge carriers (ceramic materials).

Chapter 15

Nonlinear Bisoliton Mechanism of High-T_c Superconductivity

55. Discovery of High-T_c Superconductors and their Basic Properties

In 1986, Bednortz and Müller in Switzerland [315], and, later, other researchers produced ceramic oxide materials containing copper, oxygen, lanthanum, barium, or strontium ($La_{2-\delta}BA_{\delta}CuO_4$) which pass into a superconducting state and reveal the Meissner effect under cooling below $40\,K$. In February 1987, Chu, Coen, Wu, Asburn and their collaborators from Houston University [318,324] showed that their critical temperature could be raised to $93\,K$ by substituting lanthanum by yttrium ($YBa_2Cu_3O_{7-\delta}$). Subsequent studies of these and similar materials in many laboratories confirmed these results.

The basic properties of novel high-T_c compounds have not been explained by the generally accepted theory of the superconductivity of metals proposed by Bardeen, Cooper, and Schrieffer [313] (BCS theory) thirty years ago and further developed mathematically by Bogoliubov [306]. This theory was developed within the approach of a weak electron-phonon coupling characterized by a dimensionless parameter $\lambda \ll 1$. Such a coupling results in a pairing of electrons which prevents them being scattered by acoustic phonons. The electron attraction is conditioned by their interaction with the displacements of the equilibrium positions of surrounding ions caused by electrons and leading to crystal deformation. In terms of quantum theory, these displacements are characterized by virtual phonons. This is the reason for describing this electron-deformation interaction as an electron-phonon interaction. The role of this interaction in the phenomenon of superconductivity was firstly emphasized by Fröhlich in 1950 [323].

In the BCS theory, the critical temperature T_c of the transition into the super conducting state is determined as follows:

$$T_c \sim \Theta \exp(-1/\lambda),$$

where λ is a small dimensionless parameter characterizing the intensity of the electron-phonon interaction; and Θ is the temperature ($\sim 100-500K$) corresponding to the energy of phonon vibrations actively participating in the interaction. The energy of these vibrations is inversely proportional to the square root of the mass M of the ions involved. Hence, $T_c \simeq M^{-1/2}$, which emphasizes the so-called *isotopic effect* as the most important consequence of the BCS theory. The condition that λ is small restricts the range of validity of the BCS theory to the values $\lambda < 1/3$. It was believed that such a restriction is due to the electron-phonon mechanism of superconductivity, which cannot explain the origin of critical temperatures greater than $30\,K$ even if $\Theta \sim 500\,K$. The condition that λ is small ($<1/3$) is well satisfied for many simple superconducting metals for which the temperature of superconductivity transition is less $10\,K$. All these metals are characterized by a strongly pronounced isotopic effect which confirms the BCS theory based on a weak electron-phonon interaction.

It is astonishing that novel high-T_c superconductors do not reveal an isotopic effect [314,317]. This raises some doubt as to whether or not the dominating role of the electron-phonon interaction appears in novel superconductors. For instance, Schrieffer remarked with that the absence of an isotopic effect experimentally revealed by Batlogg et al. [314] requires the search for a novel mechanism for elucidating high-T_c superconductivity. Many theoreticians have attempted to replace the electron-phonon interaction by other ones which could lead to pairing. For example, Allender, Brey, and Bardeen developed the exciton (bound state of an electron with a positively charged hole) mechanism of electron pairing. In its simplest form, the exciton mechanism of superconductivity results in replacing phonons by excitons. Anderson [311] proposed a model of magnetic interaction by using the concept of 'resonance valence bonds'. According to this, there occurs a fluctuating reverse of the direction of spins in the singlet spin state of a pair. Many other mechanisms of pairing have been suggested, too.

We suggest that all these and other similar attempts to explain high-T_c superconductivity by replacing electron-phonon interaction by other types of interaction lead to difficulties. The electron-phonon interaction is present in all cases, and it cannot be neglected. If a pairing is provided by another mechanism, the existence of a rather weak electron-phonon interaction should lead to an instability of their Bose condensate and therefore damp electron movement. For example, in good conductors (silver, copper, etc.), a rather weak electron-phonon interaction does not lead to pairing, however it is completely responsible for their nonvanishing resistance.

Some authors [312,319] have investigated the bipolaron states of a small radius localized at one or two lattice sites (electron pairs). The localization was caused by an interaction with optical phonons. The Coulomb repulsion of electrons prevents this strong localization, but this can be realized only if the interaction of electrons with optical phonons is rather strong. In this case the effective bipolaron mass should be rather large, too. Such bipolarons become immobile if the dispersion of optical phonons [320] is neglected. For example, the latter is taken into account within the second-order perturbation treatment in [305] on the basis of a jump mechanism. However, it is difficult to understand in what way these really immobile bipolarons result in superconductivity. It is worth noting that, in this reason, too, together with the interaction with optical phonons, the electrons should interact with acoustic phonons which lead to resistance. One should recall here that any type of electron pairing is far from that which leads to superconductivity. For instance, the Peierls electron pairing leads to the transition of a metal into a dielectric rather than a superconductor.

Novel high-T_c superconducting materials (ceramic oxides) possess a rather large anisotropy and a high temperature of transition into the superconducting state (above $30\,K$), unlike ordinary superconducting metals. All contain copper ions. The highest superconductivity transition temperatures T_c in ceramic superconductors are observed at comparatively small concentrations of charge carriers ($\sim 0.15\%$). Because of this property, novel superconductors differ from ordinary metallic ones whose temperature T_c increases with electron density. In his review [326], Khurana notes that no researchers have observed superconductivity of ceramic oxides with $\delta < 0.05$ and $\delta > 0.4$. The radius of correlation of paired particles appeared to be small ($\sim 30\,A$) compared with that of paired electrons ($\sim 10^{-4} cm$) in superconductors of the first kind, i.e. in metals such as mercury, lead, etc. The sharp decrease of the radius of correlation in ceramic oxides in comparison with metallic superconductors indicates a rather large interaction of quasiparticles (holes and electrons) with the acoustic branch of the lattice vibrations inherent in ceramic oxides.

In the present Chapter, we develop a new theory of high-T_c superconductivity of nonmetallic oxide ceramics, based on the concept of a moderately strong electron-phonon coupling which results in perturbation theory being invalid. This theory, developed by the present author and his co-workers, utilizes the concept of *bisolitons*, or electron (or hole) pairs coupled in a singlet spin state due to the local deformation of the chain. Spinless bisolitons, carrying a double electron charge, are Bose particles. One can demonstrate that for concentrations below and above a critical concentration, they form a Bose condensate, i.e. a collective state with a periodic arrangement of bisolitons migrating as a whole with a unique velocity and without resistance. The critical temperature of formation of bisolitons and the condensate is independent of the mass of ions providing the pairing (the isotopic effect vanishes). In Subsection 56.3, we discuss the conditions under which bisolitons are generated in a quasi-one-dimensional chain by a flow of electrons coming to it from a metal. We also show that if the energy of electrons in a flow exceeds a somewhat critical value, bisolitons are generated in the chain over equal time intervals and migrate with the same velocity, forming therefore the bisoliton condensate.

56. The Basis of the Bisoliton Model of High-T_c Superconductivity

Together with the ceramic oxides, La_2CuO_4 and $YBa_2Cu_3O_7$, there have been discovered over ten similar compounds which go into a superconducting state following cooling and the addition of admixtures such as bivalent ions of barium, strontium, etc., or under the existence of oxygen ion vacancies in a lattice. All these compounds contain copper ions which are in a bivalent state, Cu^{2+}, in the absence of alloying admixtures. Under the addition of an amount ($\delta \sim 15\%$) of alloying bivalent ions, which substitute the trivalent ions La^{3+}, to the La-type compounds, $La_{2-\delta}Ba_\delta CuO_4$, there occurs an electron transfer, e.g. $La^{3+}Cu^{2+} \rightarrow Ba^{2+}Cu^{3+}$. A small number of ions of trivalent copper having an electronic configuration $3d^8$ and integer spin then arises in the crystal. Deficient electrons (positive holes) can migrate over the Cu^{2+} ions resulting in crystal conductivity.

A small number δ of oxygen vacancies is formed in the lattice of Y-type crystals $YBa_2CuO_{7-\delta}$ after special processing. A change in the oxygen ion content leads to that of the valence states of some copper ions. In these cases, there also arise in the crystals 'excess particles' (electrons or holes) which can migrate over the lattice due to the exchange interaction J between the copper ions. Such an interaction results in a collectivizing of the states of the excess particles. The quanta of such collective states are called *quasiparticles,* which are described by plane waves having a definite wavenumber k and a quasi-momentum $\hbar k$. The spin and electric charge of a quasiparticle are the same as that of an excess particle, but its mass m_e is replaced by an effective mass m which is determined by the exchange interaction energy J and the lattice constant $\mathbf{a}$ via the relation

$$m = \frac{\hbar^2}{2\mathbf{a}^2 J}.$$

The effective mass absorbs quantitatively the periodic structure of the crystal and can drastically differ from the free electron mass m_e. If the temperature exceeds a critical volume, the conductivity of the crystal and its related properties are determined by quasiparticles.

In lanthanum crystals, the groups of CuO_4 ions are arranged in the form of layers in such a way that each oxygen ion lies in the middle of four copper ions [323]. Therefore, in the $\mathbf{a}$ and $\mathbf{b}$ directions of the crystals there exist infinite parallel chains of alternating ions of copper and oxygen. In yttrium crystals $YBa_2Cu_3O_{7-\delta}$ of the perovskite type there are two positions of the copper ions. The copper ions Cu_1 occupy the sites surrounded by a pyramid of five oxygen ions, which is well preserved over a wide range of temperatures and is less sensitive to variations in the content of the oxygen ions in the crystal. The copper ions Cu_2 lie within the planes $\mathbf{a},\mathbf{b}$ forming the chains of the ions of copper and oxygen alternating along the $\mathbf{b}$th axis. One can suggest that it is these chains which are responsible for superconductivity. The arrangement of the ions in the crystal $YBa_2Cu_3O_{7-\delta}$ is given in Fig. 30 as reproduced from Refs. [316,325]. The chains of alternating ions of copper and oxygen are designated by letters $\mathbf{b}$. A large anisotropy of the superconducting properties of the yttrium crystals along the $\mathbf{b}$-axis was observed by Vlasko-Vlasova, Indenbom, and Osipyan [307] who also gave an account of the double structure of crystals.

The assumption that the parallel chains of alternating ions of copper and oxygen, arranged in the $\mathbf{ab}$-planes, play the dominating role in the superconductivity of the yttrium crystals, is confirmed by the study of high-T_c superconductivity of $Y-Ba-Cu-O$ compounds in which the trivalent yttrium ion is replaced by the ions $Dy^{3+}, Ho^{3+}, Er^{3+}, Tm^{3+}$, and Yb^{3+} possessing permanent magnetic moments of the order of $6-10\mu_B$. In the case of such a substitution, the conducting properties are completely preserved. However, there is observed a slight lowering, by $5-10\,K$, of the transition temperature T_c. The addition of the magnetic admixtures annihilates completely superconductivity in ordinary superconducting metals. The insensitivity of the ceramics mentioned above to the substitution by Y^{3+} ions with magnetic moments seems to be due to the fact that according to Fig. 30, such a replacement occurs at local sites which are considerably far from the quasi-one-dimensional chains of the Cu and O ions which are assumed to be responsible for the superconducting properties of such ceramics. Because the exchange interaction between the ions along the chains dominates considerably over that between ions in different chains, one can consider a system of parallel chains as a quasi-one-dimensional system. We also suggest that in the absence of a magnetic field, the phases of the electron states in the cross-section of the system of chains are equal. For this reason, one can restrict study to the processes occurring in a single chain.

There exists the point of view that a quasi-one-dimensional model should not lead to superconductivity. This point is based on the works by T. Rice (1965) and P. Gohberg (1967) who demonstrated that in ideal one-dimensional systems the superconducting phase transition due to heat fluctuations is impossible. However, they neglect the fact that ideal one-dimensional systems do not exist in nature. One can consider sometimes rather strong anisotropic systems with the preferred symmetry axis being a quasi-one-dimensional one. The conclusions of Rice and Cohberg are not applicable to such systems.

The outstanding specialist in superconductivity, V. Ginzburg, in his paper [308] emphasized that '... it is completely obvious that for a rather thick wire, the probability of fluctuations which transform it into the normal state over its entire cross-section becomes negligibly small ... However, for a wire with a thickness of a few hundreds of angstroms, the fluctuating break of a chain is really impossible, and the superconducting current is not damped' ([308], p. 205).

A quasi-one-dimensional model can be applied to the plane square lattice such as in $La-Ba-Cu-O$, if the transfer of pairs of quasiparticles along some direction with a constant phase in the transverse direction is of interest. This state is realized by means of the initial conditions. Let us recall that in the form of a plane wave, a laser beam propagates over large separations in three-dimensional space, also. Therefore, based on the assumption that a superconducting current flows along parallel chains of alternating ions of oxygen and copper, we study the charge migration in one of these chains. An elementary cell of each chain contains two copper ions Cu_α^{2+} and Cu_β^{2+} in the electronic configuration $3d^9$ differing by oppositely directed α and β spins. These ions are surrounded by oxygen ions. Such elementary cells are separated from each other by a distance of $a \simeq 7.76\text{Å}$. They are coupled by chemical forces between the edge ions of copper and oxygen. The mass M of an elementary cell is about $240m_p$.

Absorbing a quasiparticle in the spin state β in a given elementary cell, the copper ion Cu_α^{2+} passes into the state Cu^{3+} via a change of charge. By virtue of the strong overlap of the electronic shells of the copper ion and the oxygen ions surrounding it, such a change should refer to the whole cell rather to an isolated copper ion. Due to the exchange interaction J between elementary cells, the changed valence state can migrate along the chain. This is one way that the collectivized excess quasiparticles with an effective mass $m = \hbar^2/2a^2J$ and carrying a certain momentum, a unit electric charge, and spin $1/2$, could arise.

A change of the valence state of a copper ion in a single elementary cell is accompanied by a change in its electrostatic interaction with neighboring elementary cells. The latter leads to a local deformation of the chain. In this way there appears a large interaction of quasiparticles with local displacements of the equilibrium positions of elementary cells. As mentioned above, such an interaction is usually recognized as an electron-phonon interaction. Due to the interaction of a quasiparticle with the resulting local chain deformation, a bound state whose energy is less than the sum of energies of a free quasiparticle and a deformation, is generated. Such a stable formation is usually called a *soliton*. In Section 26 we consider solitons in soft molecular chains which can attach electrons. The breakdown of a soliton, i.e. the detachment of a quasiparticle from a local deformation, requires the expense of energy.

One can list the four basic reasons for the high stability of isolated solitons (see Subsection 56.3):

(1) They do not interact with acoustic phonons since that interaction is taken completely into account in the coupling of a quasiparticle with a local deformation. They migrate always with a velocity which is less that of the longitudinal sound velocities and do not radiate phonons, i.e. do not transform their kinetic energy into heat energy.

(2) The envelope of the soliton wave function preserves its form and does not extend similarly to a wave packet.

(3) The breakdown of a soliton into its components (quasiparticle and local deformation) requires the expense of energy.

(4) Solitons are characterized by topological stability. Under the soliton passage through elementary cells these remain displaced from their original equilibrium positions. The breakdown of a soliton requires a return to their original positions.

To illustrate the considerable change in the lifetime of a pair of particles under their coupling into a more complex particle, one can consider the case of an elementary nucleus - deuteron consisting of a neutron and a proton. If the coupling of a proton with a neutron, whose lifetime in the free state accounts for $\sim 10^3\,s$ is comparatively small, the lifetime of the latter in a nucleus becomes infinite.

56.1. Isolated Bisolitons in a Quasi-One-Dimensional Chain

Each quasiparticle in a soliton state is in a potential well created to the moving coordinate frame $\zeta = (x - Vt)/a$. This well is created by local displacements of the equilibrium positions of elementary cells. If, in a quasi-one-dimensional chain, there exist some quasiparticles with spin 1/2, they profit in terms of energy via a coupling into singlet pairs. Due to the nonlinearity of the electron-phonon interaction, these pairs result in the total and stronger local deformation which provides their combined localization. The binding energy of two such quasiparticles, with the local deformation induced by them, exceeds the sum of the binding energies of the isolated quasiparticles (see Subsection 26.1). Two quasiparticles coupled by a local deformation into a singlet spin state, form the unique Bose particle which we call a *bisoliton*. At low temperatures, bisolitons are stable if the gain in the binding energy under their coupling exceeds the screened Coulomb repulsion of their charges. We further assume that this condition holds.

The energy of an immobile bisoliton lies beneath the conduction band of two quasiparticles which is four times deeper compared with the energy level of a single soliton. Therefore, the energy is released following the formation of an immobile bisoliton from two solitons. At small bisoliton velocities, its kinetic energy is proportional to the square of velocity. The coefficient of proportionality is just its effective mass. The effective mass of a bisoliton exceeds that of two isolated solitons since movement of the former is related to a translocation of a deeper local chain deformation.

It is worth noting that the formation of a bisoliton under the direct coupling of two quasiparticles with the resulting local deformation is energetically profitable compared with bisoliton formation from two isolated solitons. In the next Subsection we discuss the basic properties of an ensemble of bisolitons at a temperature close to absolute zero.

56.2. Bose Condensation of Bisolitons

Let us consider the movement of quasiparticles along a chain of alternating ions of copper and oxygen. These quasiparticles arise as a result of the aforementioned change of valency of the copper ions. Due to the electron-phonon interaction of quasiparticles with the resulting displacements of the positions of elementary cells, there appear local chain deformations which, in turn, lead to a coupling of quasiparticles into singlet spin pairs.

Let us assume that along the chain, the elementary cells of mass M are separated by a distance a from each other. Within the continuum approach, we characterize their positions by a continuous variable $x = na$. The equation of motion of a pair of quasiparticles in the total potential field

$$U(x,t) = -\rho(x,t) \tag{56.1}$$

caused by the local deformation of an infinite chain, takes the form

$$\left[i\hbar\frac{\partial}{\partial t}+\frac{\hbar^2}{2m}\frac{\partial^2}{\partial x_i^2}+U(x_i,t)\right]\varphi_j(x_i,t) = 0, \quad i,j = 1,2. \tag{56.2}$$

Here $\varphi_j(x_i,t)$ is the coordinate function of the ith particle which is in the spin state j in the potential field (56.1). The function $\rho(x,t)$ in eqn. (56.1) characterizes a local chain deformation conditioned by a reduction of the relative distances between elementary cells. It is defined by the following equation

$$\left[\frac{\partial^2}{\partial t^2}-V_0^2\frac{\partial^2}{\partial x^2}\right]\rho(x,t)-\frac{\sigma a^2}{m}\frac{\partial^2}{\partial x^2}\{|\varphi_1(x,t)|^2+|\varphi_2(x,t)|^2\} = 0 \tag{56.3}$$

where $V_0 = a\sqrt{\kappa/M}$ is the longitudinal sound velocity in the chain, $a\sigma$ is the energy of deformation interaction of quasiparticles with the chain, and κ is the coefficient of longitudinal elasticity.

The set of equations (56.1)-(56.3) represents the continuum approximation version of the discrete one. The latter is derived from the Hamiltonian of a discrete chain,

$$H = \sum_n \left\{ \sum_{j=1}^{2} \left[J\varphi_{j,n}^{*}(2\varphi_{j,n} - \varphi_{j,n-1}) + \right. \right.$$

$$\left. \left. \sigma \, |\varphi_{j,n}|^2 (u_{n+1} - u_{n-1}) \right] + \frac{a}{2} \left[M \left(\frac{du_n}{dt} \right)^2 + \kappa (u_n - u_{n-1})^2 \right] \right\}$$

where $\varphi_{1,n}$ and $\varphi_{2,n}$ are the fields of two quasiparticles with opposite spins, and u_n are the displacements of the equilibrium positions of the nth cells. If $\sigma = 2a\chi$, this Hamiltonian is the natural generalization of the Hamiltonian given by eqn. (25.1), for the case of two quasiparticles interacting with the chain deformations which result.

Solving eqn. (56.1)-(56.2), as a result of the translational symmetry of an infinite chain one can investigate the excitations propagating along this chain with a constant velocity $V < V_0$. In this case, the forced solution of eqn. (56.3) which satisfies the condition $\rho(x,t) \neq 0$ for all x and t under which $|\varphi_j(x,t)|^2 \neq 0$ relative to the coordinate system

$$\xi = (x - Vt)/a \tag{56.4}$$

moving with velocity V, has the form

$$\rho(\xi) = -\frac{\sigma}{\kappa(1-s^2)} \left[|\varphi_1(\xi)|^2 + |\varphi_2(\xi)|^2 \right], \quad s^2 \equiv V^2/V_0^2. \tag{56.5}$$

The energy of the local deformation of the chain is determined by the integral

$$W = \frac{1}{2}\kappa(1+s^2)\int \rho^2(\xi)d\xi. \tag{56.6}$$

Substituting the value (56.5) into eqn. (56.2) and taking eqn. (56.1) into account, one obtains the equation for the function $\Psi(x_1,x_2,t)$ which determines the motion of a pair of particles in the potential field (56.1):

$$\left\{ i\hbar\frac{\partial}{\partial t} + J\left[\frac{\partial^2}{\partial \xi_1^2} + \frac{\partial^2}{\partial \xi_2^2} \right] + G\left[|\varphi_1(\xi_1)|^2 + |\varphi_2(\xi_2)|^2 + \right. \right. \tag{56.7}$$

$$\left. \left. |\varphi_1(\xi_2)|^2 + |\varphi_2(\xi_1)|^2 \right] \right\} \Psi(x_1,x_2,t) = 0.$$

Here, we introduce the following notations

$$J = \hbar^2/2ma^2, \quad G = \sigma^2/\kappa(1-s^2). \tag{56.8}$$

We investigate further the states with a small velocity of motion ($V^2 \ll V_0^2$). In this case, the parameter G can be replaced by the constant

$$G_0 \approx \sigma^2/\kappa. \tag{56.8a}$$

The coordinate function of a pair of particles in the singlet spin state is symmetric and can be written in the form

$$\Psi(x_1,x_2,t) = \frac{1}{\sqrt{2}}\left[\varphi_1(\xi_1)\varphi_2(\xi_2)+\varphi_1(\xi_2)\varphi_2(\xi_1)\right]\exp\left[-\frac{i\mathcal{E}_p t}{\hbar}\right] \tag{56.9}$$

where $\mathcal{E}_p$ is the energy of two paired quasiparticles in the potential field $U(\xi)$.

In a chain consisting of a large number N of elementary cells and containing N_1 pairs of quasiparticles, a pairing is realized at low temperature only from those states of free quasiparticles of wavenumbers k which are close to the wavenumber $k_F = N_1\pi/2aN$ of the Fermi surface. Due to the conservation law of quasi-momentum, a paired particle, moving with velocity $V = \hbar k/m$, can be formed from two free quasiparticles with the wavenumbers

$$k_1 = 2k - k_F, \quad k_2 = k_F. \tag{56.10}$$

If the wave functions of quasiparticles in the paired state are represented by the modulated plane waves

$$\varphi_j(\xi_i) = \Phi(\xi_i)\exp(ik_j\xi_i), \tag{56.11}$$

the coordinate function (56.9) of paired quasiparticles transforms into the following form

$$\Psi(x_1,x_2,t) = \sqrt{2}\,\Phi(\xi_1)\Phi(\xi_2)\cos[(k-k_F)(\xi_1-\xi_2)] \times \\ \exp\{i[k(\xi_1+\xi_2)-\mathcal{E}_p t/\hbar]\}. \tag{56.12}$$

The appearance here of the cosine function is provided by the conditions of symmetry imposed.

Substituting the function (56.12) into eqn. (56.7), one obtains the equation for the amplitude functions $\Phi(\xi_j)$

$$\left[\frac{\partial^2}{\partial \xi_j^2}+4g\right]\Phi^2(\xi_j)+\Lambda\Phi(\xi_j) = 0, \;\; j = 1,2. \tag{56.13}$$

In eqn. (56.13), the dimensionless parameter

$$g \equiv \frac{G_0}{2J} = \frac{\sigma^2}{2\kappa J} \tag{56.13a}$$

characterizes the coupling of a quasiparticle with the deformation field. The energy $\mathcal{E}(V)$ of a pair of quasiparticles in this field is expressed in terms of the eigenvalue Λ of eqn. (56.13) via the relation

$$\mathcal{E}(V) = \mathcal{E}_p(0)+mV^2-\hbar V k_F \tag{56.14}$$

where

$$\mathcal{E}_p(0) = \Lambda J+\mathcal{E}_F \tag{56.15}$$

characterizes the position of the energy level of an immobile pair of quasiparticles beneath their Fermi level

$$\mathcal{E}_F = \hbar_F^2 k^2/m. \tag{56.15a}$$

The deformation field $\rho(\xi)$ is expressed through the function $\Phi(\xi)$ as follows

$$\rho(\xi) = -\frac{2\sigma}{\kappa(1-s^2)}\Phi^2(\xi). \tag{56.16}$$

Therefore, the energy of the local chain deformation (56.6) is defined by the integral

$$W = \frac{2G(1+s^2)}{1-s^2}\int\Phi^4(\xi)d\xi. \tag{56.17}$$

Equation (56.13) admits periodic solutions corresponding to the uniform distribution of a pair of quasiparticles over the chain. Real functions $\Phi(\zeta_i)$ of these solutions must satisfy the conditions of periodicity

$$\Phi(\zeta_i) = \Phi(\zeta_i+L) \tag{56.18}$$

and normalization

$$\int_0^L\Phi^2(\zeta)d\zeta = 1. \tag{56.19}$$

The latter requires that each pair of quasiparticles should be within each period. The exact periodic solutions of eqn. (56.13) is expressed in terms of the Jacobian elliptic function $dn(u,q)$ via the relation

$$\Phi_q(\zeta_i) = \sqrt{\frac{g}{2}}E^{-1}(q)dn(u,q), \quad u = g\zeta / E(g). \tag{56.20}$$

An explicit form of the Jacobian function $dn(u,q)$ depends on the modulus q taking continuous values in the interval $(0,1)$. The eigenvalue Λ of eqn. (56.13) is also given in terms of the modulus q by means of the relation

$$\Lambda_q = -g^2q^2E^{-1}(q). \tag{56.21}$$

The function $E(q)$ appearing in eqns. (56.20) and (56.21) is a complete elliptic integral of the second kind which is defined via eqn. (28.32) through the complete elliptic integral of the first kind,

$$K(q) = \int_0^1 [(1-t^2)(1-q^2t^2)]^{-1/2}dt. \tag{56.22}$$

The value of the modulus q of the elliptic function is determined by the product of the period L with the dimensionless parameter g of the electron-phonon interaction, and the elliptic integrals of the first, $K(q)$, and second, $E(q)$, kinds,

$$gL = 2E(q)K(q). \tag{56.23}$$

The spatial distributions of both quasiparticles within the bisoliton is characterized by the square of the function (56.20). According to eqn. (56.16), the field of the local chain deformation is also the periodic function

$$\rho_q(\xi) = \frac{\sigma g}{\kappa(1-s^2)E^2(q)}dn^2\left[\frac{g\xi}{E(q)},q\right]. \tag{56.24}$$

Substituting (56.20) into eqn. (56.17) and further the integral derived (see eqn. (28.41)), one obtains the energy of the local chain deformation per period

$$W(q) = Jg^2\frac{(1+s^2)}{(1-s^2)2E(q)}\theta_d(q) \tag{56.25}$$

where the function $\theta_d(q)$ is defined by eqn. (28.42). According to the formulas (56.14), (56.15), and (56.21), the total energy $\ni_q(V) = \mathcal{E}(V)+W$, in terms of the Fermi energy (56.15), takes the following form at small velocities

$$\ni_q(V) = -D_0(q)+\frac{1}{2}M_{bs}V^2-\hbar Vk_F. \tag{56.26}$$

In eqn. (56.26), the term

$$D_0(q) = g^2q^2E^{-1}(q)-W = g^2JF(q) \tag{56.27}$$

characterizes the energy of pairing of quasiparticles including the energy of formation of the local deformation. At absolute zero, this energy is released upon the formation of a bisoliton from two free quasiparticles close to the Fermi surface. The quantity

$$\Delta = \frac{1}{2}D_0(q) \tag{56.27a}$$

determines the energy gap in the quasiparticle spectrum resulting from a pairing.

The magnitude (56.27) is obtained regardless of the screened Coulomb repulsion between quasiparticles. Such an approximation is justified iff the spatial extension of each bisoliton, equal to $2\pi a/g$ approximately, exceeds the lattice constant a. The function $F(q)$ in eqn. (56.21) is determined by the modulus of the Jacobian elliptic function via the relation

$$F(q) = \frac{2}{E^3(q)}[2-q^2-\Xi(q)E^{-1}(q)] \tag{56.28}$$

where

$$\Xi(q) = \frac{1}{3}[2(2-q^2)E(q)-(1-q^2)K(q)]. \tag{56.28a}$$

The effective bisoliton mass within each period is

$$M_{bs} = 2m+4Jg^2\Xi(q)/V_0^2E^3(q). \tag{56.29}$$

Within each period L of an infinite chain there exists a correspondence between the energy (56.26) and the coordinate wave function $\Psi_g(\xi_1,\xi_2,t)$. The latter characterizes the bisoliton distribution given for fixed values of k and the modulus q which is determined by eqn. (56.23) with the parameters g and L characterizing the given chain. This becomes

$$\Psi_q(\xi_1,\xi_2,t) = \sqrt{2}E^{-2}(q)\cos\{(k-k_F)(\xi_1-\xi_2)\}dn\left[\frac{g\xi_1}{E(q)},q\right]\times \tag{56.30}$$

$$dn\left[\frac{g\xi_2}{E(q)},q\right]\exp\left\{i\left[\delta(k)+\frac{\mathcal{E}_p t}{\hbar}\right]\right\}.$$

Here, $\delta(k)$ is the phase which is coherent for the whole chain. It is given by the following relation

$$\delta(k) = 2k(\Re - Vt). \tag{56.31}$$

It smoothly varies under a change of the coordinate $\Re = 1/2(x_1+x_2)$. Due to coherency, the bisolitons are held as a condensate in a state of homogeneous motion. A sharp change in the velocity of one bisoliton should result in the character of motion of all bisolitons constituting the condensate. This is one of the reasons for the existence of superconductivity.

The velocity of motion of the condensate is expressed through the spatial derivative of the phase (56.31) of its total wave function (56.30). Hence,

$$V = \frac{\hbar}{2m}\frac{d\delta(k)}{d\Re} = \frac{\hbar k}{m} \tag{56.32}$$

where $2m$ is the mass of two free quasiparticles constituting a bisoliton. The coherent velocity of motion of the bisoliton condensate is upper bounded. Because the bisolitons are formed as a result of the coupling of two quasiparticles with a local chain deformation, their consistent motion becomes possible if the following condition is satisfied: the velocity of motion of the combined formation does not exceed whichever is the smallest of the following velocities:

(i) the longitudinal sound velocity V_0 which is of the order of $10^5\ cm{\cdot}s^{-1}$;

(ii) the maximum group velocity v_g of free quasiparticles within their conduction band, which is determined by the exchange integral J characterizing the width of this band,

$$v_g = \frac{2aJ}{\hbar} = \frac{\hbar}{am} \tag{56.33}$$

where m is the effective mass of quasiparticle. If $a \approx 10^{-7}\ cm$, $J \approx 0.05\ eV$, and $m = 1.2m_e$, then $v_g \approx 10^6\ cm{\cdot}s^{-1}$.

In superconducting ceramics a more important restriction on the velocity of the condensate is apparently due to the possibility of the decay of bisolitons, constituting the condensate, into free quasiparticles. This restriction is investigated in the following Subsection. Since the velocity (56.32) of the bisoliton condensate is always smaller than the longitudinal sound velocity V_0 in a given chain, it does not loose the energy required for the creation of acoustic phonons. Unlike the condensate of an ideal Bose gas as the state of Bose particles having zeroth energy, the bisolitons in the condensate migrating with a velocity V which is considerably less than V_0 and v_g, carry the energy (56.26).

The breakdown of a bisoliton at absolute zero with the further transition of two quasiparticles onto the Fermi level requires the expense of the energy D_T of pairing (56.27), which decreases under increasing temperature.

The critical temperature T_c of the breakdown of superconductivity is defined as that under which D_T vanishes. Within the present nonlinear theory of superconductivity, the explicit dependence of D_T on the temperature T has not been determined so far. A decrease of D_T with an increase in temperature is apparently and primarily related to a lowering of the exchange integral J appearing in eqn. (56.27), due to fluctuations of the distances between the elementary cells. Within the BCS theory, the relation between T_c and the energy of pairing, $2\Delta_{BCS}$, becomes rather simple at $T = 0$

$$k_B T_c = 3.54\Delta_{BCS}.$$

One can suggest that a proportionality between T_c and D_0 is also preserved for the bisoliton theory of superconductivity. Therefore, this means formally that

$$k_B T_c \sim D_0 = g^2 J F(q). \tag{56.34}$$

The energy of pairing D_0 is independent of the mass M of an elementary cell. This mass only appears in the kinetic energy of the bisoliton. For this reason, the temperature T_c is independent of the mass M in the theory of bisolitons. Therefore, the isotopic effect is absent, notwithstanding the fact that the basis of pairing is an electron-phonon interaction.

Within the framework of the model developed, the properties of bisolitons depend on the energy J of the exchange interaction, the concentration of bisolitons ($\sim 1/L$), and the dimensionless parameter g of the electron-phonon interaction. In particular, the spatial distribution of bisolitons in the chain depends on the modulus q of the elliptic function which is given by eqn. (56.23) for each value $gL > \pi/2$. If the inequality $gL \gg 1$, which corresponds to small densities of solitons, is satisfied, then $q \approx 1$. The value of q decreases with increasing density.

56.3. The Stability of the Bisoliton Condensate and Critical Superconducting Currents

The existence of a condensate of charged Bose particles is a necessary but insufficient condition of the possibility of superconductivity arising. An elucidation of the stability conditions for the bisoliton condensate necessarily requires a knowledge of the nature of other excited states with small energies. Within the model developed, the stability of the condensate of bisolitons at small velocities is provided, according to the general requirements [310], by

(i) the absence of other energetic one-particle excitations lying beneath the band of energies (56.26);

and

(ii) the fulfilment of the conduction of superfluidity by Landau

$$[d\mathcal{E}_q(V)/dV]_{V\to 0} = \hbar k_F \tag{56.35}$$

which follows from the general expression (56.26).

The first requirement in the bisoliton model as well as in the *BCS* theory is fulfilled due to the fact that the pairing of quasiparticles is provided by an electron-phonon mechanism, i.e. by the interaction of quasiparticles with displacements of the equilibrium positions of elementary cells. Under these conditions, free acoustic phonons are absent. An increase in the velocity V of the motion of the condensate results in an upper-bound constraint on the velocity. As mentioned in Subsection 56.3, the first constraint is that this velocity does not exceed the smallest of the velocities V_0 and v_g. The second constraint due to the presence of a term which is linear in velocity in the energy of the condensate (56.26). In the case of velocities which are greater than the critical one, V_{cr}, this term leads to the decay of a bisoliton into two free quasiparticles with conservation of the total quasi-momentum. Such a process refers to a class of fast processes under which the local chain deformation W_q has no time to vary, and the quasiparticles pass from the bound state in the bisoliton, moving with velocity $V = \hbar k/m$, to the free state with quasi-momenta $(2k - k_F)\hbar$ and $\hbar k_F$ under the condition of the conservation of the total quasi-momentum. Under this change with the conservation of quasimomentum $2k\hbar$ and local deformation W_q, the kinetic energy $\hbar^2 k^2/m$ of two quasiparticles within the soliton transforms into the kinetic energy of free quasiparticles

$$\ni_d(V) - W_q = \frac{1}{2m}\hbar^2(2k - k_F)^2 + \frac{1}{2m}\hbar^2 k_F^2. \quad (56.36)$$

Therefore, the new state possesses the energy

$$\ni_q(V) = W_q + \frac{\hbar^2(k-k_F)^2}{2m} + \frac{\hbar^2 k_F^2}{2m} - \frac{\hbar^2 k_F^2}{m}. \quad (56.37)$$

Comparing eqns. (56.37) and (56.36), one finds

$$\ni_q(V) - \ni_d(V) = -D_0(q)V\hbar k_F. \quad (56.38)$$

If this difference becomes negative, the decay of a bisoliton is forbidden. So, this decay of a bisoliton does not occur if the velocity V of the condensate is less than the critical velocity V_{cr} given by the following relation

$$V_{cr} = \frac{D_0(q)}{\hbar k_F}. \quad (56.39)$$

If ν is the density of bisolitons in the condensate, moving with velocity V, the density of the electric current transferred is

$$j = 2e\nu V.$$

Thus, the superconductivity of ceramic superconductors is preserved if the current density is less than the critical value,

$$j_{cr} = 2e\nu D_0(q)/\hbar k_F. \quad (56.40)$$

In ceramic superconductors, the density ν of solitons and the Fermi momentum k_F are both small, but the critical current density can be rather large. Measurements of the current density along the most favourable directions of the monocrystal $Y-Ba_2-Cu_3-O_7$ carried out by Dinger et al. [322] at $T = 4.5K$ and zero magnetic field, give the value $j_{cr} = 3 \times 10^8 A/cm^2$.

The absence of an isotopic effect in yttrium ceramics stimulated a search for other (non-electron-phonon) interactions for pairing mechanism. In particular, some authors [308,312,330] proposed a bipolaron mechanism of pairing and explained the origin of pairing by the strong interaction of electrons with local optical vibrations. In this case, the radius of paired quasiparticles accounts for an order of one or two elementary cells. However, there are no grounds to ignore completely the interaction with acoustic phonons with smaller energy which always exists. Therefore, the first requirement of stability of the condensate of paired particles is not fulfilled.

Enolskii and the present author [320] demonstrated that under the interaction with optical phonons (solitons) which move with a velocity equal or less than the maximum (v_0) group velocity of optical phonons. For optical phonons satisfying the dispersion law

$$\Omega(k) = \sqrt{\Omega_0^2 + v_0^2 k^2} \approx \Omega + \frac{k^2 v_0^2}{2\Omega_0} \tag{56.41}$$

the ratio $v_0 / 2\Omega_0 \ll 1$. It is usually assumed in theoretical investigations that $v_0 \approx 0$. Within this approximation, bipolarons should be at rest. The bipolaron theories of superconductivity [305,312,330] utilize the concept of a rather strong coupling of heavy quasiparticles (a narrow conduction band, small velocity v_g) with optical phonons which results in the localization of excitation at one or two lattice nodes. In this case, the effective mass of excitation becomes rather large. Therefore, due to the reasons mentioned in Subsection 56.3, the velocity V of the coherent movement of 'dressed' quasiparticles will be considerably small ($V < v_g$ and v_0). The migration of such particles is possible only by an incoherent jump mechanism.

Finally, the bipolaron theories ignore completely the coupling of quasiparticles with free acoustic phonons of small energies which exists in all cases. This coupling leads to strong damping. As is known, in good current conductors (silver, copper, etc.) a weak electron-phonon interaction cannot provide a pairing of electrons; however, it is completely responsible for their resistance.

Summarizing all these reasons, one concludes that bipolaron superconductivity cannot be realized in ceramic oxides with strong electron-phonon (optical) coupling. In the following Subsections we investigate the properties of the condensate of bisolitons at small and medium quasiparticle densities in a chain.

56.4. Small and Medium Quasiparticle Densities in a Chain

At rather small densities of quasiparticles, when the inequality

$$gL \gg 1 \tag{56.42}$$

is valid, the complete elliptic integrals have the following approximate values

$$E(q) \approx 1, \quad K(q) \approx \ln(4/\sqrt{1-q^2}). \tag{56.43}$$

In this case, eqn. (56.23), which determines the modulus q of Jacobian elliptic function, becomes

$$\sqrt{1-q^2} = \frac{1}{4}\exp(-gL) \ll 1. \tag{56.44}$$

This equation is satisfied if

$$q \approx 1. \tag{56.45}$$

Within this approximation, the function (56.28a) takes the value

$$\Xi(q) \approx \frac{2}{3}[1-4gL\exp(-gL)]. \tag{56.46}$$

Hence, the function $F(q)$, appearing in eqn. (56.21), has the form

$$F(q) \approx \frac{2}{3}[1+4gL\exp(-gL)]. \tag{56.47}$$

The energy of formation of the bisoliton (56.27),

$$D_0 = \frac{2}{3}g^2J[1+4gL\exp(-gL)]$$

then increases, while the effective mass (56.29),

$$M_{bs} = 2m+\frac{8Jg^2}{3V_0^2}[1-4gL\exp(-gL)] \tag{56.49}$$

decreases under an increase in the density of quasiparticles, if the condition $gL \gg 1$ holds.

The enveloping wave functions $\Phi(\xi)$ of quasiparticles in the soliton within each period are approximated by the hyperbolic functions

$$\Phi_q(\xi) = \left(\frac{g}{2}\right)^{1/2}[1+e^{-1/2gL}(u\tanh u+\sinh^2 u)]\operatorname{sech} u \tag{56.50}$$

where $u = \xi g$. Within the coordinate frame ξ, such a function possesses bell-like elevations with width

$$\Delta\xi \approx 2\pi/g, \tag{56.50a}$$

separated by a distance L from each other. Each elevation characterizes the position of a bisoliton. All elevations move with a single velocity $V < V_0$.

The energy of decay of a bisoliton is independent of the mass of ions at rather small velocities V and is given by the relation

$$D_0(q) = \frac{2}{3}g^2J[1+4gL\exp(-gL)]. \tag{56.51}$$

According to eqn. (50.26), with increasing velocity one should take into account the kinetic energy. This energy contains the effective mass M_{bs} which depends on the mass of the ions through V_0.

At large quasiparticle densities in the chain, when $gL < 5$, the complete elliptic integrals $E(q)$ and $K(q)$ have the following asymptotics

$$E(q) \approx \frac{\pi}{2}\left[1-\frac{1}{4}q^2-\frac{3}{64}q^4+...\right] \tag{56.25}$$

$$K(q) \approx \frac{\pi}{2}\left[1+\frac{1}{4}q^2+\frac{9}{64}q^4+...\right], \quad q^2 \ll 1. \tag{56.53}$$

In this case, from eqn. (56.23) there follows the equation

$$q^2 = 4\sqrt{2}\left[\frac{2gL}{\pi^2}-1\right]^{1/2} \ll 1 \tag{56.54}$$

which determines the modulus q of the Jacobian elliptic function. By virtue of eqn. (56.54), the periodic solutions (56.30) in the form of Jacobian elliptic functions exist under the condition $gL < \pi^2/2$. Therefore, the permissible maximum density of quasiparticles of the bisoliton condensate corresponds to the minimum period

$$L_{\min} = \pi^2/2g. \tag{56.55}$$

If the inequality

$$\pi^2/2 < gL < 5 \tag{56.56}$$

is obeyed, the Jacobian functions are then approximated as

$$\Phi_q(\xi) \approx \frac{1}{\pi}\sqrt{2g}\left[1-\frac{q^2}{2}\sin^2(2g\xi/\pi)\right], \quad q^2 \ll 1. \tag{56.57}$$

If the density of bisolitons attains the critical value ($q^2 \approx 0$), the modulation of the function (56.57) vanishes, and the ordered distribution of bisolitons is broken, which means a breakdown of the density of the condensate.

For $q^2 \ll 1$, the functions (56.28a) and $E(q)$ are approximated by the following expressions

$$\Xi(q) \approx \frac{\pi^2}{2}\left[1-\frac{3}{4}q^2+\frac{9}{64}q^4\right] \tag{56.58}$$

$$E(q) \approx \frac{\pi}{2}\left\{1-\frac{1}{4}q^2\right\}.$$

With the help of eqn. (56.28) one finds that

$$F(q) \approx 16/\pi^3. \tag{56.59}$$

According to eqn. (56.27), the binding energy $\Delta_9(q)$ then takes the value

$$D_0(q) \approx 2g^2J\left(\frac{2}{\pi}\right)^3. \tag{56.60}$$

One should, of course, take into account that for a large density of bisolitons, their Coulomb interaction reduces considerably the magnitude (56.60). So, according to the bisoliton model, the most favourable conditions for superconductivity correspond to the optimal concentration of quasiparticles when $F(q)$ attains its maximum.

The dependence of the superconductivity of ceramic superconductors on the concentration of charge carriers obtained on the basis of a bisoliton model is apparently confirmed by experiment. For example, Khurana [326] noted that neither group of researchers observed superconductivity of such compounds at concentrations smaller than 0.05 and greater than 0.4. The maximum of the critical temperature is attained for a concentration of charge carriers of about 0.15. In accordance with the theoretical studies presented, one should expect that the bisolitons are generated in high-T_c ceramic superconductors under cooling below the critical temperature T_c, given by eqn. (56.34). These bisolitons form a Bose condensate with their arrangement at equal distances from each other being similar to a one-dimensional crystal, if the concentration δ of quasiparticles is considerably less than the critical concentration under which the distance between bisolitons becomes equal to $L_{\min} = \pi^2/2g$. Placing a superconductor between two metallic electrodes (with an electric field applied), the whole condensate moves with the same velocity overcoming small obstacles and defects. In this case, a decrease of bisolitons at one electrode is compensated by their creation at the other electrode.

The process of such a generation of bisolitons at the metal-superconductor contact is discussed in the next Subsection.

56.5. The Generation of Bisolitons at the Boundary of a Metal with Superconductors

In Section 54 we discussed the process of the creation of solitons and bisolitons at the boundary of rigid and soft chains allowing the formation of solitons and bisolitons. We demonstrated there that in a soft chain with the exchange interaction J and the rather large dimensionless parameter g of the interaction of a quasiparticle with a chain deformation, the flow of quasiparticles which pass from the rigid chain can generate solitons if their energy exceeds the critical value

$$E_{1,cr} = \mathcal{E}_0 - \frac{1}{4}g^2 J.$$

Here $\mathcal{E}_0$ is the energy of the bottom of the conduction band of free quasiparticles in a soft chain. If the density of the incident quasiparticles is not too small, the generation of a bisoliton in a soft chain is most probable (see Subsection 54.2), since the critical energy $E_{2,cr}$ of their deformation lies much deeper below the bottom of the conduction band

$$E_{2,cr} = \mathcal{E}_0 - \frac{2}{3}g^2 J.$$

In ceramic oxides, under the absence of a current and at absolute zero, all immobile bisolitons have the energy

$$E_{bs} = \mathcal{E}_F - D_0 \tag{56.61}$$

which is underneath the Fermi surface. At medium quasiparticle densities, the quantity D_0 is given by eqn. (56.49). If a current appears, the bisolitons disappear at one end of a superconducting wire and are generated one by one at the other end in contact with a metallic electrode. The process of such a generation can be described by formulas similar to those used in Subsection 54.2 by replacing $\mathcal{E}_0$ by $\mathcal{E}_F$. In this case E_{bs} becomes the critical energy of generation of bisolitons. Therefore, if $E > E_{bs}$, bisolitons are generated at the contact over equal time intervals

$$T = \frac{3am\ln 3\sqrt{J(W+D_0)}}{4k\hbar(E-E_{bs})}. \tag{56.62}$$

They further move inside a conductor with the velocity

$$V = \frac{\hbar k(E - E_{bs})}{2m(W + D_0)}. \tag{56.63}$$

56.6. Direct Measurements of the Energy Gap in Superconductors

There exist two methods of direct measurement of the energy gap Δ in superconductors:

(i) the study of the tunnelling of electrons from a metal to a superconductor;

(ii) the study of reflection and absorption of far-infrared radiation by superconductors.

Infrared experiments in lanthanum superconductors [331-333] demonstrate that the dimensionless ratio $2\Delta/k_B T_c$ falls into the interval (1.3 - 2.7) with values which are less than the ratio 3.53 arising from the *BCS* theory. Tunnelling measurements with the same samples give the value 3.5-5 which exceed by a factor of two of those obtained in infrared studies. In a study of the superconductors $YBa_2Cu_3O_{9-y}$ (T_c=87K) Kirtley et al. [327] confirmed that the energy gap measured in experiments on tunnelling is larger than that obtained by infrared experiments. For example, the method of infrared spectroscopy gave values of the ratio $2\Delta/k_B T_c$ in the interval (1.6 - 3.4), which implies that $\Delta = (6-13) \times 10^{-3}\, eV$, while investigations of tunnelling of the same crystals gave the values in the interval (3.7 - 5.6). In the latter case, $\Delta_{\text{tunn}} = (13.7 - 20.7) \times 10^{-3}\, eV$.

Such an astonishing difference in the magnitude of the energy gap determined by different methods have received no the theoretical explanation. However, one can explain this fairly easy by using the proposed bisoliton model of high-T_c superconducting condensate are three-component formations - each pair of quasiparticles moves in the local deformation field which they create. Under the absorption of far-infrared radiation ($\sim 100 cm^{-1}$), there occurs a break in a pair of quasiparticles with the simultaneous disappearance of a local deformation (slow process). For this reason, the energy required, which is just the gap width, is determined by $D_0(q)$ as given by eqn. (56.27). The same energy is expended under the destruction by heat of a bisoliton. That is why the critical temperature for the transition into the superconducting state is determined by the binding energy $D_0(q)$ via the approximate formula (56.34).

Under tunnelling, the break of a pair of quasiparticles occurs with the preservation of the local deformation well which implies the expense of a larger energy

$$D_{\text{tunn}}(q) = D_0(q) + W = \frac{g^2 q^2}{E(q)}.$$

Taking into account the experimental value of the correlation length $a\Delta\xi = 30\,\mathring{A}$ with $a = 7.8\,\mathring{A}$, from the relation (56.60a) one obtains the dimensionless parameter $g \approx 1.6$. Hence, if $F \approx 0.8$, the energy of a pairing, D_0, takes the form $D_0 \approx 2J$. It follows from eqn. (56.34) that the critical temperature $T_c \approx 87\,K$ corresponds to the exchange integral $J \approx 8.7 \times 10^{-4}\,eV$. In this case, the energy gap Δ accounts for $0.017\,eV$, which is in good agreement with the values of the energy gap measured using infrared spectroscopy.

Bibliography

1. Abdulaev N.G., A.V. Kiselev, M.Yu. Feigina, and Yu.A. Ovchinnikov: 1977, Bioorg. Khimia 3, 709.
2. Abramowitz M. and I.A. Stegun (eds): 1964, *Handbook of Mathematical Functions,* National Bureau of Standard, New York.
3. Agranovich V.M.: 1968, *Theory of Excitons,* Nauka, Moscow.
4. Agranovich Z.S. and V.A. Marchenko: 1960, *Inverse Problem of Quantum Theory,* Kharkov State University Press, Kharkov.
5. Antonchenko V.Ya., A.S. Davydov and V.V. Illyn: 1982, *Microscopic Theory of the Characteristics of Thin Water Layers,* Preprint ITP-82-113R, Institute for Theoretical Physics, Kiev.
6. Antonchenko V.Ya., V.V. Illyn and N.N. Makovsky: 1980, *Charge Density Effects in Thin Layers of Hard sphere Systems,* Preprint ITP-80-118R, Institute for Theoretical Physics, Kiev.
7. Antonchenko V.Ya., V.V. Illyn and N.N. Makovsky: 1983, Doklady Acad. Sci. Ukrainian SSR A1, 54.
8. Artemenko S.N. and A.F. Volkov: 1981, Zh. Eksp. Teor. Fiz. 80, 2018.
9. Artemenko S.N. and A.F. Volkov: 1981, ibid. 81, 1872.
10. Atanasov B.P., G.B. Postnikova, Yu.Ch. Sadykov and M.V. Vol'kenstein: 1977, Molek. Biologia 11, 537.
11. Akhiezer A.I., V.G. Bar'yakhtar and S.V. Peletminskii: 1967, *Spin Waves,* Nauka, Moscow.
12. Akhiezer I.A. and A.E. Borovik: 1967, Zh. Eksp. Teor. Fiz. 52, 1332.
13. Bar'yakhtar B.G., B.A. Ivanov and A.A. Sukstanskii: 1980, Zh. Eksp. Teor. Fiz. 78, 1509.
14. Bateman H. and A. Erdelyi: 1954, *Higher Transcendental Functions,* McGraw-Hill Book Co., Inc., New York.
15. Belokolos E.D.: 1980, Teor. Mat. Fiz. 45, 268.
16. Berezin Yu.A. and V.I. Karpman: 1966, Zh. Eksp. Teor. Fiz. 51, 1557.

17. Bishop A.R.: 1978, *Bisolitons and Physical Perturbations,* in K. Lonngren and A. Scott (eds), *Solitons in Action,* Academic Press, New York, pp. 61-87.

18. Bogoliubov N.N.: 1950, Ukr. Mat. Zh. 2, 3.

19. Bogoliubov N.N. and Yu.A. Mitropolskii: 1974, *Asymptotic Methods in the Theory of Nonlinear Vibrations,* Nauka, Moscow.

20. Bogoliubsky L.L.: 1980, Teor. Mat. Fiz. 43, 378.

21. Brizhik L.S. and A.S. Davydov: Fiz. Nizk. Temperatur 10, 358.

22. Brizhik L.S. and A.S. Davydov: 1984, ibid. 10, 748.

23. Brodsky A.I.: 1964, *Role of Hydrogen Bonds in Proton Transfer Process,* in N.D. Sokolov and V.M. Chulanovsky (eds), *Hydrogen Bonding,* Nauka, Moscow, pp. 115-125.

24. Bulaevsky L.N.: 1975, Uspekhi Fiz. Nauk 115, 263.

25. Bullough R.K., P.J. Caudrey and H.M. Gibbs: 1980, *Double Sine-Gordon Equation: System having Physical Applications,* in R.K. Bullough and P.J. Caudrey (eds), *Solitons*, Springer, Berlin.

26. Vakhnenko A.A. and Yu.B. Gaididei: 1986, Teor. Mat. Fiz. 68, 350.

27. Vol'kenstein M.V.: 1975, *Molecular Biophysics,* Nauka, Moscow.

28. Gel'fand I.M. and B.M. Levitan: 1951, Doklady Acad. Sci. USSR, 77, 557.

29. Glauber R.J.: 1969, *The Coherence and Detection of Quanta,* in Rendiconti della Scuola Internazionale di Fisica 'E. Fermi', XLII Corso, Soc. Italiana di Fisica, Bologna.

30. Gorshkov K.A. and L.A. Ostrovskii: 1981, *Interaction of Solitons in Non-Integrable Systems, Direct Method of Perturbation Theory,* Preprint No. 12, Institute of Applied Physics Acad. Sci. USSR, Gor'ky.

31. Grigorov L.M. and D.S. Chernavskii: 1972, Biofizika 17, 195.

32. Davydov A.S.: 1968, *Theory of Molecular Excitations,* Nauka, Moscow.

33. Davydov A.S.: 1974, Biofizika 19, 670.

34. Davydov A.S.: 1975, Ukr. Fiz. Zh. 20, 179.

35. Davydov A.S.: 1976, *Theory of the Solid State,* Nauka, Moscow.

36. Davydov A.S.: 1976, *Molecular Mechanism of Contraction of Transversely Striated Muscles,* in Biophysics and Biochemistry of Muscle Contraction,

Nauka, Moscow, pp. 254-257.

37. Davydov A.S. and N.I. Kislukha: 1976, Zh. Eksp. Teor. Fiz. 71, 1090.
38. Davydov A.S.: 1981, *Biology and Quantum Mechanics,* Pergamon Press, Oxford.
39. Davydov A.S.: 1979, Teor. Mat. Fiz. 40, 408.
40. Davydov A.S.: 1979, *Elementary Excitations of the Soliton Type in Biology, in Methodological and Theoretical Problems in Biophysics*, Nauka, Moscow, pp. 243-254.
41. Davydov A.S.: 1982, Usp. Fiz. Nauk 138, 603.
42. Davydov A.S.: 1980, Zh. Eksp. Teor. Fiz. 78, 789.
43. Davydov A.S., V.Ya. Antonchenko and A.V. Zolotariuk: 1981, *Solitons and Proton Motion in Ice,* Preprint ITP-81-60R, Institute for Theoretical Physics, Kiev.
44. Davydov A.S. and A.A. Eremko: 1977, Ukr. Fiz. Zh. 22, 881.
45. Davydov A.S., A.A. Eremko and A.I. Sergienko: 1978, Ukr. Fiz. Zh. 23, 983.
46. Davydov A.S. and A.A. Eremko: 1980, Teor. Mat. Fiz. 43, 367.
47. Davydov A.S. and A.V. Zolotariuk: 1981, *Selflocalized Collective Excitations in Molecular Chains with Cubic Anharmonicity,* Preprint ITP-81-100R, Institute for Theoretical Physics, Kiev.
48. Davydov A.S. and A.V. Zolotariuk: 1982, *Electrons and Excitons in Nonlinear Molecular Chains,* Preprint ITP-82-113R, Institute for Theoretical Physics, Kiev.
49. Davydov A.S.: 1985, *Solitons in Molecular Systems, Reidel, Dordrecht, The Netherlands.*
50. Davydov A.S.: 1986, *Solitons in Bioeneregetics,* Naukova Dumka, Kiev.
51. Davydov A.S. and G.M. Pestryakov: 1981, *Excitation States of the Field with Inertial Self-Action,* Preprint ITP-81-112R, Institute for Theoretical Physics, Kiev.
52. Davydov A.S. and G.M. Pestryakov: 1981, *Space-Periodic Excitations in Molecular Chains with Strong Electron-Vibrational Interaction,* Preprint ITP-81-127R, Institute for Theoretical Physics, Kiev.

53. Davydov A.S. and A.D. Suprun: 1974, Ukr. Fiz. Zh. 19, 44.

54. Davydov A.S. and V.Z. Enol'skii: 1980, Zh. Eksp. Teor. Fiz. 79, 1888.

55. Davydov A.S. and V.Z. Enol'skii: 1981, Zh. Eksp. Teor. Fiz. 81, 1088.

56. Devyatkov, N.D.: 1973, Usp. Fiz. Nauk 110, 453.

57. Devyatkov, N.D. and M.B. Golant: 1982, Pis'ma Zh. Eksp. Teor. Fiz. 8, 39.

58. Dzyaloshinsky, I.E. and B.A. Ivanov: 1979, Pis'ma Zh. Eksp. Teor. Fiz. 29, 592.

59. Dmitrenko, I.M.: 1981, *The World of Superconductivity,* Naukova Dumka, Kiev.

60. Eremko, A.A. and A.I. Sergienko: 1982, Fiz. Tverd. Tela, 24, 3720.

61. Eremko, A.A. and A.I. Sergienko: 1980, *Exciton and Soliton Excitations in One-Dimensional Molecular Chains,* Preprint ITP-80-90R, Institute for Theoretical Physics, Kiev.

62. Eremko, A.A. and A.I. Sergienko: 1980, *Soliton Excitations in α-Helix Protein Molecules,* Preprint ITP-80-1R, Institute for Theoretical Physics, Kiev.

63. Eremko A.A.: 1984, Doklady Acad. Sci. Ukrainian SSR A3, 51.

64. Zakharov V.E.: 1972, Zh. Eksp. Teor. Fiz. 62, 1745.

65. Zakharov V.E.: 1975, *Method of Inverse Scattering Problem,* in I.A. Kunin (eds), *Theory of Elastic Media with Microstructure,* Nauka, Moscow, pp. 226-269.

66. Zakharov V.E., S.V. Manakov, S.P. Novikov and L.P. Pitaevsky: 1980, *The Soliton Theory. Method of Inverse Scattering Problem,* Nauka, Moscow.

67. Zakharov V.E. and L.D. Faddeev: 1971, Funkt. Analiz and Prilozhenie, 5, 18.

68. Zakharov V.E. and A.B. Shabat: 1971, Zh. Eksp. Teor. Fiz. 61, 118.

69. Zakharov V.E. and A.B. Shabat: 1974, Funkt. Analiz and Prilozhenie 8, 43.

70. Ivanov B.A. and A.M. Kosevich: 1978, Pis'ma Zh. Eksp. Teor. Fiz. 24, 495.

71. Ivanov B.A. and A.M. Kosevich: 1977, Zh. Eksp. Teor. Fiz. 72, 2000.

72. Indel'bom V.L. and A.N. Orlov: 1968, *Modern Aspects of Dislocation Mobility, in Dynamics of Dislocations,* Physics Technical Institute of Low Temperatures, Kharkov.

73. Kadomtsev V.I. and V.I. Karpman: 1971, Uspekhi Fiz. Nauk 109, 193.

74. Karpman V.I.: 1973, *Nonlinear Waves in Dispersive Media*, Nauka, Moscow.

75. Karpman V.I. and E.M. Maslov: 1977, Zh. Eksp. Teor. Fiz. 73, 537.

76. Kislukha N.I.: 1986, Ukr. Fiz. Zh. 31, 1323.

77. Klauder J.R. and E.C.G. Sudarshan: 1968, *Fundamentals of Quantum Optics,* W.A. Benjamin, Inc., New York.

78. Kovalev A.C., A.M. Kosevich and K.V. Maslov: 1979, Pis'ma Zh. Eksp. Teor. Fiz. 30, 321.

79. Kosevich A.M.: 1978, *Dislocations in Elasticity Theory,* Naukova Dumka, Kiev.

80. Kosevich A.M.: 1982, Fiz. Metal. and Metalowedenie 53, 420.

81. Kosevich A.M. and Yu.S. Kivzhar': 1982, Fiz. Nizk. Temperatur 8, 1270.

82. Krivnov V.Ya. and A.A. Ovchinnikov: 1976, Zh. Eksp. Teor. Fiz. 72, 1559.

83. Kulik I.O. and I.K. Yanson: 1970, *The Josephson Effect in Superconducting Tunnelling Structures,* Nauka, Moscow.

84. Kuprievich V.A. and Z.G. Kudritskaya: 1982, *Davydov's Solitons and Determination of Parameters of Exciton-Phonon Interaction,* in *Modern Problems of Solid State Physics and Biophysics*, Naukova Dumka, Kiev, pp. 96-105.

85. Kuprievich V.A.: 1985, Teor. Mat. Fiz. 64, 269.

86. Landau L.D.: 1969, Zbornik Trudov, Nauka, Moscow, vol. 1, pp. 90-91.

87. Landau L.D. and E.M. Lifshitz: 1963, Kwantowa Mekhanika, Gostekhizd., Moscow.

88. Landau L.D. and S.I. Pekar: 1948, Zh. Eksp. Teor. Fiz. 18, 419.

89. Litvak A.G. and A.M. Sergeev: 1978, Pis'ma Zh. Eksp. Teor. Fiz. 27, 549.

90. Lonngren K. and A. Scott (eds): 1978, *Solitons and Action,* Academic Press, New York.

91. Lupichev L.N., V.N. Kadanov and A.V. Savin: 1987, *Dynamics of Intramolecular Excitations in Thermolized Chain,* in Physics of Many-Particle Systems, vol. 13, Naukova Dumka, Kiev.

92. McLaughlin D.W. and A.C. Scott: 1978, *A Multisoliton Perturbation Theory,* in K. Lonngren and A. Scott (eds), *Solitons in Action*, Academic Press, New

York, pp. 201-256.

93. Malkin I.A. and V.M. Man'ko: 1979, *Dynamical Symmetries and Coherent States of Quantum Systems,* Nauka, Moscow.

94. Naberukhin Yu.I. and S.I. Shyiskii: 1970, Zh. Struct. Khim. 11, 197.

95. Ovchinnikov Yu.A., N.G. Abdulaev, M.Yu. Feigina, A.V. Kiselev, N.A. Lobanov and I.V. Nasimov: 1978, Bioorg. Khimia 4, 979.

96. Onsager L. and M. Dupuis: 1960, *Electric Properties of Ice,* in *Termodinamika dei Processi Irreversibili* (Scuola Internazionale di Fisica 'Enrico Fermi'), Soc. Italiana di Fisica, Bologna.

97. Parmentier R.D.: 1978, *Fluxons in Long Josephson Junctions* in K. Lonngren and A. Scott (eds), *Solitons in Action*, Academic Press, New York, pp. 173-199.

98. Pekar S.I.: 1946, Zh. Eksp. Teor. Fiz. 16, 335.

99. Pekar S.I.: 1951, *Studies in Electronic Theory of Crystals,* Gostekhizd., Moscow.

100. Pekar S.I.: 1982, *Crystalloptics and Additional Light Waves,* Naukova Dumka, Kiev.

101. Petviashvili V.I.: 1979, *Inhomogeneous Solitons,* in *Nonlinear Waves,* Nauka, Moscow, pp. 5-19.

102. Rashba E.I.: 1957, Zh. Opt. Spectrosk. 2, 75.

103. Rashba E.I.: 1957, ibid. 3, 568.

104. Sagdeev R.Z.: 1958, in *Problems in Plasma Theory and the Problem of Controlled Thermonuclear Reactions,* Publ. by Acad. Sci. USSR, vol. 4, p. 384.

105. Sagdeev R.Z.: 1964, in *Problems in Plasma Theory,* Atomizdat, Moscow, vol. 4, pp. 20-81.

106. Sergienko A.I.: 1982, *Soliton-Like Excitations in Closed Molecular Chains,* Preprint ITP-82-29E, Institute for Theoretical Physics, Kiev.

107. Sergienko A.I.: 1986, Ukr. Fiz. Zh. 31, 1637.

108. Sergienko A.I.: 1986, *Motion of Bjerrum Defects in Ordered Chains of Water Molecules,* Preprint ITP-86-110R, Institute for Theoretical Physics, Kiev.

109. Sergienko A.I.: 1986, *Interaction of Bjerrum Defects in Ordered Chain of Water Molecules,* Preprint ITP-86-124R, Institute for Theoretical Physics,

Kiev.

110. Sikorsky Yu.S.: 1936, *Elements of Elliptic Function Theory,* Nauchn.-Tekhn. Izd. NKTP USSR, Moscow.

111. Simenog I.V.: 1977, Teor. Mat. Fiz. 30, 408.

112. Scott A.: 1982, *Davydov's Solitons and α-Helix Protein Molecules,* in *Modern Problems of Solid State Physics and Biophysics,* Naukova Dumka, Kiev, pp. 176-179.

113. Sokolov N.D.: 1955, Usp. Fiz. Nauk 57, 205.

114. Sokolov N.D.: 1964, *Some Problems of Hydrogen Bonding,* in N.D. Sokolov and V.M. Chulanovsky (eds), *Hydrogen Bonding,* Nauka, Moscow, pp. 7-49.

115. Solodovnikov E.P., A.N. Tavkhelidze and O.A. Krustalev: 1972, Teor. Mat. Fiz. 10, 162.

116. Solodovnikova E.P., A.N. Tavkhelidze and O.A. Krustalev: 1972, ibid. 11, 317.

117. Takhtadzian L.A. and L.D. Faddeev: 1974, ibid. 21, 160.

118. Tyablikov S.V.: 1951, Zh. Eksp. Teor. Fiz. 21, 377.

119. Tyablikov S.V.: 1952, ibid. 22, 153.

120. Tyablikov S.V.: 1954, ibid. 23, 381.

121. Ukrainskii I.I. and S.I. Mironov: 1979, Teor. Eksp. Khimia 15, 144.

122. Faddeev L.D.: 1959, Usp. Mat. Nauk 14, 57.

123. Frenkel, Ya.I.: 1972, *Introduction to the Theory of Metals,* Nauka, Moscow.

124. Frenkel Ya.I. and T.A. Kontorova: 1938, Zh. Eksp. Teor. Fiz. 8, 89.

125. Chernavskaya N.M. and D.S. Chernavskii: 1978, *Physical Aspects of Bacteriorhodopsin Function,* Preprint No. 61, Physical Institute Acad. Sci. USSR, Moscow.

126. Eilbeck J.C., P.S. Lomdahl and A.C. Scott: 1984, *Abstracts Third Intern. Conf. on Problem of Statistical Mechanics,* Dubna.

127. Yanke E. and F. Emde: 1945, *Tables of Functions,* Dover Publication, New York.

128. Yanke E., F. Emde and F. Lösch: 1960, *Tafeln Höherer Funktionen,* B.G. Teubner Verlagsgesellschaft, Stuttgart.

129. Ablowitz M.J.: 1978, Stud. Appl. Math. 58, 17.

130. Ablowitz M.J., D.J. Kaup, A.C. Newell and H. Segur: 1973, Phys. Rev. Lett. 30, 1262.

131. Ablowitz M.J., D.J. Kaup, A.C. Newell and H. Segur: 1974, Stud. Appl. Math. 53, 249.

132. Alexander D.M.: 1985, Phys. Rev. Lett. 54, 138.

133. Anderson D.L.T.: 1971, J. Math. Phys. 12, 945.

134. Antonchenko V.Ya., S.A. Davydov and A.V. Zolotariuk: 1983, Phys. Stat. Sol. (b) 115, 631.

134a. Antonchenko V.Ya., V.V. Illyin and Makovskii: 1986, *Molecular Properties of Water near a Surface,* Preprint ITP-86-71E, Institute for Theoretical Physics, Kiev.

135. Arrecchi F.T. and R. Bonifacio: 1965, IEEE J. Quant. Electr. 1, 169.

136. Aubry S.A.: 1975, J. Chem. Phys. 62, 3217.

137. Bardeen J.: 1973, Solid State Comm. 13, 357.

138. Barone A., F. Esposito, C.J. Magee and A.C. Scott: 1971, Riv. Nuovo Cim. 1, 227.

139. Bachgaard K. and D. Jerome: 1981, J. Am. Chem. Soc. 103, 2440; 1982, Sci. Amer. 247, 50, N1.

140. Bethe H.Z.: 1931, Z. Phys. B71, 205.

141. Bishop A.R., J.A. Krumhansl and J.K. Schrieffer: 1975, Solid State Comm. 17, 1515.

142. Bishop A.R. and T. Schneider: 1978, *Solitons and Condensed Matter Physics,* Springer-Verlag, New York.

143. Bjerrum M.K.: 1951, Danske. Vid. Selsk. Math. Fys. Medd. 27, 3; 1952, Science 115, 385.

144. Blumenfeld L.A. and D.S. Chernavskii: 1973, J. Theor. Biol. 39, 1.

145. Bondenson A., M. Lisak and D.L.T. Anderson: 1979, Phys. Scr. 20, 479.

146. Bogan A.C., C.R. Hanines and A.F.G. Stuart: 1980, Nuovo Cim. ser. II, 58B, 1.

147. Brizhik L.S. and A.S. Davydov: 1983, Phys. Stat. Sol. (b) 115, 615.

148. Careri G.: 1973, *Search for Cooperative Phenomena in Hydrogen-Bonded Amide Structures,* in H. Haken and W. Wagner (eds), *Cooperative Phenomena,* Springer, Berlin, pp. 391-394.

149. Careri G., V. Buontempo, B. Carta and A.C. Scott: 1983, Phys. Rev. Lett. 51, 304.

150. Careri G., V. Buontempo, F. Galuzzi, A.C. Scott, E. Gaitton and E. Shyamsunder: 1984, Phys. Rev. B30, 4689.

151. Calogero F. and A. Degasperis: 1976, Nuovo Cim. 32B, 201.

152. Calogero F.: 1978, *Nonlinear Evolution Equations Solvable by the Spectral Transform,* Pitman, London.

153. Davydov A.S. and N.I. Kislukha: 1973, Phys. Stat. Sol. (b) 59, 456.

154. Davydov A.S.: 1969, ibid. 36, 211.

155. Davydov A.S.: 1974, Studia Biophysika 47, 221.

156. Davydov A.S.: 1976, Phys. Stat. Sol. (b) 75, 735.

157. Davydov A.S.: 1977, Studia Biophysika 62, 1.

158. Davydov A.S.: 1977, J. Theor. Biol. 66, 379.

158a. Davydov A.S.: 1986, Physicochem. Biol. Rev. 6, 201.

159. Davydov A.S.: 1978, Phys. Stat. Sol. (b) 90, 457.

160. Davydov A.S.: 1979, Intern. J. Quantum Chem. 26, 5.

161. Davydov A.S.: 1979, Phys. Scripta 20, 387.

162. Davydov A.S.: 1982, *Biology and Quantum Mechanics,* Pergamon Press, Oxford.

163. Davydov A.S.: 1983, *Energy Transfer along Alpha-Helical Proteins,* in E. Clementi and R. Sarma (eds), *Structure and Dynamics: Nucleic Acids and Proteins,* Adenine Press, New York, pp. 377-388.

164. Davydov A.S.: 1980, Phys. Stat. Sol. (b) 102, 275.

165. Davydov A.S.: 1983, *Energy Transfer along Alpha-Helical Proteins,* in P.-O. Löwdin and B. Pullman (eds), *New Horizons of Quantum Chemistry,* D. Reidel, Dordrecht, pp. 65-75.

166. Davydov A.S. and A.V. Zolotariuk: 1983, Phys. Stat. Sol. (b) 115, 115.

167. Davydov A.S. and A.V. Zolotariuk: 1983, Phys. Lett. 94A, 49.

168. Davydov A.S. and A.V. Zolotariuk: 1984, Phys. Scr. 30, 426.

169. Davydov A.S.: 1985, *Solitons in Molecular Systems,* Reidel, Dordrecht.

170. Eilbeck J.C.: 1979, *Davydov's Solitons,* 16mm silent film (available from swift film productions), I Wool Road Wimbledon, London.

171. Enz V: 1961, Helv. Phys. Acta 37, 245.

172. Eremko A.A., Yu.B. Gaididei and A.A. Vaknenko: 1985, Phys. Stat. Sol. (b) 127, 703.

173. Eigen M. and L. De Maeyer: 1958, Proc. Roy. Soc. London A247, 505.

174. Eremko A.A.: 1984, *Photodissociation of Davydov Solitons,* in *Nonlinear and Turbulent Processes in Physics,* Vol. 2, Harward, pp. 757-761.

175. Eilbeck J.C., P.S. Lomdahl and A.C. Scott: 1984, Phys. Rev. B30, 4703.

176. Faddeev L.D. and V.E. Korepin: 1978, Phys. Reports 42C, 1.

177. Fedyanin V.K. and L.V. Yakushevich: 1982, Intern. J. Quantum Chem. 21, 1019.

178. Feldtkeller E.: 1968, Phys. Stat. Sol. (b) 27, 161.

179. Fogel M.B., S.E. Trullinger, A.R. Bishop and J.A. Krumhansl: 1976, Phys. Rev. Lett. 36, 1411.

180. Fogel M.B., S.E. Trullinger and A.R. Bishop: 1976, Phys. Lett. 59A, 81.

181. Fogel M.B., S.E. Trullinger, A.R. Bishop and J.A. Krumhansl: 1977, Phys. Rev. B15, 1578.

182. Frenkel J. and T. Kontorova: 1939, J. Phys. USSR 1, 137.

183. Fröhlich H.: 1954, Proc. Roy. Soc. London A223, 296.

184. Fröhlich H.: 1954, Adv. Phys. 3, 325.

185. Fröhlich H., H. Peizer and S. Zienau: 1950, Phil. Mag. 41, 221.

186. Gardner C.S., J.M. Green, M.D. Kruskal and R.M. Miura: 1967, Phys. Rev. Lett. 19, 1095.

187. Del Giudice E., S. Doglia and M. Milani: 1981, Phys. Scripta 23, 307.

188. Gibions J., S.G. Thornhill, M.J. Wardrop and D. ter Haar: 1977, J. Plasma Phys. 17, 153.

189. Glauber R.L.: 1963, Phys. Rev. 131, 2766.

190. Glauber R.J.: 1963, Phys. Rev. Lett. 10, 84.

191. Green D.E.: 1973, Science 181, 583.

192. Gross E.: 1976, Ann. Phys. (USA), 99, 1.

193. Hasegawa A. and Tappert: 1973, Appl. Phys. Lett. 23, 142.

194. Hasegawa A. and Y. Kodama: 1981, Proc. IEEE 69, 1145.

195. Heeger A.J. and A.F. Garito: 1974, *The Electronic Properties of TTE-TCNQ,* in H.J. Keller (ed.), *Low Dimensional Cooperative Phenomena,* Plenum Press, New York, pp. 89-123.

196. Henderson R. and P. Unwin: 1975, Nature 257, 28; 1975, J. Mol. Biol. 94, 425.

197. Henderson R.: 1977, Ann. Rev. Biophys. Bioenerg. 6, 87.

198. Henderson R.: 1978, J. Mol. Biol. 93, 123.

199. Hinkle P.C. and R.E. McCarty: 1978, Sci. Amer. 238, 104, N3.

200. Hol W.G., P.T. van Duijnen and H.J.C. Berendsen: 1978, Nature 273, 443.

201. Holstein T.: 1959, Ann. Phys. (USA) 8, 325.

202. Hyman J.M., D.W. McLaughlin and A.C. Scott: 1981, Physica D3, 23.

203. Ichikawa T.H.: 1979, Phys. Scripta 20, 296.

204. Jaccard C.: 1973, *Structure of Water and Aqueous Solutions,* in Proc. Int. Symp. (Marburg, 1973) Verl. Chem. Weinheim, 1974, pp. 409-424.

205. Karpman V.I.: 1967, Phys. Rev. Lett. 25A, 708.

206. Karpman V.I.: 1979, Phys. Scripta 20, 462.

207. Kawahara J.: 1969, J. Phys. Soc. Japan 27, 1331.

208. Kay I. and U.E. Moses: 1956, J. Appl. Phys. 27, 1503.

209. Keener J.R. and D.W. McLaughlin: 1977, Phys. Rev. A16, 777.

210. Keller H.J. (ed.): 1974, *Low Dimensional Cooperative Phenomena,* Plenum Press, New York.

211. Kodama Y.: 1978, J. Phys. Soc. Japan 45, 311.

212. Koehler J.R., A.R. Bishop, J.A. Krumhansl and J.R. Schrieffer: 1975, Solid State Comm. 17, 1515.

213. Korteweg D.J. and G. de Vries: 1895, Phil. Mag. 39, 422.

214. Krumhansl J.A. and J.R. Schrieffer: 1975, Phys. Rev. B11, 3535.

215. Kruskal M.D., R.M. Miura, C.S. Gardner and N.I. Zabusky: 1970, J. Math. Phys. 11, 952.

216. Kuprievich V.A.: 1985, Physica D14, 395.

217. Laedke E.W., K.H. Spatschek, M. Wilkens and A.V. Zolotariuk: 1985, Phys. Rev. A32, 1161.

218. Lamb G.L.: 1969, Phys. Lett. 29A, 207.

219. Lamb G.L.: 1971, Rev. Mod. Phys. 43, 99.

220. Landau L.: 1933, Phys. Z. Sowjetunion Bd.3, 664.

221. Lax P.D.: 1968, Comm. Pure and Appl. Math. 21, 467; 1975, ibid. 38, 141.

222. Little W.A.: 1964, Phys. Rev. A134, 1416.

223. Little W.A.: 1965, Sci. Amer. 212, 21, N2.

224. Lomdahl P.S., L. MacNeil, A.C. Scott, M.E. Stoneham and S.J. Webb: 1982, Phys. Lett. 92A, 207.

225. Lomdahl P.S. and W.C. Kerr: 1985, Phys. Rev. Lett. 55, 1235.

226. Lomdahl P.S., S.P. Layne and J. Bigo: 1984, Los Alamos Sci. N10, 3.

227. Layne S.P.: 1984, ibid. N10, 23.

228. Lomdahl P.S.: 1984, ibid. N10, 27.

229. Miura P.M., C.S. Gardner and M.D. Kruskal: 1968, J. Math. Phys. 9, 1204.

230. McCall S.L. and E.L. Hahn: 1967, Phys. Rev. Lett. 18, 908.

231. McCall S.L. and E.L. Hahn: 1969, Phys. Rev. 183, 457.

232. Mollenauer L.E., R.H. Stolen and J.P. Gordon: 1980, Phys. Rev. Lett. 45, 1095.

233. Nagle J.F., M. Mille and H.J. Morowitz: 1980, J. Chem. Phys. 72, 3959.

234. Nagle J.F. and H.J. Morowitz: 1978, Proc. Natl. Acad. Sci. USA, 75, 298.

235. Nakamura A.: 1977, J. Phys. Soc. Japan 42, 1892.

236. Nevskaya N.A. and Yu.N. Chirgadze: 1976, Biopolymers 15, 639.

237. Onodera Y.: 1970, Progr. Theor. Phys. 44, 1477.

238. Osborne A. and A.E. Stuart: 1978, J. Math. Phys. 19, 1573.

239. Petrov E.G., I.I. Ukrainskii and V.N. Kharkyanen: 1978, J. Theor. Biol. 73, 29.

240. Ragulskis K.M., R.J. Bansevichuns, R.B. Didzgalvis and V.I. Raguleskiene: 1976, *Oblique Inpacts and their Transformation into Continuous Motion. Theory of Machines and Mechanisms,* Nauka, Moscow.

241. Rashba E.I.: 1982, *Self-Trapping of Excitons,* in E.I. Rashba and M.D. Sturge (eds), *Excitons,* North-Holland, Dordrecht, pp. 547-597.

242. Rebbi C.: 1979, Sci. Am. 240, 76.

243. Rice M.J., A.R. Bishop, J.A. Krumhansl and S.E. Trullinger: 1976, Phys. Rev. Lett. 36, 432.

243a. Rice T.M., P.A. Lee and M.C. Cross: 1979, Phys. Rev. B20, 1345.

244. Rubinstein J.: 1970, J. Math. Phys. 11, 258.

245. Russell J.S.: 1844, *Report on Waves,* 14th Meeting of the British Assoc., Adv. Sci., London, p. 311.

246. Schneider T. and E. Stoll: 1973, Phys. Rev. Lett. 31, 1254.

247. Schuster H. (ed.): 1975, *One-Dimensional Conductors,* Springer-Verlag, Berlin.

248. Scott A.C., F.Y.E. Chu and D.W. McLaughlin: 1973, Proc. IEEE 61, 1443.

249. Scott A.C., F.Y.E. Chu and S.A. Reible: 1976, J. Appl. Phys. 47, 3272.

250. Scott A.C. and W.J. Josephson: 1969, Appl. Phys. Lett. 14, 316.

251. Scott A.C., A. Barone, F. Esposito and C.J. Magee: 1971, Rivesta del Nuovo Cimento 1, 227.

252. Scott A.C.: 1979, Phys. Scripta 20, 509.

253. Scott A.C.: 1981, Phys. Lett. 86A, 60.

254. Scott A.C.: 1982, Phys. Rev. A26, 578.

255. Scott A.C.: 1982, Phys. Scripta 25, 651.

256. Scott A.C.: 1983, *Solitons in the Alpha-Helix Protein,* in E. Clementi and R. Sarma (eds), *Structure and Dynamics of Nucleic Acids and Proteins,* Adenine Press, New York, pp. 389-404.

257. Scott A.C.: 1981, *Introduction to Nonlinear Waves,* in R.H. Enns, B.L. Jones, R.M. Miura and S.S. Rangnekar (eds), *Nonlinear Phenomena in Physics and Biology,* Plenum Press, New York, pp. 9-82.

258. Scott J.F.: 1974, Rev. Mod. Phys. 46, 83.

258a. Scott A.C.: 1984, Phys. Scripta 33.

259. Scott A.C., E. Gratton, E. Shyamsunder and G. Careri: 1985, Phys. Rev. B32, 5551.

260. Scott A.C.: 1987, *On Davydov Solitons at 300 K, in Physics of Many-Particle Systems,* Vol. 12, Naukova Dumka, Kiev.

261. Scott-Russell J.: 1844, *Report on Waves,* Proc. Roy. Soc. Edinburgh 319.

262. Skulachev V.P.: 1976, FEBS Lett. 64, 23.

263. Skulachev V.P.: 1980, Biochem. Biophys. Acta 604, 297.

264. Stoeckenius W., R.H. Lozier and R.A. Bogomolni: 1979, ibid. 505, 215.

265. Toda M.: 1967, J. Phys. Soc. Japan 23, 501.

266. Toda M.: 1967, ibid. 22, 431.

267. Toda M.: 1970, Progr. Theor. Phys. Suppl. 45, 174.

268. Toda M.: 1975, Phys. Repts. 18C, 1.

269. Toda M.: 1979, Phys. Scripta 20, 424.

270. Toda M. and M.A. Waddati: 1973, J. Phys. Japan 34, 18.

271. Tappert F.D.: 1972, Phys. Fluids 15, 2446.

272. Toyozawa Y.: 1961, Progr. Theor. Phys. 26, 29.

273. Toyozawa Y.: 1980, Tech. Report of ISSP Secr. A No. 1036.

274. Turner I.E., V.E. Anderson and K. Fox: 1968, Phys. Rev. 174, 81.

275. Vault D., J.H. Parkas and B. Chance: 1967, Nature 215, 642.

276. Zabusky N.J.: 1967, *A Synergetic Approach to Problems of Nonlinear Dispersive Wave Propagation and Interaction in Nonlinear Partial Differential Equation,* Academic Press, New York, pp. 203-222.

277. Zabusky N.J.: 1973, Comp. Phys. Commun. 5, 1.

278. Zabusky N.J. and M.D. Kruskal: 1965, Phys. Rev. Lett. 15, 240.

279. Webb S.J.: 1980, Phys. Repts. 60, 201.

280. Webb S.J., P.S. Lomdahl, L. MacNeil, A.C. Scott and M.E. Stoneham: 1982, Phys. Lett. 92, 207.

281. Zmuidzinas J.S.: 1978, Phys. Rev. B17, 3919.

282. Davydov A.S.: 1985, *Solitons in Molecular Systems,* Reidel, Dordrecht.

283. Schegolev I.F., E.B. Yagubskii, and V.N. Laukhin: 1984, Pis'ma Zh. Eksp. Teor. Fiz. 39, 12.

284. Yagubskii E.B., V.N. Ginodman and A.V. Gudenko: 1985, Pis'ma Zh. Eksp. Teor. Fiz. 42, 384.

285. Yagubskii E.B., I.F. Schegolev and A.V. Zvarykina: 1986, Pis'ma Zh. Eksp. Teor. Fiz. 43, 257.

286. Yagubskii E.B., K.I. Pokhodnya and M.K. Sheykman: 1986, Pis'ma Zh. Eksp. Teor. Fiz. 44, 293.

287. Brizhik L.S. and A.S. Davydov: 1987, *Soliton Mechanism of Superconductivity in Organic Quasi-One-Dimensional Crystals,* Preprint ITP-87-53E, Institute for Theoretical Physics, Kiev.

288. Chu C.W., P.H. Hor, R.L. Meng, L. Gao, Z.J. Huang and Y. Wang: 1987, Phys. Rev. Lett. 58, 405.

289. Cava R.J., R.B. van Dover, B. Batlogg and E.A. Rietman: 1987, Phys. Rev. Lett. 58, 408.

290. Davydov A.S. and V.I. Ermakov: 1987, *Generation of Solitons and Bisolitons and the End of Molecular Chain,* Preprint ITP-87-62E, Institute for Theoretical Physics, Kiev.

291. Bardeen J., L.N. Cooper and J.R. Schrieffer: 1957, Phys. Rev. 108, 1175.

292. Batlogg B. et al.: 1987, Phys. Rev. Lett. 58, 2333.

293. Bourue L.C.: 1987, Phys. Rev. Lett. 58, 2337.

294. Peyrard M., St. Pnevmatikos and N. Flytzanis: 1987, *Dynamics of Two-Component Solitons in Hydrogen Bonded Chains,* Preprint Lab. ORC, Univ. de Bourgogne, France.

295. Davydov A.S. and V.Z. Enol'skii: 1987, *Towards the Effective Mass of Pekar's Polaron,* Preprint ITP-87-44E, Institute for Theoretical Physics, Kiev.

296. Kashimori Y., F. Chien and K. Nishimoto: 1986, Chem. Phys. 107, 389.

297. Pnevmatikos S.: 1987, Phys. Lett. A122, 249; 1988, Phys. Rev. Lett. 60, 1534.

298. Weberpals H. and K.H. Spatschek: 1987, Phys. Rev. A36, 2946.

299. Halding J. and P.S. Lomdahl: 1988, Phys. Rev. A37, 2608.

300. Kryachko E.S.: 1987, Chem. Phys. Lett. 141, 346.

301. Kryachko E.S.: 1988, Solid State Commun. 65, 1609.

302. Yanovitskii O.E. and E.S. Kryachko: 1988, Phys. Stat. Sol. (b) 147, 69.

303. Kryachko E.S.: 1988, *Collective Proton Transfer in System* $(A-H...)_\infty$ *with Double-Morse Symmetric Potential,* Preprint ITP-88-173E, Institute for Theoretical Physics, Kiev; 1990, Chem. Phys. 143, 359.

304. Kryachko E.S.: 1989, *Proton Transfer in Water and Ice: Solitonic Models for Structural Defects in the Hydrogen-Bonded Networks,* in D. Mukherjee (ed.), *Aspects of Many-Body Effects in Molecules and Extended Systems* (Lecture Notes in Chemistry), Springer, Heidelberg, pp. 301-318.

305. Alexandrov A.S.: 1987, Pis'ma Zh. Eksp. Teor. Fiz. 46, 128;
Alexandrov A.S. and A.B. Krebbs: 1988, *Ground State of Continuum Polaron in Model with Local Electron-Phonon Interaction,* Preprint 018-88, Engineering Physical Institute, Moscow.

306. Bogoliubov N.N.: 1958, Zh. Eksp. Teor. Fiz. 34, 58, 71.

307. Vlasko-Vlasov V.K., M.V. Indenbom and Yu.A. Osipyan: 1988, Pis'ma Zh. Eksp. Fiz. 47, 312.

308. Ginzburg V.L.: 1970, Usp. Fiz. Nauk 101, 185.

309. Pokrovski V.L.: 1988, Pis'ma Zh. Eksp. Teor. Fiz. 47, 539.

310. Tilli D.R. and J. Tilli: 1977, in V.L. Ginzburg (ed.), *Superfluidity and Superconductivity,* Mir, Moscow.

311. Anderson P.W.: 1987, Science 235, 1196.

312. Alexander A.S. and J. Ranniger: 1982, Phys. Rev. B23, 1796.

313. Bardeen J., L.N. Cooper and J.R. Schrieffer: 1957, Phys. Rev. 108, 1175.

314. Batlog B., R.J. Cava, A. Jayaraman et al.: 1987, Phys. Rev. Lettt. 58, 2333.

315. Bednorts I.G. and K.A. Müller: 1986, Z. Phys. 64, 189.

316. Beno M.A. et al.: 1987, App. Phys. Lett. March.

317. Bourne L.C., M.F. Crommie, A. Zettl et al.: 1987, Phys. Rev. Lett. 58, 2337.

318. Chu C.W., P.H. Hor, P.L. Meng et al.: 1987, Phys. Rev. Lett. 58, 405.

319. Conwell E.M, and I.A. Howard: 1985, Phys. Rev. B31, 7835.

320. Davydov A.S. and V.E. Enol'skii: 1987, Phys. Stat. Sol. (b) 143, 167.

321. Davydov A.S. and V.N. Ermakov: 1987, *Soliton Generation at the Boundary of Molecular Chains,* Preprint ITP-87-104E, Institute for Theoretical Physics, Kiev.

322. Dinger T.E., T.K. Worthington, W.J. Gallagher et al.: 1987, Phys. Rev. Lett. 58, 2637.

323. Fröhlich H.: 1950, Phys. Rev. 79, 845.

324. Gomes R. et al.: 1987, Phys. Rev. B36, 7226.

325. Jorgensen J.D. et al.: 1987, Phys. Rev. Lett. 58, 1024.

326. Khurana A.: 1987, Physics Today 40, 17.

327. Kirtley , J.K. et al.: 1987, Phys. Rev. B35, 8846.

328. Massida S., A.J. Freeman, D.D. Koelling: 1987, Phys. Lett. 132A, 198.

329. Pauling L.: 1987, Phys. Rev. Lett. 59, 225.

330. Rice M.J. and S. Jegalev: 1988, Synth. Metals 22, 209.

331. Schlesinger Z. et al.: 1987, Phys. Rev. B35, 5334.

332. Sklewski P.E. et al.: 1987, ibid. B35, 5930.

333. Wlater V. et al.: 1987, ibid. B35, 5327.

334. Wu M.K. et al.: 1987, Phys. Rev. Lett. 58, 908.

Index

MIX
Papier aus verantwortungsvollen Quellen
Paper from responsible sources
FSC® C105338

If you have any concerns about our products,
you can contact us on
ProductSafety@springernature.com

In case Publisher is established outside the EU,
the EU authorized representative is:
Springer Nature Customer Service Center GmbH
Europaplatz 3, 69115 Heidelberg, Germany

Printed by Libri Plureos GmbH
in Hamburg, Germany